Topology

McGraw-Hill Higher Mathematics Titles

Ahlfors, Lars
Complex Analysis

Brown, James Ward and Ruel V. Churchill
Complex Variables and Applications, 7th edition

Brown, James Ward and Ruel V. Churchill
Fourier Series and Boundary Value Problems, 6th edition

Burton, David M.
The History of Mathematics, 5th edition

Burton, David M.
Elementary Number Theory, 5th edition

Falmagne, Jean-Claude
Lectures in Elementary Probability Theory and Stochastic Processes

Rudin, Walter
Real and Complex Analysis, 3rd edition

Rudin, Walter
Principles of Mathematical Analysis, 3rd edition

Rudin, Walter
Functional Analysis, 2nd edition

Walter Rudin Series in Advanced Mathematics

Editor-in-Chief: Steven Krantz, Washington Univeristy

David Barrett
Univeristy of Michigan

Steven Bell
Purdue University

John P. D'Angelo
University of Illinois at Urbana-Champaign

Robert Ferrerman
University of Chicago

William McCallum
University of Arizona

Bruce Palka
University of Texas at Austin

Harold R. Parks
Oregon State University

Jean-Pierre Rosay
University of Wisconsin

Jonathan Wahl
University of North Carolina

Lawrence Washington
University of Maryland

C. Eugene Wayne
Boston University

Michael Wolf
Rice University

Hung-His Wu
University of California-Berkeley

Topology

Sheldon W. Davis
Miami University

Boston   Burr Ridge, IL   Dubuque, IA   Madison, WI   New York   San Francisco   St. Louis
Bangkok   Bogotá   Caracas   Kuala Lumpur   Lisbon   London   Madrid   Mexico City
Milan   Montreal   New Delhi   Santiago   Seoul   Singapore   Sydney   Taipei   Toronto

The McGraw·Hill Companies

Higher Education

TOPOLOGY

Published by McGraw-Hill, a business unit of The McGraw-Hill Companies, Inc., 1221 Avenue of the Americas, New York, NY 10020. Copyright © 2005 by The McGraw-Hill Companies, Inc. All rights reserved. No part of this publication may be reproduced or distributed in any form or by any means, or stored in a database or retrieval system, without the prior written consent of The McGraw-Hill Companies, Inc., including, but not limited to, in any network or other electronic storage or transmission, or broadcast for distance learning.

Some ancillaries, including electronic and print components, may not be available to customers outside the United States.

This book is printed on acid-free paper.

1 2 3 4 5 6 7 8 9 0 QPF/QPF 0 9 8 7 6 5 4

ISBN 0–07–291006–2

Publisher, Mathematics and Statistics: *William K. Barter*
Executive editor: *Robert E. Ross*
Senior marketing manager: *Nancy Anselment*
Senior project manager: *Vicki Krug*
Lead production supervisor: *Sandy Ludovissy*
Senior designer: *David W. Hash*
Cover image: *Rokusek Design*
Typeface: *10/12 Computer Modern*
Printer: *Quebecor World Fairfield, PA*

Library of Congress Cataloging-in-Publication Data

Davis, Sheldon (Sheldon W.)
    Topology / Sheldon Davis. — 1st ed.
        p.   cm.
    Includes index.
    ISBN 0–07–291006–2 (acid-free paper)
    1. Topology. I. Title.

QA611.D38   2005
514—dc22   2003068896
                                CIP

www.mhhe.com

Preface

This text has evolved from topics covered in the two-semester topology course that we have given at Miami University for the past twenty-five years. Part one is the material that we have typically covered in the first semester. It is intended for advanced undergraduates and beginning graduate students. I have found that most of this material can usually be covered in a one-semester, three–credit hour course. The entirety of part one could easily be covered in two quarter-long, three–credit hour courses. This book contains what I consider to be the bare essentials every student should know before beginning the study of continuum theory, set-theoretic topology, algebraic topology, functional analysis, or any of the many areas of mathematics that use topology in an important way.

Part two is the material that we usually cover in the second semester (which is a four–credit hour course for graduate students). I regard this section as the beginning of the study of general topology, or set-theoretic topology. The students who take these courses at Miami have usually had at least a semester of abstract algebra and a semester of analysis. These are not required for this course, but there are occasional remarks in the text that refer to this prior knowledge.

Part two of the book can be used following part one, and that is what we customarily do at Miami. However, I have tried to build in some flexibility since there is variation in opinion among schools about what the first course in topology should be. I have included in part one a very brief chapter introducing continuum theory for those interested in going that way in succeeding courses and a very brief chapter introducing homotopy theory for those who would go into algebraic topology in the next course.

For those who do not teach topology at the undergraduate level, part two could serve as the text for a first course in general topology for graduate students. This would require referring to part one for proofs of some theorems, but the statements are included in part two. With that course in mind, the material, especially at the beginning of part two, does refer to some portions of part one; we revisit

viii

many of the concepts introduced in part one from a deeper point of view. Many concepts are defined again, and much of the information developed in part one is mentioned again, but not generally proved, in part two. I have not found this to be awkward for the students who have just completed the course from part one, as the discussion of some of these ideas a second time often helps to firm up their understanding (which was perhaps only partially gained the first time through). One structural feature in part two is that for each new concept after the constructions of subspace, quotient, and product, we will prove the preservation theorem. That is, we will gather together the results about preservation of the new concept by subspaces, products, and quotients into a single theorem. This structure was introduced (to me) in the text by Willard, and both I, as a student, and my students have always found it to be helpful.

Almost all of the material is what would be called general topology or set-theoretic topology. However, in part one, the use of set theory is kept to an intuitive level. For example, we deal with countable and uncountable sets, but we do not distinguish cardinality more than that. In part two, we use the ordinals, especially the countable ordinals, and a small amount of cardinal arithmetic, at least to the point of being able to consider the continuum hypothesis. If the project in chapter 7 is not covered in the first semester, I would usually begin the second semester by covering that material.

Some readers may consider it to be an odd thing that we don't use the phrase *point-set topology*. For many years this was a commonly used phrase denoting the area that is now usually called *general topology*. The origin of this is probably R. L. Moore's famous book, *Foundations of Point Set Theory*. Because of the tremendous impact that set theory has had on this field in the last forty years, the name that is used by the people actually working in the field today is *set-theoretic topology*, although one does hear the phrase *general topology* used especially when there is no fancy set theory involved.

Many people need to be thanked for their contributions to this manuscript. First and foremost, my wife and daughter, Brenda and Kim Davis, have been understanding, and occasionally prompting, that I have needed to spend much time in the summer and fall of 2003 writing the manuscript. My colleagues Zoltan Balogh, Dennis Burke, Roman Pol, and Dimitrina Stavrova have used part one in our course at Miami University, and they have made many valuable suggestions. David Lutzer has also twice used part one at the College of William and Mary, and his suggestions have been excellent. One of my students, Jennifer Secor, took very careful notes in the second-semester course in 2002, which was taught without a text, and her notes have been the outline for much of part two. Dennis Burke has also contributed many valuable comments and suggestions about the content of part two.

In addition, many reviewers contributed greatly to this project. They are Robert Andersen, University of Wisconsin, Eau Claire; Fernanda Botelho, University of Memphis; Mark Brittenham, University of Nebraska; Murray Eisenberg, University of Massachusetts; Michael Friedberg, University of Houston; Stavros Garoufalidis, Georgia Institute of Technology; Robert Ghrist, University of Illinois, Urbana-Champaign; James Keesling, University of Florida; Wayne Lewis, Texas Tech University; Charles Livingston, Indiana University; P.K. Subramanian, California State University, Los Angeles; Neil Stoltzfus, Louisiana State University; Alex Suciu, Northeastern University; and Arthur Wasserman, University of Michigan.

Finally, I would like to thank Howard H. Wicke for teaching me topology.

Part I

Chapter 1

Sets, Functions, Notation

In this chapter we introduce the basic terminology and background from set theory that will be used throughout this book. We will introduce the notions of countable and uncountable sets, which we will use in part one, but we defer a detailed description of the countable ordinals and more discussion of such things as cardinal arithmetic until part two.

We assume that the reader has some familiarity with ideas of sets as collections of objects. We remind the reader that a set is determined by its elements (*the axiom of extensionality*). That is, if two sets have the same elements, then the two sets are equal. We say a set A is a *subset* of a set B if every element of A is also an element of B, and we write $A \subset B$. In particular, we see that $A = B$ if and only if both $A \subset B$ and $B \subset A$ are true. This is the basis of the so-called *pointwise argument*, which is probably the most commonly used way to show that two sets are equal. In a pointwise argument, we first prove the implication $x \in A \implies x \in B$, which gives the result that $A \subset B$, and then we prove that $x \in B \implies x \in A$, which gives the result that $B \subset A$.

We first introduce the notation we will need to discuss functions. We follow the custom that a function is a set of ordered pairs; that is, a subset of a certain Cartesian product.

Definition 1.1 *Suppose X and Y are sets. The **Cartesian product** of X and Y is the set*

$$X \times Y = \{(x, y) : x \in X \ and \ y \in Y\} .$$

Certain special subsets of Cartesian products are given the name *function*.

Definition 1.2 *A **function** is a set of ordered pairs with the property that no two of the pairs have the same first term.*

Definition 1.3 *If f is a function, the set of all first terms of f is called the **domain** of f, and we denote this by $dom(f)$. If f is a function, the set of all second terms of f is called the **range** of f, and we denote this by $ran(f)$. If f is a function such that $dom(f) = X$ and $ran(f) \subset Y$, then we say f **is a function from** X **into** Y, and we write $f : X \to Y$.*

One common usage is to use the word *map* interchangeably with the word function, although some authors reserve the word map to mean a *continuous* function. We now establish some additional terminology regarding functions.

Notation 1.4 *If $f : X \to Y$ and $x \in X$, then we denote by $f(x)$ the unique element of Y such that $(x, f(x)) \in f$.*

Definition 1.5 *Suppose $f : X \to Y$. If $ran(f) = Y$, then we say f is **onto**. Some authors call such a function **surjective**. If, for every two distinct points x_1 and x_2 in $dom(f)$, we have $f(x_1) \neq f(x_2)$, then we say f is **one-to-one**, and we often write, "f is $1 - 1$." Some authors call such functions **injective**. For authors who use this terminology, a function which is both an injection and a surjection is called a **bijection**. In this case, we will usually say, "f is $1 - 1$ and onto."*

Definition 1.6 *Suppose $f : X \to Y$ and $A \subset Y$. The **inverse image** of A under the function f is the set*

$$f^{-1}(A) = \{x \in X : f(x) \in A\}.$$

When $A = \{y\}$, we usually write $f^{-1}(y)$ rather than $f^{-1}(\{y\})$. The sets $f^{-1}(y)$, for all $y \in Y$, are called the **fibers** of the function f.

Definition 1.7 *Suppose $f : X \to Y$ and $A \subset X$. The **direct image** of A under the function f is the set*

$$f(A) = \{f(x) : x \in A\}.$$

We often just refer to $f(A)$ as the **image** of A. Note that $f(X) = ran(f)$.

Both the notations for inverse and direct images have some logical difficulties. For example, the exponent of -1 is thought in most of mathematics to mean reciprocal, and here it doesn't. Even if the reader is sophisticated enough to know that we mean inverse function, we are *not* requiring the inverse function to exist. The set defined by the inverse image of a subset A of Y is defined for any function $f : X \to Y$ and $A \subset Y$. The problem with the notation for direct image is that for $f(A)$ to exist, A should really be an element of the $dom(f)$, rather than a subset of it. The context will usually remove any confusion related to these difficulties, but notation has been proposed, and is used by some authors, to remove these problems; that is, to denote the image and inverse image by $f^{\rightarrow}(A)$ and $f^{\leftarrow}(A)$, respectively. We only mention this alternative notation. The notation we have used in definitions 1.6 and 1.7 is quite standardly used, and we will use it throughout this book.

When we have sets that are labeled as G_α for each α in some index set A, we refer to the collection $\{G_\alpha : \alpha \in A\}$ as an *indexed family of sets*. There are some logical difficulties with this notation since we are distinguishing between G_α and G_β when $\alpha \neq \beta$ even though G_α and G_β may actually be the same set. So it is conceivable that $\{G_\alpha : \alpha \in A\}$ could be a singleton. The way to be careful about this is to use a notation that more clearly refers to the indexing, such as $(G_\alpha : \alpha \in A)$. However, it is rare that any confusion results, and the notation we have chosen is commonly used.

Definition 1.8 *Suppose that $\mathcal{G} = \{G_\alpha : \alpha \in A\}$ is an indexed family of sets. The **union** of the family is the set*

$$\bigcup \mathcal{G} = \bigcup_{\alpha \in A} G_\alpha = \{x : \exists \alpha \in A \text{ such that } x \in G_\alpha\}.$$

*The **intersection** of the family is the set*

$$\bigcap \mathcal{G} = \bigcap_{\alpha \in A} G_\alpha = \{x : \forall \alpha \in A \text{ we have } x \in G_\alpha\}.$$

Notation 1.9 *The notation that we have just used may not be familiar to all readers. So let us formally introduce these standard shorthands: "$\forall$" means "for all" or "for every" and "$\exists$" means "there exists" or "there is."*

Somewhat more casually, the union of a family is the collection of all points that are elements of *at least one* member of the family. In contrast, the intersection of a family is the collection of all points that are elements of *every* member of the family. In certain special families, we modify these notations somewhat.

Notation 1.10 *When $\mathcal{G} = \{A, B\}$, we usually write $A \cup B$ for $\bigcup \mathcal{G}$ and $A \cap B$ for $\bigcap \mathcal{G}$, and when the family is indexed by the natural numbers, i.e. $\mathcal{G} = \{A_1, A_2, A_3, ...\}$, we often write $\bigcup_{n=1}^{\infty} A_n$ and $\bigcap_{n=1}^{\infty} A_n$ for the union and intersection, respectively, of the family $\mathcal{G}$.*

Definition 1.11 *If X is a set and $A \subset X$, we define the* **complement** *of A relative to X to be the set given by*

$$X \backslash A = \{x : x \in X \text{ and } x \notin A\}.$$

Even if $A \not\subset X$, we still write $X \backslash A = \{x : x \in X \text{ and } x \notin A\}$. This set is sometimes called the *difference* of the two sets and is then read "*X* minus *A*."

There are many notations for complement in the literature. This one is probably the least likely to be misinterpreted, so it is the one we will use. In particular, the notation $\overline{A}$, which is used in many books, especially probability books, will be reserved for another purpose in this book.

The effect that complement has on unions and intersections is given in the following theorem.

Theorem 1.12 *(DeMorgan's laws) Suppose that $\mathcal{G} = \{G_\alpha : \alpha \in A\}$ is an indexed family of subsets of a set X.*

1. $X \backslash (\bigcap_{\alpha \in A} G_\alpha) = \bigcup_{\alpha \in A} (X \backslash G_\alpha)$.

2. $X \backslash (\bigcup_{\alpha \in A} G_\alpha) = \bigcap_{\alpha \in A} (X \backslash G_\alpha)$.

To prove DeMorgan's laws, we will use a pointwise argument.

Proof. We will prove part 1. (We leave part 2 as an exercise.) Suppose that $x \in X \backslash (\bigcap_{\alpha \in A} G_\alpha)$. There must exist some $\alpha_0 \in A$ with $x \notin G_{\alpha_0}$. Hence $x \in X \backslash G_{\alpha_0}$, and consequently, $x \in \bigcup_{\alpha \in A} (X \backslash G_\alpha)$. Thus we have $X \backslash (\bigcap_{\alpha \in A} G_\alpha) \subset \bigcup_{\alpha \in A} (X \backslash G_\alpha)$.

Conversely, suppose that $x \in \bigcup_{\alpha \in A} (X \backslash G_\alpha)$. Choose $\alpha_1 \in A$ such that $x \in X \backslash G_{\alpha_1}$. Since $x \notin G_{\alpha_1}$, $x \notin \bigcap_{\alpha \in A} G_\alpha$, and thus $x \in X \backslash (\bigcap_{\alpha \in A} G_\alpha)$. Hence $\bigcup_{\alpha \in A} (X \backslash G_\alpha) \subset X \backslash (\bigcap_{\alpha \in A} G_\alpha)$, and thus $X \backslash (\bigcap_{\alpha \in A} G_\alpha) = \bigcup_{\alpha \in A} (X \backslash G_\alpha)$. ∎

We now turn to the effects of direct and inverse images under functions upon the operations on the sets that we have just defined.

Theorem 1.13 *Suppose $f : X \to Y$. The following statements are true.*

1. $f(\varnothing) = \varnothing$.

2. *If A and B are subsets of X, and $A \subset B$, then $f(A) \subset f(B)$.*

3. *If $\{G_\alpha : \alpha \in A\}$ is an indexed family of subsets of X, then $f(\bigcup_{\alpha \in A} G_\alpha) = \bigcup_{\alpha \in A} f(G_\alpha)$.*

4. If $\{G_\alpha : \alpha \in A\}$ is an indexed family of subsets of X, then $f(\bigcap_{\alpha \in A} G_\alpha) \subset \bigcap_{\alpha \in A} f(G_\alpha)$.

Proof. Statement 1 is clear. To see 2, we will again use a pointwise argument, that is, we will show that every point of $f(A)$ is also a point of $f(B)$. Suppose that $y \in f(A)$. Choose $x \in A$ such that $f(x) = y$. Since $A \subset B$, we must have $x \in B$. Thus $y = f(x) \in f(B)$. Hence we have $f(A) \subset f(B)$.

To see 3, we will use pointwise arguments to show that each of these sets is a subset of the other. Suppose $y \in f(\bigcup_{\alpha \in A} G_\alpha)$. Choose $x \in \bigcup_{\alpha \in A} G_\alpha$ with $y = f(x)$. Now we can choose $\alpha_0 \in A$ such that $x \in G_{\alpha_0}$. We then have $y = f(x) \in f(G_{\alpha_0})$. Hence $y \in \bigcup_{\alpha \in A} f(G_\alpha)$. So we have $f(\bigcup_{\alpha \in A} G_\alpha) \subset \bigcup_{\alpha \in A} f(G_\alpha)$.

Conversely, suppose $y \in \bigcup_{\alpha \in A} f(G_\alpha)$. We can choose $\alpha_0 \in A$ such that $y \in f(G_{\alpha_0})$. Choose $x \in G_{\alpha_0}$ such that $y = f(x)$. Since $x \in G_{\alpha_0}$, we have that $x \in \bigcup_{\alpha \in A} G_\alpha$, and, consequently, $y = f(x) \in f(\bigcup_{\alpha \in A} G_\alpha)$. Thus $\bigcup_{\alpha \in A} f(G_\alpha) \subset f(\bigcup_{\alpha \in A} G_\alpha)$. Since each is a subset of the other, we then have $f(\bigcup_{\alpha \in A} G_\alpha) = \bigcup_{\alpha \in A} f(G_\alpha)$.

To see 4, we will again use a pointwise argument. Suppose $y \in f(\bigcap_{\alpha \in A} G_\alpha)$. Choose $x \in \bigcap_{\alpha \in A} G_\alpha$ such that $y = f(x)$. Now for each $\alpha \in A$, we have $x \in G_\alpha$, and thus $f(x) \in f(G_\alpha)$. So $y \in f(G_\alpha)$ for every $\alpha \in A$, that is $y \in \bigcap_{\alpha \in A} f(G_\alpha)$. Hence $f(\bigcap_{\alpha \in A} G_\alpha) \subset \bigcap_{\alpha \in A} f(G_\alpha)$. This completes the proof. ∎

None of the results in the theorem above are particularly surprising. In fact, probably the most surprising thing at a casual glance is the lack of equality in 4. We see below that this is the best we can hope to achieve.

Example 1.14 *Consider the function $f : \mathbb{R} \to \mathbb{R}$ defined by $f(x) = x^2$. Notice that $f([0, \infty)) = [0, \infty)$, and $f((-\infty, 0]) = [0, \infty)$. Now $[0, \infty) \cap (-\infty, 0] = \{0\}$, so that $f([0, \infty) \cap (-\infty, 0]) = \{0\}$. On the other hand, $f([0, \infty)) \cap f((-\infty, 0]) = [0, \infty)$. So the inclusion in 4 above can be proper.*

In the next theorem, we see that this slightly disagreeable behavior by direct images is not present in inverse images.

Theorem 1.15 *Suppose $f : X \to Y$. The following statements are true.*

1. $f^{-1}(\varnothing) = \varnothing$.

2. $f^{-1}(Y) = X$.

3. If A and B are subsets of Y and $A \subset B$, then $f^{-1}(A) \subset f^{-1}(B)$.

4. If $\{G_\alpha : \alpha \in A\}$ is an indexed family of subsets of Y, then $f^{-1}(\bigcup_{\alpha \in A} G_\alpha) = \bigcup_{\alpha \in A} f^{-1}(G_\alpha)$.

5. If $\{G_\alpha : \alpha \in A\}$ is an indexed family of subsets of Y, then $f^{-1}(\bigcap_{\alpha \in A} G_\alpha) = \bigcap_{\alpha \in A} f^{-1}(G_\alpha)$.

6. If $A \subset Y$, then $f^{-1}(Y \backslash A) = X \backslash f^{-1}(A)$.

Proof. This is another good example of pointwise arguments, so we leave the proof to the reader as an exercise. ∎

Definition 1.16 *A nonempty set is A called **finite** if there is a natural number n such that there exists a one-to-one correspondence between the points of A and the numbers $\{1, 2, 3, ..., n\}$. The empty set $\varnothing$ is also defined to be finite. A set which is not finite is called **infinite**.*

We are all familiar with the number of points in a finite set, but what is the number of points in an infinite set? The idea is really the same idea that we use when we count a set with five elements by pointing with each finger (including the thumb) of one hand; that is, we assign a $1-1$ and *onto* function from the digits on a hand to the elements of the set.

Definition 1.17 *Two sets A and B are called **equipotent** if there exists a $1-1$ and onto function $f : A \to B$. In this case, we often say that A and B have the same **cardinality**, and we write $|A| = |B|$.*

The next result is very useful in determining when two sets have the same cardinality. We will follow the many authors who call this result the Cantor-Bernstein theorem. (There are some authors who call this the Schröder-Bernstein theorem.)

Theorem 1.18 *(Cantor-Bernstein theorem) If X is equipotent with a subset of Y, and Y is equipotent with a subset of X, then $|X| = |Y|$.*

Proof. Suppose $f : X \to Y$ is $1-1$, and $g : Y \to X$ is also $1-1$. We want to find a mapping $\phi : X \to Y$ which is $1-1$ and *onto*. First suppose $x \in g(Y)$. We will call $g^{-1}(x)$ the first ancestor of x. Note that g^{-1} is actually a function in this case, as is f^{-1}. Similarly, if $y \in f(X)$, then we will call $f^{-1}(y)$ the first ancestor of y. Continuing in this way, we call $f^{-1}(g^{-1}(x))$ the second ancestor of x, and $g^{-1}(f^{-1}(y))$ the second ancestor of y, and so on, as long as this continues to be possible. For a given x, there are three possibilities. First, x could have infinitely many ancestors. Second, x could have an even number of ancestors, which includes 0 ancestors in case $x \notin g(Y)$. Finally, the last possibility is that x could have an odd number of ancestors.

We define subsets of X according to these possibilities. Let

$$
\begin{aligned}
X_i &= \{x \in X : x \text{ has infinitely many ancestors}\}, \\
X_e &= \{x \in X : x \text{ has an even number of ancestors}\}, \text{ and} \\
X_o &= \{x \in X : x \text{ has an odd number of ancestors}\}.
\end{aligned} \tag{1.1}
$$

We define Y_i, Y_e, and Y_o similarly. It is easy to see that f maps X_i onto Y_i, that f maps X_e onto Y_o, and that g^{-1} maps X_o onto Y_e. It is also clear that X_i, X_e, and X_o are pairwise disjoint, and their union is X. Similarly, Y_i, Y_e, and Y_o are pairwise disjoint, and their union is Y.

Now we define the function $\phi : X \to Y$ by

$$
\phi(x) = \begin{cases} f(x), & \text{if } x \in X_i \cup X_e, \\ g^{-1}(x), & \text{if } x \in X_o. \end{cases}
$$

It can be checked that ϕ is $1-1$ and *onto*. ∎

We now want to define symbols for certain special sets. The use of the bold block letters for special subsets of the real numbers is sometimes called the Bourbaki notation. The use of ω for the natural numbers including 0 stems from set theory and the fact that this set is the first infinite ordinal. We will use these notations throughout this book. In particular, we will use both notations for the natural numbers depending on whether we include 0 or not. It is clearly possible to get along with just one or the other notation for the natural numbers, but since some students will be going on to analysis or algebraic topology, we will sometimes use $\mathbb{N}$, which is what is customary in those fields, and since some students will be going on to set theory or set-theoretic topology, we will sometimes use ω, which is customary in those fields.

Notation 1.19 $\mathbb{N}$ *is the set of natural numbers (not including 0), i.e.,* $\mathbb{N} = \{1, 2, 3, ...\}$. ω *is the set of natural numbers including 0, i.e.* $\omega = \{0, 1, 2, 3, ...\}$. $\mathbb{Z}$ *is the set of integers.* $\mathbb{Q}$ *is the set of rational numbers.* $\mathbb{R}$ *is the set of real numbers.*

Definition 1.20 *A set C is called **countable** if C is equipotent with some subset of $\mathbb{N}$. Any set which is not countable is called **uncountable**.*

Example 1.21 *The following sets are countable*

1. $\mathbb{N}$

2. *Every finite set*

3. ω

4. $\mathbb{Z}$

Proof. For 3, $f : \omega \to \mathbb{N}$ defined by $f(x) = x + 1$ is $1-1$ and *onto*. For 4, define $f : \mathbb{Z} \to \mathbb{N}$ by

$$f(x) = \begin{cases} 2x, & \text{if } x \geq 1, \\ 2(-x) + 1, & \text{if } x \leq 0. \end{cases}$$

It is easy to check that f is both $1-1$ and *onto*. We have simply mapped the positive integers to the even natural numbers and the negative integers and 0 to the odd natural numbers. ∎

Theorem 1.22 *Every subset of a countable set is also countable.*

Proof. Left to the reader. ∎

Our next result is a useful result in its own right, but the method of proof is perhaps more important for readers who have not seen such an argument. We will construct a set by induction, or using correct terminology, by recursion, but we will leave the distinction between these words to the set theorists. We will begin the construction at some point, and then, assuming we have the set constructed up to the nth point, we will describe how to get the next point. This method describes a set indexed by the set of all natural numbers since there cannot be a first natural number where the construction is not defined.

Theorem 1.23 *If A is an infinite countable set, then $|A| = |\mathbb{N}|$.*

Proof. Since A is countable, we know that A is equipotent with a subset of $\mathbb{N}$. We will construct by induction a subset of A that is equipotent with $\mathbb{N}$. Since A is infinite, we know that A is nonempty. Choose a point $a_1 \in A$. Suppose that distinct points $a_1, a_2, ..., a_n$ have been chosen. Since A is infinite, $A \backslash \{a_1, a_2, ...a_n\} \neq \varnothing$. Let $a_{n+1} \in A \backslash \{a_1, a_2, ...a_n\}$. Note that a_{n+1} is distinct from all the points $a_1, a_2, ...a_n$. By induction we define the set $\{a_n : n \in \mathbb{N}\}$. Now we define a function $f : \mathbb{N} \to A$ by $f(n) = a_n$. Since the points chosen in the construction are all distinct, f is a $1-1$ mapping. Thus $\mathbb{N}$ is equipotent with a subset of A. Now by the Cantor-Bernstein theorem, we have the result. ∎

Corollary 1.24 *Every infinite set has a subset which is equipotent with* $\mathbb{N}$.

Proof. This is the same construction as above, and is left to the reader. ∎

Remark 1.25 *Infinite countable sets are often called "countably infinite" and sometimes called "denumerable."*

The next theorem uses an argument that is due to Cantor, and so is generally called the "Cantor diagonalization argument." This very clever construction has inspired similar arguments in a variety of seemingly unrelated areas.

Theorem 1.26 $\mathbb{R}$ *is uncountable.*

Proof. If the theorem were not true, then there would have to exist a function $f : \mathbb{N} \to \mathbb{R}$ which is $1-1$ and *onto*. We will show that no such function can exist. Suppose f were such a function. We will define a point of $\mathbb{R}$ which is not an element of $ran(f)$. For each $n \in \mathbb{N}$, we define

$$a_n = \begin{cases} 4, \text{ if the } n^{th} \text{ place to the right of the decimal in } f(n) \text{ is } 5, \\ 5, \text{ if the } n^{th} \text{ place to the right of the decimal in } f(n) \text{ is not } 5. \end{cases}$$

Now let $x = 0.a_1a_2a_3... = \sum_{n=1}^{\infty} \frac{a_n}{10^n}$. Certainly, x is a real number, but the decimal representation for x differs from the decimal representation for $f(n)$ at the nth place. Since the decimal representation for x consists only of 4's and 5's, the representation is unique. Hence the function f cannot be *onto*, and this is a contradiction. So the result is established. ∎

The next result is usually referred to as "any countable union of countable sets is countable."

Theorem 1.27 *If A_n is a countable set for each $n \in \mathbb{N}$, then $\bigcup_{n=1}^{\infty} A_n$ is a countable set.*

Proof. For each n, choose a subset $P_n \subset \mathbb{N}$ such that $|A_n| = |P_n|$. Now, we can index A_n in a $1-1$ way by $A_n = \{a_{n,m} : m \in P_n\}$. We define $f : \bigcup_{n=1}^{\infty} A_n \to \mathbb{N}$ as follows. If $x \in \bigcup_{n=1}^{\infty} A_n$, then let k be the first n such that $x \in A_n$. Now choose the unique $m \in P_k$ such that $x = a_{k,m}$. Define $f(x) = 2^k 3^m$. By the uniqueness of the factorization of integers, f is a $1-1$ function, and the theorem is proved. ∎

Corollary 1.28 $\mathbb{Q}$ *is countable.*

Proof. Note that $\mathbb{Q}$ can be written as $\left\{\frac{p}{q} : p \in \mathbb{Z} \text{ and } q \in \mathbb{N}\right\}$. For each $n \in \mathbb{N}$, let $A_n = \left\{\frac{p}{n} : p \in \mathbb{Z}\right\}$. Since $\mathbb{Z}$ is countable, so is A_n, for each $n \in \mathbb{N}$. Hence $\mathbb{Q} = \bigcup_{n=1}^{\infty} A_n$ is countable. ∎

Corollary 1.29 $\mathbb{N} \times \mathbb{N}$ *is countable.*

Proof. Left to the reader. ∎

We now turn to the definition of a sequence. This is the same idea that we recall from calculus with the minor modification that the terms of our sequences will be in an abstract set, rather than the real numbers. There are many notations for sequences, and we will use the one that makes a sequence look like a generalization of ordered pair.

Definition 1.30 *Suppose A is a set. A **sequence** in A is a function $x : \mathbb{N} \to A$.*

Rather than using the function notation, we generally denote $x(n)$ by x_n, and we call x_n the nth *term of the sequence.* We often denote the sequence whose n^{th} term is x_n by $(x_n : n \in \mathbb{N})$ or by $(x_n)_{n=1}^{\infty}$.

Many notations for sequences are popular, including $\{x_n\}_{n=1}^{\infty}$, $\langle x_n \rangle_{n=1}^{\infty}$, $\langle x_n \rangle_{n \in \mathbb{N}}$, $\langle x_n : n \in \mathbb{N} \rangle$, and others. The set theorists all use ω for the natural numbers, and so they would say that a sequence is a function from ω into A. Hence the notation would be $(x_n : n \in \omega)$; we will use this notation as often as the other.

One last set-theoretic tool will be introduced in this chapter. This is Zorn's lemma, and it is equivalent to the *axiom of choice*. We will have more to say about this in chapter 13 (when we will have more to say about several set theory topics). Zorn's lemma is a statement about when a partially ordered set will have a maximal element. A *maximal element* is an element for which there is no larger element.

In the picture opposite, we will say that $a < b$ if b is above a and there is a sequence of upward pointing lines from a to b. Notice that there are several (how many?) *maximal* elements but no *largest* element.

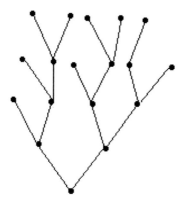

Definition 1.31 *A set* $\mathbb{P}$ *is **partially ordered** by a relation* $\leq$ *on* $\mathbb{P}$ *(we call* $\leq$ *a **partial order**) provided the following are true:*

1. *($\leq$ is reflexive) for each* $x \in \mathbb{P}$, $x \leq x$,

2. *($\leq$ is transitive) if* $x \leq y$ *and* $y \leq z$, *then* $x \leq z$, *and*

3. *($\leq$ is antisymmetric) if* $x \leq y$ *and* $y \leq x$, *then* $x = y$.

Definition 1.32 *A **linear order** (or **total order**) is a partial order such that any two points are comparable.*

We are now ready to state Zorn's lemma.

Theorem 1.33 *(Zorn's lemma) If* $\mathbb{P}$ *is a nonempty partially ordered set, and* $\mathbb{P}$ *has the property that every linearly ordered subset of* $\mathbb{P}$ *has an upper bound in* $\mathbb{P}$, *then* $\mathbb{P}$ *has a maximal element.*

Exercises

1. Prove theorem 1.15.

2. Prove part 2 of DeMorgan's laws, theorem 1.12.

3. Show that the function constructed in the proof of the Cantor-Bernstein theorem is well defined, and show that it is $1 - 1$ and *onto*.

4. Prove theorem 1.22.

5. Prove corollary 1.24.

6. Suppose that $|A| = |B|$ and that A is countable. Prove that B is also countable.

7. Prove corollary 1.29.

8. Prove that if X is an infinite set, then $|A| = |X|$ for some proper subset $A \subset X$. Note: by a *proper* subset of X, we simply mean a subset of X that does not equal X.

Chapter 2

Metric Spaces

Topology can be described as the study of those structures on abstract spaces that will support a theory of limits and continuity. If you think about the definitions of limit and continuity from calculus, you will notice that they are not really statements about absolute values and ε's and δ's. They are statements about when points are close together. Our goal in this book is to develop and study structures that will allow us to tell when points are close together.

It turns out that this is a very rich subject. It gives rise to several areas of mathematics that have developed far beyond these meager roots. We will first introduce the spaces where we actually know the distance from each point to each other point. These not only are the closest to the familiar setting in calculus, they also give us a template for what will happen in more general spaces. The spaces where we can measure the distance from one point to another are called metric spaces.

Definition 2.1 *Suppose X is a set. A **metric** on X is a function $d : X \times X \to [0, \infty)$ such that the following conditions are satisfied:*

1. $d(x, y) = 0$ if and only if $x = y$.

2. $d(x, y) = d(y, x)$, $\forall x, y \in X$.

3. $d(x, y) \leq d(x, z) + d(z, y)$, $\forall x, y, z \in X$.

We sometimes say "distance function" in place of "metric," and we refer to $d(x, y)$ as the *distance* from x to y. Property 2 is called the **symmetry property**, or one sometimes says d is **symmetric**. Property 3 is called the *triangle inequality* because of the picture this represents in $\mathbb{R}^2$ (see figure 2.1).

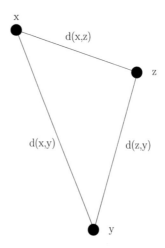

Figure 2.1: The triangle inequality.

Definition 2.2 *If X is a set, and d is a metric on X, we say that (X, d) is a **metric space**. When the metric d is understood from the context, we often simply refer to X as a metric space.*

Example 2.3 *Suppose $X = \mathbb{R}$, and $d_1(x, y) = |x - y|$. This is called the **usual metric** or **Euclidean metric** on $\mathbb{R}$.*

Let us prove that this function is a metric.

Proof. For any real number z, $|z| \geq 0$. So d_1 is indeed a function from $\mathbb{R} \times \mathbb{R}$ into $[0, \infty)$. For any $x \in \mathbb{R}$, $d_1(x, x) = |x - x| = |0| = 0$. If $d_1(x, y) = |x - y| = 0$, then $x - y = 0$, and thus $x = y$. So property 1 is true. For any $x, y \in \mathbb{R}$, $d_1(y, x) = |y - x| = |(-1)(x - y)| = |-1|\,|x - y| = |x - y| = d_1(x, y)$. So property 2 is true. Using the triangle inequality for absolute value from calculus, for any $x, y, z \in \mathbb{R}$, $d_1(x, y) = |x - y| = |x - z + z - y| \leq |x - z| + |z - y| = d_1(x, z) + d_1(z, y)$. So property 3 is true, and the proof is complete. ∎

Example 2.4 *Suppose $X = \mathbb{R}$, and $d_2(x, y) = \frac{|x-y|}{1+|x-y|}$. This is also a metric on $\mathbb{R}$, and it has the interesting property that 1 is an upper bound for the distances between points with this metric.*

Example 2.5 *Suppose $X = \mathbb{R}$, and $d_3(x,y) = \min\{1, |x-y|\}$. This is also a metric on $\mathbb{R}$, and it also has the property that 1 is an upper bound for the distances between points.*

Remark 2.6 *We have shown that d_1 is a metric, and we leave it to the reader to verify that d_2 and d_3 satisfy the properties required of metrics. We also note that these three are all distinct, for instance $d_1(0,2) = 2$, $d_2(0,2) = \frac{2}{3}$, and $d_3(0,2) = 1$. So it is possible to define many different metrics on the same set.*

It is natural to wonder if there any sets that do not have any metrics defined on them. We see below that every set has at least one metric defined on it.

Definition 2.7 *Suppose X is a set. The **discrete metric** on X is defined by*

$$d(x,y) = \begin{cases} 1, & \text{if } x \neq y, \\ 0, & \text{if } x = y. \end{cases}$$

It is easy to verify, and we leave it to the reader, that the discrete metric on X is a metric no matter what the set X may be.

Definition 2.8 *If (X,d) is a metric space, $x \in X$, and $\varepsilon > 0$, then the **open ball** centered at x of radius ε is the set $B(x,\varepsilon) = \{y \in X : d(x,y) < \varepsilon\}$.*

If there are two or more metrics under consideration, we will denote $B(x,\varepsilon)$ by $B_d(x,\varepsilon)$ to indicate that d is the metric being used.

Example 2.9 *Let $X = \mathbb{R}^2 = \mathbb{R} \times \mathbb{R}$. We define three metrics on X.*

1. $d_e =$ *the Euclidean metric (sometimes called the l_2-metric) is defined by*

$$d_e((x_1,x_2),(y_1,y_2)) = \sqrt{(x_1-y_1)^2 + (x_2-y_2)^2}.$$

   The usual metric on $\mathbb{R}^2$ is d_e.

2. $d_1 =$ *the l_1-metric (sometimes called the taxi cab metric, think of a city laid out so that all intersections are at right angles) is defined by*

$$d_1((x_1,x_2),(y_1,y_2)) = |x_1-y_1| + |x_2-y_2|.$$

3. $d_\infty =$ *the sup metric (sometimes called the l_∞-metric) is defined by*

$$d_\infty((x_1,x_2),(y_1,y_2)) = \max\{|x_1-y_1|, |x_2-y_2|\}.$$

We leave it to the reader to check that these really are metrics.

To get an idea of the appearance of the balls, let us draw pictures of the unit ball at the origin for the first two of these metrics. (We leave it as an exercise for the student to draw the third.)

$$
\begin{aligned}
B_{d_e}((0,0),1) &= \{(x,y) : x^2 + y^2 < 1\}, \\
B_{d_1}((0,0),1) &= \{(x,y) : |x| + |y| < 1\}, \\
B_{d_\infty}((0,0),1) &= \{(x,y) : \max\{|x|,|y|\} < 1\}.
\end{aligned}
$$

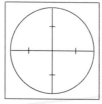

$$B_{d_e}((0,0),1)$$

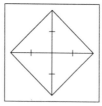

$$B_{d_1}((0,0),1)$$

The fact that some of these balls have flat spots on their surfaces is important in certain areas of mathematics that involve uniqueness of best approximations or nearest points in subsets. For example, in the d_1 metric, the subset $\{(x, x+1) : x \in \mathbb{R}\}$ has an entire interval of points at distance 1 from the origin, and all of these points are the "nearest" to the origin, while in the d_e metric, there is only one nearest point to the origin in this subset. (By the way, how far is it from the origin to the nearest point?)

Definition 2.10 *Suppose that (X,d) is a metric space and that $U \subset X$. A point $x \in U$ is called an **interior point** of U if and only if $\exists \varepsilon > 0$ such that $B(x,\varepsilon) \subset U$. The **interior** of U, denoted $Int(U)$ or $IntU$, is the set of all interior points of U.*

Remark 2.11 *It is clear that $IntU \subset U$.*

Definition 2.12 *In the event that $IntU = U$, we say that U is an **open** subset of X. In other words, a subset $U \subset X$ is open in X provided that $\forall x \in U \exists \varepsilon_x > 0$ such that $B(x, \varepsilon_x) \subset U$.*

We have used "ε_x" in this definition intentionally, even though it makes the writing look more messy, to emphasize that the choice of the radius of the ball that witnesses the fact that $x \in IntU$ depends on the point x.

   The next theorem gives us several properties of the collection of all open subsets of metric spaces. When we discuss more general spaces, this theorem will be the model for the properties that we will want the collection of open sets to have.

Theorem 2.13 *Suppose X is a metric space. The following subsets of X are open:*

1. *the entire space X.*

2. *the empty set $\varnothing$.*

3. *$B(x, \varepsilon)$, for any $x \in X$ and any $\varepsilon > 0$.*

4. *$IntA$, for any $A \subset X$.*

5. *$U \cap V$, for any open subsets $U \subset X$ and $V \subset X$.*

6. *$\bigcup_{\alpha \in I} U_\alpha$, for any indexed family $\{U_\alpha : \alpha \in I\}$ of open subsets of X.*

Proof. 1 and 2 are obvious. We now prove 3. Suppose $x \in X$ and $\varepsilon > 0$. Let $y \in B(x, \varepsilon)$, and let $\eta = \varepsilon - d(x, y)$. Notice that $\eta > 0$. We claim that $B(y, \eta) \subset B(x, \varepsilon)$. Suppose that $z \in B(y, \eta)$. Hence $d(y, z) < \eta$. Hence, by the triangle inequality, we have that $d(x, z) \leq d(x, y) + d(y, z) < d(x, y) + \eta = d(x, y) + \varepsilon - d(x, y) = \varepsilon$. Since $d(x, z) < \varepsilon$, we have that $z \in B(x, \varepsilon)$. Thus we know our claim is true, and y is an interior point of $B(x, \varepsilon)$. Since y was arbitrary, it follows that $B(x, \varepsilon)$ is an open set.

   Now we prove 4. Suppose that $A \subset X$ and $x \in IntA$. Choose $\varepsilon > 0$ such that $B(x, \varepsilon) \subset A$. By the construction in the proof of 3, for each $y \in B(x, \varepsilon)$ we can find $\eta > 0$ such that $B(y, \eta) \subset B(x, \varepsilon)$, and thus $B(y, \eta) \subset A$. Hence, each such y is an interior point of A, that is, $B(x, \varepsilon) \subset IntA$. Hence $IntA$ is open.

   Now we prove 5. Suppose that U and V are open subsets of X, and $x \in U \cap V$. Choose $\varepsilon_1 > 0$ and $\varepsilon_2 > 0$ such that $B(x, \varepsilon_1) \subset U$ and $B(x, \varepsilon_2) \subset V$. Let $\varepsilon = \min\{\varepsilon_1, \varepsilon_2\}$. Now suppose $y \in B(x, \varepsilon)$. Since $d(x, y) < \varepsilon$, we have that $y \in B(x, \varepsilon_1) \cap B(x, \varepsilon_2) \subset U \cap V$. Hence $B(x, \varepsilon) \subset U \cap V$. Therefore, $U \cap V$ is open.

Finally, to see 6, suppose that $\{U_\alpha : \alpha \in I\}$ is an indexed family of open subsets of X. Let $x \in \bigcup_{\alpha \in I} U_\alpha$. Choose $\beta \in I$ such that $x \in U_\beta$. Since U_β is open, we can choose $\varepsilon > 0$ such that $B(x, \varepsilon) \subset U_\beta$. Now since $U_\beta \subset \bigcup_{\alpha \in I} U_\alpha$, we have that $B(x, \varepsilon) \subset \bigcup_{\alpha \in I} U_\alpha$. Hence $\bigcup_{\alpha \in I} U_\alpha$ is an open set. ∎

We come now to the definition of the word that you see on the front cover of the book. The *topology* generated by a metric, or rather the properties of this topology, which we proved in theorem 2.13, will be the prototype for topologies on general spaces.

Definition 2.14 *The collection of all open subsets of the metric space (X, d) is called the **topology** on X **generated by the metric** d.*

We have seen that it is possible to have many different metrics on the same set. If two metrics generate the same topology, then we call them *equivalent*.

Definition 2.15 *We say that two metrics d_1 and d_2 on the same set X are **equivalent** if they generate the same topology.*

Example 2.16 *The metrics d_e, d_1, and d_∞ defined on $\mathbb{R}^2$ in example 2.9 are all equivalent.*

Before proving that these three metrics are equivalent, we will establish a helpful lemma that is also independently interesting.

Lemma 2.17 *If d and ρ are metrics on a set X, and there is a positive real number k such that $d(x, y) \leq k\rho(x, y)$ for all x, y in X, then the topology generated by d is a subcollection of the topology generated by ρ.*

Proof. Suppose U is an open set in (X, d). We wish to show that U is an open set in (X, ρ). Suppose $x \in U$. Choose $\varepsilon > 0$ such that $B_d(x, \varepsilon) \subset U$. We will show that $B_\rho(x, \frac{\varepsilon}{k}) \subset B_d(x, \varepsilon)$. Suppose that $y \in B_\rho(x, \frac{\varepsilon}{k})$. We know that $\rho(x, y) < \frac{\varepsilon}{k}$. Hence we have that $d(x, y) \leq k\rho(x, y) < k\frac{\varepsilon}{k} = \varepsilon$. Thus $y \in B_d(x, \varepsilon)$, and since y was arbitrary, we know that $B_\rho(x, \frac{\varepsilon}{k}) \subset B_d(x, \varepsilon)$. This shows that U contains a ρ-ball around each of its points. So U is an open set in (X, ρ). ∎

We now use this lemma to prove the equivalence of the metrics mentioned in the example.

Proof. (of example 2.16) Suppose $\overline{x}$ and $\overline{y}$ are elements of $\mathbb{R}^2$. We adopt the usual custom that $\overline{x} = (x_1, x_2)$ and $\overline{y} = (y_1, y_2)$. Note that both $|x_1 - y_1|$ and $|x_2 - y_2|$

are less than $\sqrt{(x_1 - y_1)^2 + (x_2 - y_2)^2}$. Hence we have that

$$\text{(i) } d_\infty(\overline{x}, \overline{y}) \le d_e(\overline{x}, \overline{y}).$$

Since d_e satisfies the triangle inequality,

$$\begin{aligned} d_e(\overline{x}, \overline{y}) &\le d_e(\overline{x}, (y_1, x_2)) + d_e((y_1, x_2), \overline{y}) \\ &= \sqrt{(x_1 - y_1)^2 + (x_2 - x_2)^2} + \sqrt{(y_1 - y_1)^2 + (x_2 - y_2)^2} \\ &= \sqrt{(x_1 - y_1)^2} + \sqrt{(x_2 - y_2)^2} \\ &= |x_1 - y_1| + |x_2 - y_2| = d_1(\overline{x}, \overline{y}). \end{aligned}$$

Thus we have

$$\text{(ii) } d_e(\overline{x}, \overline{y}) \le d_1(\overline{x}, \overline{y}).$$

Also note that it is trivial that

$$\text{(iii) } d_1(\overline{x}, \overline{y}) \le 2d_\infty(\overline{x}, \overline{y}).$$

From (i), (ii), and (iii), we have that $d_\infty(\overline{x}, \overline{y}) \le d_e(\overline{x}, \overline{y}) \le d_1(\overline{x}, \overline{y}) \le 2d_\infty(\overline{x}, \overline{y})$. Hence, by the lemma, the topology generated by each of these metrics is contained in the topology generated by each of the others. ∎

Our next theorem is often useful, and it basically says that only the balls of small radius matter in determining which sets are open.

Remark 2.18 *When we say a metric ρ is* bounded, *we mean that $\exists M > 0$ such that $\rho(x, y) \le M \ \forall x, y \in X$.*

Theorem 2.19 *If (X, d) is a metric space, then there is a bounded metric on X which is equivalent to d.*

Proof. Left to the student. Hint: consider $\min\{1, d(x, y)\}$. ∎

We will now develop the idea of a closed subset of a metric space. This generates an equivalent structure to the one generated by the idea of an open set, but it has its origins in the idea of a convergent sequence. Thus it may feel a little more comfortable to people with stronger backgrounds in analysis.

Definition 2.20 *Suppose X is a metric space and $A \subset X$. A point $x \in X$ is called a* **cluster point** *of A if and only if for every $\varepsilon > 0$, $B(x, \varepsilon)$ contains at least one point of A distinct from x, i.e. $(B(x, \varepsilon) \cap A) \backslash \{x\} \ne \varnothing$. The set of all cluster points of A is called the* **derived set** *of A and is denoted A'. The* **closure** *of A is the set $\overline{A} = A \cup A'$.*

Cluster points are called *limit points* in many texts, but the phrase "limit point" is rather overused. So we will reserve this for the situation where we are working with sequences (or later on, nets and filters), and we will use cluster point for the more general situation. There are also other notations for closure of a set, and we will use them from time to time. Specifically, the other notations for $\overline{A}$ that will be handy once in a while are clA and $cl(A)$. (Some authors also use $cls(A)$.) When we have more than one space containing the set A, we will use $cl_X(A)$ to denote the closure of A in the space X.

Definition 2.21 *A subset A of a metric space X is called* **closed** *if and only if* $A = \overline{A}$.

Lemma 2.22 *If X is a metric space and $A \subset X$, then A is closed iff $A' \subset A$.*

Proof. Left to the reader. ∎

The next result shows that the notion of a closed set is really dual to the notion of an open set.

Theorem 2.23 *If X is a metric space and $A \subset X$, then A is closed if and only if $X \backslash A$ is open.*

Proof. Suppose that A is closed and $x \in X \backslash A$. Since $x \notin A'$ and $x \notin A$, there exists $\varepsilon > 0$ such that $B(x, \varepsilon) \cap A = \varnothing$. Thus $B(x, \varepsilon) \subset X \backslash A$. Hence $X \backslash A$ is open.
Conversely, suppose that $X \backslash A$ is open. We need only show that $A' \subset A$. Suppose $x \in A'$. If $x \notin A$, then $x \in X \backslash A$, but $X \backslash A$ is open. So, there must exist $\varepsilon > 0$ such that $B(x, \varepsilon) \subset X \backslash A$. Consequently, $B(x, \varepsilon) \cap A = \varnothing$, and so $x \notin A'$, which is a contradiction. Hence we must have $x \in A$. Therefore $A' \subset A$, and thus $\overline{A} = A$. ∎

The duality between open sets and closed sets is made a bit more clear through the next theorem. The reader should compare this with theorem 2.13.

Theorem 2.24 *Suppose X is a metric space. The following subsets of X are closed:*

1. *the empty set $\varnothing$.*

2. *the entire space X.*

3. *$\{y : d(x, y) \leq \varepsilon\}$ (often called the **closed ball** centered at x), for any $x \in X$ and any $\varepsilon > 0$.*

4. $\overline{A}$, for any $A \subset X$.

5. $F \cup G$, for any closed subsets $F \subset X$ and $G \subset X$.

6. $\bigcap_{\alpha \in I} F_\alpha$, for any indexed family $\{F_\alpha : \alpha \in I\}$ of closed subsets of X.

Proof. To see 1, simply note that $\varnothing = X \backslash X$. To see 2, note that $X = X \backslash \varnothing$. Let us prove 3. Suppose $x \in X$ and $\varepsilon > 0$. Let $B = \{y : d(x,y) \le \varepsilon\}$. Suppose $z \in X \backslash B$. Since $d(x,z) > \varepsilon$, we let $\eta = d(x,z) - \varepsilon$ and note that $\eta > 0$. We claim that $B(z,\eta) \subset X \backslash B$. Suppose $y \in B(z,\eta)$, and note that $d(x,z) \le d(x,y) + d(y,z)$. Hence we have $d(x,y) \ge d(x,z) - d(y,z) > d(x,z) - \eta = \varepsilon$. Thus $y \in X \backslash B$. So our claim is justified, and $X \backslash B$ is open.

To prove 4, we will again show that the complement is open. Suppose $x \in X \backslash \overline{A}$. We know then that $x \notin A$ and $x \notin A'$. Hence there exists $\varepsilon > 0$ such that $B(x,\varepsilon) \cap A = \varnothing$. We will show that we actually have $B(x,\varepsilon) \cap \overline{A} = \varnothing$. Suppose $y \in B(x,\varepsilon)$. Clearly $y \notin A$. Also, since $B(x,\varepsilon)$ is open, there is $\eta > 0$ such that $B(y,\eta) \subset B(x,\varepsilon)$, but then $B(y,\eta) \cap A = \varnothing$. Hence $y \notin A'$. Thus $y \notin A \cup A' = \overline{A}$, that is $B(x,\varepsilon) \subset X \backslash \overline{A}$. Hence we have that $X \backslash \overline{A}$ is open, i.e. $\overline{A}$ is closed.

We now prove 5. Suppose H and K are closed subsets of X. Then $X \backslash H$ and $X \backslash K$ are open. Hence $X \backslash (H \cup K) = (X \backslash H) \cap (X \backslash K)$ is open. Hence $H \cup K$ is closed.

To prove 6, we will again use DeMorgan's laws. Suppose $\{F_\alpha : \alpha \in I\}$ is an indexed family of closed subsets of X. We have then that $\{X \backslash F_\alpha : \alpha \in I\}$ is an indexed family of open subsets of X. Hence $X \backslash (\bigcap_{\alpha \in I} F_\alpha) = \bigcup_{\alpha \in I} (X \backslash F_\alpha)$ is open, i.e., $\bigcap_{\alpha \in I} F_\alpha$ is closed. ∎

We will now give three more characterizations of the closure of a set in a metric space. The first is often given as the definition. The second is the one that most easily extends to nonmetric spaces, and the third shows the historical link with analysis.

Theorem 2.25 *If X is a metric space, and $A \subset X$, then a point $x \in \overline{A}$ if and only if whenever $U \subset X$ is open and $x \in U$, we have $U \cap A \neq \varnothing$.*

Proof. Suppose $x \in \overline{A}$ and U is open with $x \in U$. If $x \in A$, then $U \cap A \neq \varnothing$. If $x \in A'$, then we choose $\varepsilon > 0$ such that $B(x,\varepsilon) \subset U$. Since x is a cluster point of A, $B(x,\varepsilon) \cap A \neq \varnothing$, and thus $U \cap A \neq \varnothing$.

Conversely, suppose that whenever $U \subset X$ is open and $x \in U$, we have $U \cap A \neq \varnothing$. If $x \in A$, then we are done since $A \subset \overline{A}$. If $x \notin A$, then for each $\varepsilon > 0$ the open set $B(x,\varepsilon)$ must intersect A, but $x \notin A$. So $(B(x,\varepsilon) \cap A) \backslash \{x\} \neq \varnothing$. Hence $x \in A'$. So in either case, $x \in A \cup A' = \overline{A}$. ∎

Lemma 2.26 *If X is a metric space and A and B are subsets of X with $A \subset B$, then $\overline{A} \subset \overline{B}$.*

Proof. Left to the reader. ∎

Theorem 2.27 *If X is a metric space, and $A \subset X$, then*

$$\overline{A} = \bigcap \{F : A \subset F \text{ and } F \text{ is closed subset of } X\}.$$

Proof. Since $\overline{A}$ is closed and $A \subset \overline{A}$, we clearly have that

$$\overline{A} \supset \bigcap \{F : A \subset F \text{ and } F \text{ is closed subset of } X\}.$$

To see the other inclusion, suppose

$$x \notin \bigcap \{F : A \subset F \text{ and } F \text{ is closed subset of } X\}.$$

Choose a closed set F with $A \subset F$ and $x \notin F$. Since $X \backslash F$ is open, $x \in X \backslash F$, and $A \cap (X \backslash F) = \varnothing$, by theorem 2.25, $x \notin \overline{A}$. Hence

$$\overline{A} \subset \bigcap \{F : A \subset F \text{ and } F \text{ is closed subset of } X\},$$

and the result is proved. ∎

Corollary 2.28 *If X is a metric space, $A \subset X$, and F is a closed subset of X with $A \subset F$, then $\overline{A} \subset F$.*

Definition 2.29 *Suppose (X, d) is a metric space and $(x_n : n \in \mathbb{N})$ is a sequence in X. We say the sequence $(x_n : n \in \mathbb{N})$ **converges** to a point $x \in X$ if and only if for every $\varepsilon > 0$ there exists $N \in \mathbb{N}$ such that if $n \geq N$, then $d(x, x_n) < \varepsilon$ (or equivalently, if $n \geq N$, then $x_n \in B(x, \varepsilon)$).*

We denote this by $x_n \to x$, by $\lim_{n \to \infty} x_n = x$, or by $\lim x_n = x$, and we call x the *limit point* of the sequence $(x_n : n \in \mathbb{N})$. The use of the phrase "the limit point" would seem to imply that there can be only one limit point. We will see, in theorem 2.32, that that is indeed the case.

From time to time, we will need to express the negation of this idea. The symbol $x_n \nrightarrow x$ means that the sequence $(x_n : n \in \mathbb{N})$ *does not converge to x.*

Theorem 2.30 *If X is a metric space, and $A \subset X$, then a point $x \in \overline{A}$ if and only if there is a sequence $(x_n : n \in \mathbb{N})$ in A with $x_n \to x$.*

Proof. First, suppose there is a sequence $(x_n : n \in \mathbb{N})$ in A with $x_n \to x$, and let U be an open set with $x \in U$. Choose $\varepsilon > 0$ such that $B(x, \varepsilon) \subset U$ and $N \in \mathbb{N}$ such that $n \geq N \implies d(x, x_n) < \varepsilon$. Now $x_N \in B(x, \varepsilon) \cap A$, so $U \cap A \neq \varnothing$. Since this is true for any open set, $x \in \overline{A}$.

Now suppose that $x \in \overline{A}$. For each $n \in \mathbb{N}$, the ball $B(x, \frac{1}{n})$ must intersect A. So for each $n \in \mathbb{N}$ we choose $x_n \in A \cap B(x, \frac{1}{n})$. So $(x_n : n \in \mathbb{N})$ is a sequence in A. For any $\varepsilon > 0$, we can choose $N \in \mathbb{N}$ such that $\frac{1}{N} < \varepsilon$. Now if $n \geq N$, then $d(x, x_n) < \frac{1}{n} \leq \frac{1}{N} < \varepsilon$. Hence $x_n \to x$, as desired. ∎

The definition of the limit of a sequence that we have given is drawn directly from calculus. We now extend it slightly using the idea of an open set.

Theorem 2.31 *If (X, d) is a metric space and $(x_n : n \in \mathbb{N})$ is a sequence in X, then the sequence $(x_n : n \in \mathbb{N})$ converges to a point $x \in X$ if and only if for every open set $U \subset X$ with $x \in U$ there exists $N \in \mathbb{N}$ such that if $n \geq N$, then $x_n \in U$.*

Proof. Suppose $x_n \to x$ and U is open with $x \in U$. First choose $\varepsilon > 0$ with $B(x, \varepsilon) \subset U$. Now choose $N \in \mathbb{N}$ such that $n \geq N \implies d(x, x_n) < \varepsilon$. This N works, that is, if $n \geq N$, then $x_n \in U$.

To see the converse, suppose that for every open set $U \subset X$ with $x \in U$ there exists $N \in \mathbb{N}$ such that if $n \geq N$, then $x_n \in U$. Let $\varepsilon > 0$ be given. Since $B(x, \varepsilon)$ is open, choose $N \in \mathbb{N}$ such that if $n \geq N$, then $x_n \in B(x, \varepsilon)$. But then if $n \geq N$, we have $d(x, x_n) < \varepsilon$, i.e., $x_n \to x$. ∎

Our next theorem is that a sequence in a metric space can have only one limit. In particular, it makes sense to talk about *the* limit of a convergent sequence.

Theorem 2.32 *(Uniqueness of limits) Suppose (X, d) is a metric space and $(x_n : n \in \mathbb{N})$ is a sequence in X. If $x_n \to x$ and $x_n \to y$, then $x = y$.*

Proof. Suppose $x \neq y$. Let $\varepsilon = \frac{1}{2} d(x, y)$ and note that $\varepsilon > 0$. By convergence, we choose N_1 and N_2 in $\mathbb{N}$ such that if $n \geq N_1$, then $d(x, x_n) < \varepsilon$, and if $n \geq N_2$, then $d(y, x_n) < \varepsilon$. Now take $m = \max\{N_1, N_2\}$, and we have $d(x, y) \leq d(x, x_m) + d(x_m, y) < \varepsilon + \varepsilon = d(x, y)$. This is a contradiction, so we must have $x = y$. ∎

The completeness of the real numbers is of fundamental importance in analysis. We would like to extend the concept of completeness to metric spaces. Of course, the real line is a very special set, and completeness can be described in several ways. The most common is probably that every bounded set of real numbers has a least upper bound. This depends on the fact that the real line has an order on

it. In general, we will not have an order present in metric spaces. Another way
to describe completeness of the reals is to say that every Cauchy sequence of real
numbers converges. This depends only on the distance function, and it is the
concept we will now extend to metric spaces.

Definition 2.33 *A sequence* $(x_n : n \in \omega)$ *in a metric space* (X, d) *is called a*
Cauchy sequence *if and only if for every* $\varepsilon > 0$ *there exists* $N \in \omega$ *such that if*
$n \geq N$ *and* $m \geq N$, *then* $d(x_n, x_m) < \varepsilon$.

Theorem 2.34 *If a sequence* $(x_n : n \in \omega)$ *in a metric space* (X, d) *is convergent,*
then it is Cauchy.

Proof. Suppose $x_n \to x$ in the metric space (X, d), and suppose $\varepsilon > 0$ is given.
By convergence, we choose $N \in \omega$ such that if $n \geq N$, then $d(x, x_n) < \frac{\varepsilon}{2}$. Now if
$n \geq N$ and $m \geq N$, then we have $d(x_n, x_m) \leq d(x_n, x) + d(x, x_m) < \frac{\varepsilon}{2} + \frac{\varepsilon}{2} = \varepsilon$.
Hence $(x_n : n \in \omega)$ is a Cauchy sequence. $\blacksquare$

It is the converse to this result, or rather the question of for which spaces the
converse is true, which makes the concept of Cauchy sequences interesting.

Example 2.35 *Let* $X = (0, 1)$ *and* $d(x, y) = |x - y|$. *The sequence* $(\frac{1}{n} : n \in \mathbb{N})$
is a Cauchy sequence in this space, but it does not converge in this space since
$0 \notin (0, 1)$, *and each point of* $(0, 1)$ *has a neighborhood containing at most one*
point of $(\frac{1}{n} : n \in \mathbb{N})$.

Example 2.36 *Let* $X = \mathbb{Q}$, *the set of rational numbers, and* $d(x, y) = |x - y|$.
Choose a sequence $(x_n : n \in \omega)$ *of rationals which converges to* $\sqrt{2}$, *or whatever*
irrational you like. The sequence $(x_n : n \in \omega)$ *will be a Cauchy sequence in* X
which does not converge in the space X.

Intuitively, what goes wrong in these two examples is that a limit point, which
should be present in the space, is not there, i.e., the space is *incomplete*.

Definition 2.37 *A metric space* (X, d) *is said to be* ***complete*** *if and only if every*
Cauchy sequence in (X, d) *converges in* (X, d).

Example 2.38 *We list several examples. Assuming the completeness of the reals,*
it is relatively easy to check that they have the properties claimed. We assume that
all have the Euclidean metric.

    1. $\mathbb{R}$ *is complete.*

2. $\mathbb{R}^n$ *is complete for any natural number n.*

3. $(0, 1)$ *is not complete.*

4. $[0, 1]$ *is complete.*

5. $\mathbb{Q}$ *is not complete.*

6. $\mathbb{R} \backslash \mathbb{Q}$ *is not complete.*

If you tried to provide proofs for the examples listed above, then you have probably already discovered the next result.

Theorem 2.39 *Suppose (X, d) is a complete metric space, $Y \subset X$, and d_Y is the restriction of d to $Y \times Y$. The space (Y, d_Y) is a complete metric space if and only if Y is a closed subset of X.*

Proof. First, suppose Y is closed in X and $(x_n : n \in \omega)$ is a Cauchy sequence in (Y, d_Y). Since d_Y is the restriction of d, the sequence $(x_n : n \in \omega)$ is also a Cauchy sequence in (X, d). Now X is complete, so there is a point $x \in X$ such that $x_n \to x$. Hence $x \in \overline{Y}$. Since Y is closed, it follows that $x \in Y$. Hence $(x_n : n \in \omega)$ has a limit in Y. Thus Y is complete.

To see the converse, we suppose that (Y, d_Y) is complete. We wish to show that Y is closed. Suppose $x \in \overline{Y}$. We can choose a sequence in Y which converges to x, say $x_n \to x$ with $x_n \in Y$ $\forall n$. Since $x_n \to x$, the sequence $(x_n : n \in \omega)$ is a Cauchy sequence in X, it is also a Cauchy sequence in (Y, d_Y). Now, since (Y, d_Y) is complete, there is a point $y \in Y$ such that $x_n \to y$. This means that y is also a point of X, so, by uniqueness of limits, $x = y$. Hence $x \in Y$, and we have shown that Y is closed. ∎

There is a stronger result about subspaces that is true, although much more difficult to prove. If we look at $(-\frac{\pi}{2}, \frac{\pi}{2})$, then we readily see that this space is not complete when given the Euclidean metric. However, if we define a new metric by $d(x, y) = |\tan x - \tan y|$, then it is not very hard to show that this metric is equivalent to the Euclidean metric, and, with this metric, $(-\frac{\pi}{2}, \frac{\pi}{2})$ is complete. In a somewhat similar way, one can show that any open subset of a complete metric space has an equivalent metric with respect to which the open subset is complete. Even more is true, and we state the following theorem without proof, leaving the proof for a more advanced course in general topology.

Theorem 2.40 *If (X, d) is a complete metric space, $A \subset X$, and $A = \bigcap_{n \in \omega} G_n$ where each G_n is open in X, then there is a metric ρ on A, which is equivalent to the restriction of d to A, such that (A, ρ) is a complete metric space.*

Corollary 2.41 *The space of irrationals* $\mathbb{R}\backslash\mathbb{Q}$ *has a complete metric which generates the usual topology.*

Before stating the next theorem, we need the idea of the diameter of a set.

Definition 2.42 *Suppose* (X, d) *is a metric space and* $A \subset X$*. If* A *is nonempty, the **diameter** of* A *is given by* $diam A = \sup \{d(x, y) : x, y \in A\}$*.*

Note that it is entirely possible that $diam A = +\infty$. From the definition of sup, it would follow that $diam \varnothing = -\infty$, however we *define* $diam \varnothing = 0$.

Theorem 2.43 *(Cantor intersection theorem) If* (X, d) *is a complete metric space and* $(F_n : n \in \omega)$ *is a decreasing sequence of nonempty closed subsets of* X *such that* $diam F_n \to 0$*, then* $\bigcap_{n \in \omega} F_n$ *contains exactly one point.*

Proof. Let $F = \bigcap_{n \in \omega} F_n$. We first show that F cannot contain two distinct points. If $x \neq y$ and $\{x, y\} \subset F$, then $d(x, y) > 0$ and $\{x, y\} \subset F_n$ for all n. Since $diam F_n \to 0$, choose $k \in \omega$ such that $diam F_k < d(x, y)$. Now, since $\{x, y\} \subset F_k$, this contradicts the definition of diameter. So F is at most one point.

We will be finished once we have shown that $F \neq \varnothing$. For each $n \in \omega$, choose $x_n \in F_n$, and consider the sequence $(x_n : n \in \omega)$. Suppose $\varepsilon > 0$ is given. Since $diam F_n \to 0$, we choose $N \in \omega$ such that if $n \geq N$, then $diam F_n < \varepsilon$. If $n \geq N$ and $m \geq N$, then $x_n \in F_n \subset F_N$ and $x_m \in F_m \subset F_N$ since $(F_n : n \in \omega)$ is decreasing. Hence $d(x_n, x_m) \leq diam F_N < \varepsilon$, and thus we have that $(x_n : n \in \omega)$ is a Cauchy sequence. Since X is complete, $x_n \to x$ for some $x \in X$. Now for each n, $(x_k : k \geq n)$ is a sequence in F_n which converges to x. Thus $x \in \overline{F_n}$, but F_n is closed. So $x \in F_n$ for every n. Hence $x \in F$. ∎

Remark 2.44 *The following examples show that the hypotheses of the Cantor intersection theorem are all needed.*

1. *If* $F_n = [n, \infty)$*, then* $(F_n : n \in \omega)$ *is a decreasing sequence of closed subsets of* $\mathbb{R}$*, but* $\bigcap_{n \in \omega} [n, \infty) = \varnothing$*. So the condition that* $diam F_n \to 0$ *is needed.*

2. *If* $F_n = (0, \frac{1}{2^n})$*, then* $(F_n : n \in \omega)$ *is a decreasing sequence of subsets of* $\mathbb{R}$ *and* $diam F_n \to 0$*, but* $\bigcap_{n \in \omega} (0, \frac{1}{2^n}) = \varnothing$*. So the condition that the sets be closed is needed.*

3. *The condition that* $(F_n : n \in \omega)$ *is a decreasing sequence is not really essential provided that we have* $\bigcap_{k=1}^{n} F_k \neq \varnothing$ *for each* n*. If we let* $H_n = \bigcap_{k=1}^{n} F_k$ *for each* n*, and* $(F_n : n \in \omega)$ *has this property and satisfies all the hypotheses except "decreasing," then* $(H_n : n \in \omega)$ *satisfies all the hypotheses, and* $\bigcap_{n \in \omega} F_n = \bigcap_{n \in \omega} H_n$*.*

We close out this chapter with the Baire category theorem. This is a result that has many applications in topology, as well as both pure and applied analysis. It is typically used to show the existence of some special type of object, like solutions to a class of differential equations, or an everywhere continuous, nowhere differentiable function on the reals. It is also a motivating idea for some fancy set theory, like Martin's axiom. Unfortunately, most of those applications are a little beyond the material that we are covering, so we will see only a few in this book. Nevertheless, we are in a position to state and prove this theorem, and we do so now.

Definition 2.45 *Suppose X is a metric space. A subset $D \subset X$ is called **dense** provided $\overline{D} = X$. A subset $F \subset X$ is called **nowhere dense** provided $Int\overline{F} = \varnothing$.*

Remark 2.46 *It is easy to show that a set $D \subset X$ is dense if and only if $U \cap D \neq \varnothing$ whenever U is a nonempty open subset of X.*

Proof. Left to the reader. ∎

Notice that for a set to be nowhere dense its closure can contain no nonempty open sets; so, the set is not dense in any nonempty open set. In particular, it is not dense in any ball centered at any point.

Example 2.47 *The following are easy to verify.*

1. *$\mathbb{Z}$ is nowhere dense in $\mathbb{R}$.*

2. *$\mathbb{Q}$ and $\mathbb{R} \backslash \mathbb{Q}$ are both dense in $\mathbb{R}$.*

3. *$[0, 1]$ is neither dense nor nowhere dense in $\mathbb{R}$.*

Definition 2.48 *Suppose X is a metric space. A subset $A \subset X$ is called **first category** (also called meager or category I) provided $A = \bigcup_{n \in \omega} F_n$ for some sequence $(F_n : n \in \omega)$ nowhere dense sets in X. Any subset which is NOT first category is called **second category** (also called nonmeager or category II).*

Example 2.49 *The following are easy to verify.*

1. *A singleton set $\{x\}$ is nowhere dense unless $\{x\}$ is itself an open set.*

2. *$\mathbb{Q}$ is first category in $\mathbb{R}$.*

3. *Any nowhere dense set is first category.*

    4. The union of any countable family of first category sets is also first category.

Theorem 2.50 (*Baire category theorem*) *Every complete metric space is a second-category subset of itself.*

Proof. Suppose that X is a complete metric space. We need to show that X cannot be written as $\bigcup_{n \in \omega} F_n$ where each F_n is nowhere dense in X. Suppose to the contrary that we do have X written in this way. We will do an inductive construction to show that this is not possible. Since X is open, $\overline{F_0} \neq X$. So choose $x_0 \in X \backslash \overline{F_0}$ and $r_0 > 0$ such that $r_0 < 1$ and $B(x_0, r_0) \cap F_0 = \varnothing$. Let $H_0 = \{y : d(x_0, y) \leq \frac{r_0}{2}\}$. Note that H_0 is closed, $diam H_o < 1$, $Int H_0 \supset B(x_0, \frac{r_0}{2})$, and $H_0 \cap F_0 = \varnothing$. Now since $Int \overline{F_1} = \varnothing$, choose $x_1 \in B(x_0, \frac{r_0}{2}) \backslash \overline{F_1}$. Also choose $r_1 > 0$ such that $r_1 < \frac{1}{2}$, $B(x_1, r_1) \cap F_1 = \varnothing$, and $B(x_1, r_1) \subset B(x_0, \frac{r_0}{2})$. Let $H_1 = \{y : d(x_1, y) \leq \frac{r_1}{2}\}$. Note that H_1 is closed, $H_1 \subset H_0$, $diam H_1 < \frac{1}{2}$, $Int H_1 \supset B(x_1, \frac{r_1}{2})$, and $H_1 \cap (F_0 \cup F_1) = \varnothing$. We continue inductively, for the $n + 1$ step, choose $x_{n+1} \in B(x_n, \frac{r_n}{2}) \backslash \overline{F_{n+1}}$, $r_{n+1} > 0$ such that $r_{n+1} < \frac{1}{2^{n+1}}$, $B(x_{n+1}, r_{n+1}) \cap F_{n+1} = \varnothing$, and $B(x_{n+1}, r_{n+1}) \subset B(x_n, \frac{r_n}{2})$. Let $H_{n+1} = \{y : d(x_{n+1}, y) \leq \frac{r_{n+1}}{2}\}$. Note that H_{n+1} is closed, $H_{n+1} \subset H_n$, $diam H_{n+1} < \frac{1}{2^{n+1}}$, $Int H_{n+1} \supset B(x_{n+1}, \frac{r_{n+1}}{2})$, and $H_1 \cap (F_0 \cup F_1 \cup \cdots \cup F_{n+1}) = \varnothing$. By induction we construct the sequence $(H_n : n \in \omega)$, and note that H_n is a decreasing sequence of closed sets and $diam H_n < \frac{1}{2^n}$ for each n. Hence $diam H_n \to 0$, and by the Cantor intersection theorem, there is a point $x \in \bigcap_{n \in \omega} H_n$. For each n, we have $x \notin F_n$, and thus $x \in X \backslash \bigcup_{n \in \omega} F_n$, a contradiction. ∎

Corollary 2.51 *If $(U_n : n \in \omega)$ is a sequence of open dense subsets of a complete metric space X, then $\bigcap_{n \in \omega} U_n \neq \varnothing$.*

Proof. If U_n is open and dense, the $X \backslash U_n$ is closed and nowhere dense. Moreover, $X \backslash (\bigcap_{n \in \omega} U_n) = \bigcup_{n \in \omega} (X \backslash U_n)$. Thus if $\bigcap_{n \in \omega} U_n = \varnothing$, then $\bigcup_{n \in \omega} (X \backslash U_n) = X$, contradicting the Baire category theorem. ∎

In fact, using DeMorgan's law, as above, it is easily seen that the corollary is an equivalent statement to the Baire category theorem. With just a little more work, we can show that in the corollary, the set $\bigcap_{n \in \omega} U_n$ is actually dense in X. This is often how the Baire category theorem is stated.

Exercises

    1. Prove that the discrete metric (see definition 2.7) on any set satisfies the definition of a metric.

2. Prove that d_2 as defined in example 2.4 is a metric.

3. Prove that d_3 as defined in example 2.5 is a metric.

4. Prove that d_e, d_1, and d_∞ (see example 2.9) are metrics.

5. Prove theorem 2.19.

6. Prove lemma 2.22.

7. Prove that a subset U of a metric space (X, d) is open if and only if U can be written as the union of some collection of open balls in X.

8. Prove that if X is a metric space and $x \in X$, then $\{x\}$ is a closed set in X. Deduce then that every finite subset of a metric space is closed.

9. Suppose X is a metric space and $A \subset X$. Prove that $\overline{A} = X \backslash Int(X \backslash A)$ and $Int A = X \backslash (\overline{X \backslash A})$.

10. For a metric space (X, d), $A \subset X$, and $x \in X$, define the distance from x to A by $d(x, A) = \inf \{d(x, y) : y \in A\}$. Prove that $\overline{A} = \{x : d(x, A) = 0\}$.

11. Find an example of a family $\{U_n : n \in \omega\}$ of open subsets of $\mathbb{R}$ such that $\bigcap_{n \in \omega} U_n$ is not open.

12. Suppose that X is a metric space, and A and B are subsets of X.

    (a) Prove that $\overline{A \cup B} = \overline{A} \cup \overline{B}$.
    (b) Prove that $\overline{A \cap B} \subset \overline{A} \cap \overline{B}$.
    (c) Find an example to show that we do not generally get equality in part b.

13. For a metric space (X, d), and $A \subset X$, we define the boundary of A to be the set $Bdry(A) = \overline{A} \cap \overline{X \backslash A}$.

    (a) Prove that $Bdry(A)$ is a closed set.
    (b) Prove that A is closed if and only if $Bdry(A) \subset A$.
    (c) Prove that $Int(A) \cup Bdry(A) = \overline{A}$.

14. Describe the boundaries of the following subsets of $\mathbb{R}$.

    (a) $\mathbb{Q}$.
    (b) $\mathbb{R} \backslash \mathbb{Q}$.

(c) $[0, 1]$.

(d) $(0, 1)$.

(e) $\mathbb{Z}$.

15. Suppose $(n_k : k \in \omega)$ is a strictly increasing sequence of natural numbers and $(x_n : n \in \omega)$ is a sequence in the metric space X. We say the sequence given by $(x_{n_k} : k \in \omega)$ is a *subsequence* of the sequence $(x_n : n \in \omega)$.

    Suppose that $(x_n : n \in \omega)$ is a Cauchy sequence in the metric space (X, d) which has a convergent subsequence. Prove that the sequence $(x_n : n \in \omega)$ converges.

16. Show that a set $D \subset X$ is dense in the metric space X iff $U \cap D \neq \varnothing$ whenever U is a nonempty open subset of X.

17. Suppose that U is an open dense subset of the metric space X. Prove that $X \backslash U$ is nowhere dense.

18. Prove that the set $\mathbb{Q}$ of rational numbers cannot be written as $\mathbb{Q} = \bigcap_{n \in \omega} U_n$ where $(U_n : n \in \omega)$ is a sequence of open subsets of $\mathbb{R}$. Hint: Use the Baire category theorem.

19. Prove that the set $\mathbb{Z}$ of integers with the restriction of the usual metric on $\mathbb{R}$ is a complete metric space, and yet it is the union of countably many singletons. Why doesn't this contradict the Baire category theorem?

Chapter 3

Continuity

In this chapter, we explore continuous functions between metric spaces. Continuity for metric spaces is extracted directly from the ε-δ definition given in first-semester calculus. We will prove the characterizations which will motivate the definitions that will be given for continuity in general spaces, as well as some classical results with an analysis flavor.

Definition 3.1 *Suppose (X, d) and (Y, ρ) are metric spaces and $f : X \to Y$. We say that f is **continuous at the point** $c \in X$ if and only if for every $\varepsilon > 0$ there exists $\delta > 0$ such that if $d(x, c) < \delta$, then $\rho(f(x), f(c)) < \varepsilon$.*

The definition is easily seen to be equivalent to the following characterization in terms of open balls.

Theorem 3.2 *If (X, d) and (Y, ρ) are metric spaces and $f : X \to Y$, then f is continuous at the point $c \in X$ if and only if for every $\varepsilon > 0$ there exists $\delta > 0$ such that $f(B_d(c, \delta)) \subset B_\rho(f(c), \varepsilon)$.*

Definition 3.3 *We say f is a **continuous function** from X into Y if f is continuous at every point of the space X.*

In analysis, as well as other applications, we often like to discuss continuity in terms of convergent sequences.

Theorem 3.4 *If (X, d) and (Y, ρ) are metric spaces and $f : X \to Y$, and $c \in X$, then f is continuous at c if and only if whenever $(x_n : n \in \omega)$ is a sequence in X with $x_n \to c$, we have $f(x_n) \to f(c)$.*

33

Proof. Suppose that f is continuous at c and that $(x_n : n \in \omega)$ is a sequence in X with $x_n \to c$. We wish to show that $(f(x_n) : n \in \omega)$ converges to $f(c)$ in the space Y. Suppose $\varepsilon > 0$ is given. By continuity, choose $\delta > 0$ such that $d(x,c) < \delta \implies \rho(f(x), f(c)) < \varepsilon$. Since $x_n \to c$, we choose $N \in \omega$ such that $d(x_n, c) < \delta$ whenever $n \geq N$. Now suppose $n \geq N$. Since $d(x_n, c) < \delta$, $\rho(f(x_n), f(c)) < \varepsilon$ by the choice of δ. Hence we have that $\rho(f(x_n), f(c)) < \varepsilon$ whenever $n \geq N$. That is $f(x_n) \to f(c)$.

To prove the converse, suppose that f is not continuous at c. Choose some $\varepsilon > 0$ such that for every choice of δ, there exists at least one x with $d(x, c) < \delta$, but $\rho(f(x), f(c)) \geq \varepsilon$. Hence, for each $n \in \omega$ we choose x_n such that $d(x_n, c) < \frac{1}{2^n}$ but $\rho(f(x_n), f(c)) \geq \varepsilon$. Since $\frac{1}{2^n} \to 0$, we have that $x_n \to c$, but since $\rho(f(x_n), f(c)) \geq \varepsilon$ for every n, $f(x_n) \not\to f(c)$. ∎

This result is very useful in showing that particular functions are not continuous.

Example 3.5 *Define $f : \mathbb{R} \to \mathbb{R}$ by $f(x)$ where*

$$f(x) = \begin{cases} 1, \text{if} x \in \mathbb{Q}, \\ 0, \text{if} x \notin \mathbb{Q}. \end{cases}$$

We show that f is not continuous at any point.

Proof. If $c \in \mathbb{Q}$, then we may choose a sequence $(x_n : n \in \omega)$ of irrationals such that $x_n \to c$. Now $f(c) = 1$, but $f(x_n) = 0$ for every n, and clearly the sequence that is constantly 0 does not converge to 1. Similarly, if $c \in \mathbb{R} \backslash \mathbb{Q}$, then we can choose a sequence $(x_n : n \in \omega)$ of rationals such that $x_n \to c$. Now $f(c) = 0$, but $f(x_n) = 1$ for every n, and clearly the sequence that is constantly 1 does not converge to 0. ∎

A word of caution is in order. Theorem 3.4 is a theorem about metric spaces, and the corresponding result will not be true in all spaces once we get general spaces defined. However, the next result will generalize.

Theorem 3.6 *Suppose (X, d) and (Y, ρ) are metric spaces and $f : X \to Y$. The following are equivalent:*

1. f is a continuous function.

2. for any open set $U \subset Y$, $f^{-1}(U)$ is open in X.

3. for any closed set $H \subset Y$, $f^{-1}(H)$ is closed in X.

4. for any $A \subset X$, $f(\overline{A}) \subset \overline{f(A)}$.

Proof. We will establish the result by showing that $1 \implies 2 \implies 3 \implies 4 \implies 1$.

Suppose that 1 is true, i.e., that f is continuous on X, and suppose that U is an open subset of Y. Let $x \in f^{-1}(U)$. Since $f(x) \in U$, there exists $\varepsilon > 0$ such that $B(f(x), \varepsilon) \subset U$. Now using continuity, we can find $\delta > 0$ such that $d(x, y) < \delta \implies \rho(f(x), f(y)) < \varepsilon$. Thus if $y \in B(x, \delta)$, then $f(y) \in B(f(x), \varepsilon) \subset U$, and so $y \in f^{-1}(U)$. Hence $B(x, \delta) \subset f^{-1}(U)$. Thus $f^{-1}(U)$ is open. So we have proved that $1 \implies 2$

Now suppose 2 is true. Suppose that H is a closed subset of Y. So $Y \backslash H$ is an open subset of Y. By 2, $X \backslash f^{-1}(H) = f^{-1}(Y \backslash H)$ is open, so $f^{-1}(H)$ is closed. Hence $2 \implies 3$ is true.

Suppose that 3 is true, and let $A \subset X$. Since $\overline{f(A)}$ is closed and 3 is true, we know that $f^{-1}(\overline{f(A)})$ is closed in X. Now $A \subset f^{-1}(f(A)) \subset f^{-1}(\overline{f(A)})$, so $\overline{A} \subset f^{-1}(\overline{f(A)})$. Hence $f(\overline{A}) \subset f(f^{-1}(\overline{f(A)})) \subset \overline{f(A)}$. So $3 \implies 4$ is true.

Now suppose that 4 is true, and let $c \in X$. We want to show that f is continuous at c. Suppose $(x_n : n \in \omega)$ is a sequence in X such that $x_n \to c$. If $f(x_n) \not\to f(c)$, then we may choose $\varepsilon > 0$ such that for every $k \in \omega$ there is $n_k \geq k$ with $\rho(f(x_{n_k}), f(c)) \geq \varepsilon$. Consider $A = \{x_{n_k} : k \in \omega\}$. Since $x_n \to c$, we know $x_{n_k} \to c$ as well. So $c \in \overline{A}$. Since 4 is true, $f(c) \in \overline{f(A)}$, but $B(f(c), \varepsilon) \cap f(A) = \emptyset$. This is a contradiction, so the statement that $f(x_n) \not\to f(c)$ is impossible. Thus f is continuous at c. So $4 \implies 1$, and the proof is complete. ∎

A useful result in the theory of continuous functions, especially in counting the number of continuous functions on a particular space, is that continuous functions are determined on dense subsets.

Theorem 3.7 *If (X, d) and (Y, ρ) are metric spaces, $f : X \to Y$ and $g : X \to Y$ are continuous functions, and $\{x : f(x) = g(x)\}$ is dense in X, then $f = g$.*

Proof. Suppose $f \neq g$. We choose $c \in X$ such that $f(c) \neq g(c)$. Let $\varepsilon = \frac{1}{2}\rho(f(c), g(c))$. Note that $B(f(c), \varepsilon) \cap B(g(c), \varepsilon) = \emptyset$. Consider $U = f^{-1}(B(f(c), \varepsilon)) \cap g^{-1}(B(g(c), \varepsilon))$. Since $c \in U$, we see that $U \neq \emptyset$, and U is open by the continuity, thus we have $U \cap \{x : f(x) = g(x)\} \neq \emptyset$. Choose $x \in U$ with $f(x) = g(x)$. Now we have $f(x) \in f(U) \subset B(f(c), \varepsilon)$ and $f(x) = g(x) \in g(U) \subset B(g(c), \varepsilon)$. Hence $f(x) \in B(f(c), \varepsilon) \cap B(g(c), \varepsilon) = \emptyset$, a contradiction. ∎

We turn now to a stronger form of continuity, which is possessed by many familiar functions from analysis. Indeed, the fact that continuous functions on

closed bounded intervals satisfy this property is the key element of the proof that
those functions are Riemann integrable.

Definition 3.8 *Suppose (X, d) and (Y, ρ) are metric spaces and $f : X \to Y$.
We say that f is **uniformly continuous** provided that for every $\varepsilon > 0$ there
exists $\delta > 0$ such that for any points x_1 and x_2 in X, if $d(x_1, x_2) < \delta$, then
$\rho(f(x_1), f(x_2)) < \varepsilon$.*

The difference between this definition and definition 3.1 is that in definition 3.1
the choice of δ depends on both ε and c; here, the choice of δ depends only on ε.
That is, for a given ε, the same δ works at every point c.

Example 3.9 *The following are easy to verify.*

1. *Define $f : \mathbb{R} \to \mathbb{R}$ by $f(x) = 3x - 4$. For a given $\varepsilon > 0$, the choice of $\delta = \frac{\varepsilon}{3}$
   will satisfy the definition of continuity at c for any real number c. Thus f is
   uniformly continuous on $\mathbb{R}$.*

2. *Define $f : (0, \infty) \to \mathbb{R}$ by $f(x) = \frac{1}{x}$. The function f is continuous, but not
   uniformly continuous.*

Theorem 3.10 *Suppose (Y, ρ) is a complete metric space, (X, d) is a metric
space, and A is a dense subset of X. If $f : A \to Y$ is uniformly continuous,
then f has a unique continuous extension $\widehat{f} : X \to Y$, where $\widehat{f}$ is also uniformly
continuous.*

When we say $\widehat{f}$ is an **extension** of f, we mean that $\widehat{f}(x) = f(x)$ for all $x \in dom f$.

Proof. Since A is dense, the uniqueness follows from theorem 3.7. We prove
existence. We define $\widehat{f} : X \to Y$ as follows. Since A is dense in X, it is true that
each point of $X \backslash A$ must be the limit of some sequence of points of A. For each
$x \in X \backslash A$, we choose a particular fixed sequence $(a_n(x) : n \in \omega)$ in A such that
$a_n(x) \to x$. If $x \in A$, then we let $a_n(x) = x$ for all $n \in \omega$. We first show that for
any $x \in X$, the sequence $(f(a_n(x)) : n \in \omega)$ is a Cauchy sequence in Y. Suppose
$\varepsilon > 0$ is given. Choose a uniform $\delta > 0$ corresponding to ε as in the definition
of uniform continuity. Since $(a_n(x) : n \in \omega)$ is convergent, it is also Cauchy.
So choose $N \in \omega$ such that if $n \geq N$ and $m \geq N$, then $d(a_n(x), a_m(x)) < \delta$.
Hence we have that if $n \geq N$ and $m \geq N$, then $\rho(f(a_n(x)), f(a_m(x))) < \varepsilon$. Thus
$(f(a_n(x)) : n \in \omega)$ is a Cauchy sequence in Y. Since Y is complete, this sequence
must have a limit. We define $\widehat{f}(x) = \lim_{n \to \infty} f(a_n(x))$. It is clear that $\widehat{f}$ is an
extension of f.

We now show that $\widehat{f}$ is uniformly continuous. Suppose $\varepsilon > 0$ is given. By the uniform continuity of f on A, we choose $\eta > 0$ such that if $a, b \in A$ and $d(a,b) < \eta$, then $\rho(f(a), f(b)) < \frac{\varepsilon}{3}$. Let $\delta = \frac{\eta}{3}$. Note that $\delta > 0$. We show that this δ satisfies the definition. Suppose $x, y \in X$ and $d(x,y) < \delta$. Consider the sequences $(a_n(x) : n \in \omega)$ and $(a_n(y) : n \in \omega)$, which were the ones used to define $\widehat{f}(x)$ and $\widehat{f}(y)$ respectively, and note that if $x \in A$ we took $a_n(x) = x$ for all n. So we have $a_n(x) \to x$, $a_n(y) \to y$, $f(a_n(x)) \to \widehat{f}(x)$, and $f(a_n(y)) \to \widehat{f}(y)$. Choose $N_1 \in \omega$ such that

$$\text{if } n \geq N_1, \text{ then } \rho(f(a_n(x)), \widehat{f}(x)) < \frac{\varepsilon}{3}.$$

Choose $N_2 \in \omega$ such that

$$\text{if } n \geq N_2, \text{ then } \rho(f(a_n(y)), \widehat{f}(y)) < \frac{\varepsilon}{3}.$$

Choose $N_3 \in \omega$ such that

$$\text{if } n \geq N_3, \text{ then } d(a_n(x), x) < \frac{\eta}{3}.$$

Choose $N_4 \in \omega$ such that

$$\text{if } n \geq N_4, \text{ then } d(a_n(y), y) < \frac{\eta}{3}.$$

Note that if $n \geq \max\{N_3, N_4\}$, then

$$d(a_n(x), a_n(y)) \leq d(a_n(x), x) + d(x, y) + d(y, a_n(y)) < \frac{\eta}{3} + \frac{\eta}{3} + \frac{\eta}{3} = \eta.$$

Let $n = \max\{N_1, N_2, N_3, N_4\}$, and we have

$$\begin{aligned}
\rho(\widehat{f}(x), \widehat{f}(y)) &\leq \rho(\widehat{f}(x), f(a_n(x))) + \rho(f(a_n(x)), f(a_n(y))) + \rho(f(a_n(y)), \widehat{f}(y)) \\
&< \frac{\varepsilon}{3} + \frac{\varepsilon}{3} + \frac{\varepsilon}{3} = \varepsilon.
\end{aligned}$$

Hence $\widehat{f}$ is uniformly continuous, and the proof is complete. ∎

We note that in the proof that we have just completed the construction of $\widehat{f}$ seems to depend on the particular choice of the sequences $(a_n(x) : n \in \omega)$, but since the extension is actually unique, the function is independent of the choice of the sequences which were used to construct it. We also note that we used the axiom of choice in the construction above to choose, for each point $x \in X \backslash A$, *one* of the sequences in A which converge to x.

One type of uniformly continuous function that is especially useful in analysis is the *contractive mapping*, or *contraction*.

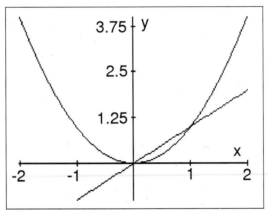

0 and 1 are fixed points of $f(x) = x^2$

Definition 3.11 *Suppose (X, d) and (Y, ρ) are metric spaces, and $f : X \to Y$. We say that f is a **contraction**, or **contractive mapping**, if and only if there exists $\alpha \in (0, 1)$ such that $\rho(f(x), f(y)) \leq \alpha d(x, y)$ for all $x, y \in X$.*

Notice that the mapping is moving points closer together. Certainly, such a mapping is uniformly continuous. However more is true. The next theorem has applications in analysis, applied mathematics, and computer science. This result is sometimes called the contractive mapping theorem, but it is more commonly called the Banach fixed-point Theorem. Before stating the theorem, we need one more definition.

Definition 3.12 *Suppose $f : X \to X$. We say a point $c \in X$ is a **fixed point** of f if $f(c) = c$.*

A fixed point is a point that is not "moved" by the function. These have arisen in other places in your mathematical training. They are the eigenvectors corresponding to the eigenvalue $\lambda = 1$ in linear algebra. They are the points where the graph of a function crosses the line $y = x$ in calculus.

Example 3.13 *Consider the function $f : \mathbb{R} \to \mathbb{R}$ defined by $f(x) = x^2$ (see the figure above). The points 0 and 1 are fixed points since $f(0) = 0$ and $f(1) = 1$. All other points are not fixed points.*

Theorem 3.14 *(Banach fixed-point theorem) If (X, d) is a complete metric space and $f : X \to X$ is a contraction, then f has a unique fixed point.*

Proof. Choose α with $0 < \alpha < 1$ such that $d(f(x), f(y)) \leq \alpha d(x, y)$ for all $x, y \in X$. Now choose a point $x_0 \in X$. We inductively define a sequence $(x_n : n \in \omega)$ by $x_{n+1} = f(x_n)$ for all $n \in \omega$. If $f(x_n) = x_n$ for some n, then we have our fixed point. So assume $x_{n+1} \neq x_n$ for all $n \in \omega$. We claim this sequence is Cauchy. Suppose $\varepsilon > 0$ is given. Since $\sum_{n=0}^{\infty} \alpha^n$ is a convergent geometric series, we choose $N \in \omega$ such that $\sum_{k=N}^{\infty} \alpha^k < \frac{\varepsilon}{d(x_0, x_1)}$. We show by induction that for $n \in \omega$, $d(x_n, x_{n+1}) \leq \alpha^n d(x_0, x_1)$. Certainly this is true when $n = 0$. Suppose the statement is true when $n = k$, that is $d(x_k, x_{k+1}) \leq \alpha^k d(x_0, x_1)$. Now we prove the $k + 1$ case. Since $d(x_{k+1}, x_{k+2}) = d(f(x_k), f(x_{k+1})) \leq \alpha d(x_k, x_{k+1}) \leq \alpha \cdot \alpha^k d(x_0, x_1) = \alpha^{k+1} d(x_0, x_1)$. Thus, by the principle of mathematical induction, we have shown that $d(x_n, x_{n+1}) \leq \alpha^n d(x_0, x_1)$ for all $n \in \omega$.

Now suppose $n \geq N$ and $m \geq N$, and without loss of generality $n < m$. We see that by triangle inequality

$$
\begin{aligned}
d(x_n, x_m) &\leq d(x_n, x_{n+1}) + d(x_{n+1}, x_{n+2}) + \cdots + d(x_{m-1}, x_m) \\
&= \sum_{k=n}^{m-1} d(x_k, x_{k+1}) \leq \sum_{k=N}^{\infty} d(x_k, x_{k+1}) \\
&\leq \sum_{k=N}^{\infty} \alpha^k d(x_0, x_1) = d(x_0, x_1) \sum_{k=N}^{\infty} \alpha^k < d(x_0, x_1) \frac{\varepsilon}{d(x_0, x_1)} = \varepsilon.
\end{aligned}
$$

Hence $(x_n : n \in \omega)$ is a Cauchy sequence.

Since X is complete, choose $c \in X$ such that $x_n \to c$. Since $(x_{n+1} : n \in \omega)$ is a subsequence of $(x_n : n \in \omega)$, we also have $x_{n+1} \to c$. Now by the continuity of f, $f(x_n) \to f(c)$ and $f(x_n) = x_{n+1} \to c$. Thus by uniqueness of limits, $f(c) = c$. We have produced a fixed point for f.

To see that there can be only one fixed point, note that if x and y are both fixed points, then $d(x, y) = d(f(x), f(y)) \leq \alpha d(x, y) < d(x, y)$ unless $d(x, y) = 0$. Hence the fixed point is unique. ∎

One of the satisfying things about this proof is that we have given a construction that produces the fixed point and can be carried out by computation. Moreover, since the difference between x_n and c at stage n in the construction is given by the sum of a geometric series, this gives a method for approximating the fixed point and an efficient method for computing the error.

The Banach fixed-point theorem is a key element in proving the existence and uniqueness of solutions to ordinary differential equations. It is also used in showing that certain recursively defined objects in computer science do not create infinite loops.

Exercises

1. Consider the metric space $\mathbb{Z}$, the integers with the usual metric, that is $d(n, m) = |n - m|$.

   (a) Describe the topology on $\mathbb{Z}$.

   (b) Suppose X is an arbitrary metric space. Which functions $f : \mathbb{Z} \to X$ are continuous?

2. Suppose D is any set with the discrete metric and X is an arbitrary metric space. Which functions $f : D \to X$ are continuous?

3. Suppose X, Y, and Z are metric spaces, and $f : X \to Y$ and $g : Y \to Z$ are continuous functions. Prove that $g \circ f : X \to Z$ is also continuous.

4. Suppose X and Y are metric spaces, and $f : X \to Y$ is a continuous *onto* function. Prove that if D is a dense subset of X, then $f(D)$ is a dense subset of Y. What happens if we do not require f to be onto?

5. Prove that if X and Y are metric spaces, and $f : X \to Y$ is a uniformly continuous function, then for any Cauchy sequence $(x_n : n \in \omega)$ in X, the sequence $(f(x_n) : n \in \omega)$ is Cauchy in Y.

6. Use the previous exercise to show that the function defined by $f(x) = \frac{1}{x}$ for $x \in (0, \infty)$ is not uniformly continuous. Also, prove that it *is* continuous.

7. Show that in the proof of the Banach fixed-point theorem, for each n , $d(x_n, c) \leq d(x_0, x_1) \sum_{k=n}^{\infty} \alpha^k$.

8. Find an example of a function $f : \mathbb{R} \to \mathbb{R}$ such that $|f(x) - f(y)| < |x - y|$ for any distinct points $x, y \in \mathbb{R}$, and yet $f(x) \neq x$ for every real number x Why doesn't this contradict the Banach fixed-point theorem?

9. Define $f : \mathbb{R} \to \mathbb{R}$ by $f(x) = \sqrt[3]{x}$. Is f uniformly continuous? Show that f is one-to-one. Consider $f^{-1} : \mathbb{R} \to \mathbb{R}$. Is f^{-1} uniformly continuous?

10. Suppose that $f : X \to Y$ is a continuous mapping of the space X onto the space Y, and X has the property that every sequence of points of X has a convergent subsequence. Prove that Y also has this property, that is, every sequence in Y has a convergent subsequence.

Chapter 4

Topological Spaces

In this chapter we introduce general topological spaces. We have already seen how to generate a topology on a metric space. Indeed, the motivation for the definition of a topology is theorem 2.13. Thus all metric spaces will be topological spaces. The question of when the converse holds, that is which topologies are generated by metrics, was a guiding force in the field for the first half of the twentieth century. We will return to this, the general metrization problem, in some depth in part two.

Counterexamples have played a major role in the development of topology, as well as all of mathematics. In set-theoretic topology, probably more than any other area of mathematics, natural conjectures have turned out to be false. As a result, it is extremely important for the student of topology to have a collection of *pathological* examples. In this chapter, we begin to help the reader build such a collection by introducing several of the important examples in set-theoretic topology.

Definition 4.1 *Suppose X is a set. A **topology** on X is a collection τ of subsets of X such that the following are true:*

1. *$\emptyset \in \tau$.*

2. *$X \in \tau$.*

3. *If $U \in \tau$ and $V \in \tau$, then $U \cap V \in \tau$.*

4. *If $\mathcal{U} \subset \tau$, then $\cup \mathcal{U} \in \tau$; that is, if $U_i \in \tau$ for all $i \in I$, then $\bigcup_{i \in I} U_i \in \tau$.*

From condition 3, we get by induction that any finite intersection of elements of τ is an element of τ. Condition 4 says that any union of elements of τ is again an element of τ. We often combine these statements by saying that τ is closed under finite intersections and arbitrary unions.

Definition 4.2 *If X is a set and τ is a topology on X, we call the pair (X, τ) a* **topological space** *. When the topology is clear from the context, we often refer to the topological space X.*

Definition 4.3 *The elements of τ are called the* **open** *subsets of the topological space X.*

Henceforth, if we use the word "space" we will always mean "topological space" unless some other modifier is attached.

Example 4.4 *We list several examples of topological spaces below. Some of these will be referred to by name in the future, so it is important to become familiar with these and to fill in any steps that are omitted.*

1. *If (X, d) is a metric space, then the topology generated by d is a topology (see theorem 2.13 and definition 2.14). We say a topological space (X, τ) is a* **metrizable** *space if and only if there is some metric on X that generates the topology τ.*

2. *If X is any set, $\tau = \{\varnothing, X\}$ is a topology on X. This is called the* **trivial topology** *, or sometimes the* **indiscrete topology**, *on X.*

3. *If X is any set, $\tau = \mathcal{P}(X) = \{A : A \subset X\}$, the power set of X, is a topology on X. This is called the* **discrete topology** *on X. When X has the discrete topology, we often say X is a* **discrete space**.

4. *Suppose $X = \{0, 1\}$ and $\tau = \{\varnothing, X, \{0\}\}$. It is easy to check that τ is a topology on X. This space is called the* **Sierpiński space**, *or, sometimes, the two-point connected space. (W. Sierpiński and K. Kuratowski were leaders of a Warsaw-based group that was one of the earliest and strongest set theory and topology groups in the world.)*

5. *Let $X = \mathbb{R}$, let*

$$\tau_M = \{U \cup F : U \text{ is open in the usual topology on } \mathbb{R} \text{ and } F \subset \mathbb{R} \backslash \mathbb{Q}\}.$$

This is a topology on $\mathbb{R}$. The space $(\mathbb{R}, \tau_M)$ is called the **Michael line**, *which we will denote by $\mathbb{M}$. (E. A. Michael, recently retired from the University of*

Washington, Seattle, is probably best known for his early work on paracompact spaces.)

6. *Let X be any set, and $\tau = \{A \subset X : A = \varnothing$ or $X\backslash A$ is finite$\}$. This is a topology on X which is called the* **cofinite topology** *on X.*

7. *Let $X = \mathbb{R}$. Let $\tau = \{A \subset \mathbb{R} : 0 \notin A\} \cup \{A \subset \mathbb{R} : 0 \in A$ and $\mathbb{R}\backslash A$ is finite$\}$. This is a topology on $\mathbb{R}$ that is obviously related to the cofinite topology, but we will see that it is quite different in some ways.*

8. *Here is another topology similar to the one in 7. Let $X = \mathbb{R}$, and let $\tau = \{A \subset \mathbb{R} : 0 \notin A\} \cup \{A \subset \mathbb{R} : 0 \in A$ and $\mathbb{R}\backslash A$ is countable$\}$. We will explore some of the differences between 7 and 8 in the exercises.*

We now define interior in general topological spaces. We leave it as an exercise to show that this is the same concept that we defined for metric spaces.

Definition 4.5 *Suppose X is a topological space, and $A \subset X$. The* **interior** *of A is the set given by $IntA = \cup \{U \subset X : U \subset A$ and U is open$\}$.*

We also denote the interior of A by $Int(A)$ or perhaps $Int_X A$ if there is the possibility of confusion about which space X we are discussing. Some books use $\overset{\circ}{A}$ or A° for $IntA$.

It is immediate from the definition that $IntA$ is the largest open set contained inside A.

Theorem 4.6 *Suppose X is a topological space, and $A \subset X$. For any open set $U \subset A$, we have that $U \subset IntA$.*

Definition 4.7 *Suppose X is a topological space, and $A \subset X$. We say A is a* **neighborhood** *of the point x if and only if $x \in IntA$.*

We sometimes abbreviate the word neighborhood by *nbd*.

A word of caution is probably in order here. This use of the word *neighborhood* is not universal, although it is convenient and common. Some books *require* neighborhoods to be open sets. We will not do this since we want to have the option of having neighborhoods with other special properties, such as compactness, which may preclude the set being open, even though the point we are considering will be in the interior.

Following the pattern established for metric spaces, we define a closed set to be the complement of an open set.

Definition 4.8 *Suppose X is a topological space. A subset $F \subset X$ is **closed** if and only if $X \backslash F$ is open.*

The next result is the analogue of theorem 2.24.

Theorem 4.9 *Suppose X is a topological space. The following subsets of X are closed.*

1. *the empty set $\varnothing$.*

2. *the entire space X.*

3. *$H \cup K$, for any closed subsets $H \subset X$ and $K \subset X$.*

4. *$\cap \mathcal{F} = \bigcap_{\alpha \in I} F_\alpha$, for any indexed family $\mathcal{F} = \{F_\alpha : \alpha \in I\}$ of closed subsets of X.*

5. *$\overline{A}$, for any $A \subset X$.*

Proof. For parts $1, 2, 3$, and 4, simply take complements and use parts $2, 1, 3$, and 4 of definition 4.1. For 5, just use 4. ∎

Definition 4.10 *Suppose X is a topological space, and $A \subset X$. A point $x \in X$ is a **cluster point** of A if and only if for every neighborhood U of x, U contains a point of A distinct from x, i.e., $(U \cap A) \backslash \{x\} \neq \varnothing$. The **derived set** of A is the set of all cluster points of A. It is denoted by A', just as it was in metric spaces.*

The comment made in the section on metric spaces is true here as well. Cluster points are also called limit points, accumulation points, and adherent points by various authors. We will again reserve the phase "limit point" to describe the limit of a convergent sequence or of a convergent net or filter (defined in chapter 16).

Definition 4.11 *If A is a subset of a space X, then the **closure** of A is the set $\overline{A} = \cap \{F \subset X : F$ is closed and $A \subset F\}$.*

Compare this definition with theorem 2.27. From the definition, it is clear that $\overline{A}$ is the smallest closed set containing A.

Theorem 4.12 *Suppose X is a topological space, and $A \subset X$. For any closed set $F \supset A$, we have that $\overline{A} \subset F$.*

Theorem 4.13 *Suppose X is a topological space, and $A \subset X$. The following are true.*

1. $\overline{A} = \{x : every\ nbd\ of\ x\ intersects\ A\}.$

2. $\overline{A} = A \cup A'.$

Proof. To see 1, suppose $x \in \overline{A}$ and U is a neighborhood of x. If $U \cap A = \varnothing$, then $X \backslash IntU$ is a closed set, and $A \subset X \backslash IntU$. Thus $\overline{A} \subset X \backslash IntU$, but $x \in \overline{A} \cap IntU$, a contradiction. Hence $U \cap A \neq \varnothing$. Now suppose every neighborhood of x intersects A. If $x \notin \overline{A}$, then there exists some closed set F with $A \subset F$ and $x \notin F$. Then we have that $X \backslash F$ is open and $x \in X \backslash F$, so that $X \backslash F$ is a neighborhood of x, but $(X \backslash F) \cap A = \varnothing$, a contradiction. Hence $x \in \overline{A}$, and part 1 is proved.

To see 2, first note that if $x \in A \cup A'$, then every neighborhood of x must intersect A. Thus, by part 1, $x \in \overline{A}$. So we have that $A \cup A' \subset \overline{A}$. Now suppose $x \in \overline{A}$. If $x \in A$, then clearly $x \in A \cup A'$. Suppose $x \notin A$. Let U be a neighborhood of x. By 1, we must have $U \cap A \neq \varnothing$, but $x \notin A$, so $(U \cap A) \backslash \{x\} \neq \varnothing$. Thus $x \in A'$. Hence we have $\overline{A} \subset A \cup A'$, completing the proof. ∎

Theorem 4.14 *Suppose X is a topological space, and A and B are subsets of X. The following are true.*

1. $\overline{\varnothing} = \varnothing.$

2. $A \subset \overline{A}.$

3. $\overline{(\overline{A})} = \overline{A}.$

4. $\overline{A \cup B} = \overline{A} \cup \overline{B}.$

Proof. We first note that $\overline{A}$ is a closed set, and if F is any closed set, then $\overline{F} = F$. Now 1 is true since $\varnothing$ is closed, 2 is clear from the definition of $\overline{A}$, and 3 is true since $\overline{A}$ is closed. We wish to prove 4. Since $\overline{A} \cup \overline{B}$ is closed, and $A \cup B \subset \overline{A} \cup \overline{B}$, we have $\overline{A \cup B} \subset \overline{A} \cup \overline{B}$. Moreover, $\overline{A \cup B}$ is closed, and $A \subset \overline{A \cup B}$, so $\overline{A} \subset \overline{A \cup B}$. Similarly, $\overline{B} \subset \overline{A \cup B}$. Hence $\overline{A} \cup \overline{B} \subset \overline{A \cup B}$. Thus we have $\overline{A \cup B} = \overline{A} \cup \overline{B}$. ∎

The properties given above are called the *Kuratowski closure axioms*. They can be used as the fundamental concept to *define* topological spaces. If $\mathcal{C}$ is an operation on the subsets of X such that (1) $\mathcal{C}(\varnothing) = \varnothing$, (2) $A \subset \mathcal{C}(A)$, (3) $\mathcal{C}(\mathcal{C}(A)) = \mathcal{C}(A)$, and (4) $\mathcal{C}(A \cup B) = \mathcal{C}(A) \cup \mathcal{C}(B)$, then $\{U : \mathcal{C}(X \backslash U) = X \backslash U\}$ is a topology on X and $\mathcal{C}$ is the closure operator for this topology.

We saw in our study of metric spaces that many of the things we want to do involve the small open sets with very special form: the open balls. We now define a notion which is analogous to describing the topology of a metric space in terms of open balls.

Definition 4.15 *Suppose* (X, τ) *is a topological space. A* **base** *for the topology* τ *is a subcollection* $\mathcal{B} \subset \tau$ *such that every element of* τ *is the union of some subcollection of* $\mathcal{B}$*. The elements of a base are called* **basic open sets** *. When the topology is understood, we say* $\mathcal{B}$ *is a base for* X*.*

It is immediate from the definition that $\mathcal{B}$ is a base for X provided that $\mathcal{B}$ is a collection of open sets with the property that a subset $U \subset X$ is open if and only if $\forall x \in U \; \exists B_x \in \mathcal{B}$ such that $x \in B_x \subset U$.

We also note that some authors use the term "basis" where we are using "base." The terminology which we are using is the one used by most set-theoretic topologists today.

Example 4.16 *We list several examples below of bases for topologies.*

1. *Any topology* τ *is a base for itself. Usually, however, we are interested in finding a "nice" base, where "nice" usually refers to the sets having some prescribed form or the base itself having some desirable characteristic (like being countable, for instance).*

2. *If* X *is a metric space, then* $\mathcal{B} = \{B(x, \varepsilon) : x \in X \text{ and } \varepsilon > 0\}$ *is a base for* X*.*

3. *Consider* $\mathbb{R}$ *with the usual Euclidean metric topology.*

$$\mathcal{B}_1 = \{(a, b) : a < b\}$$

   is a base for this topology. Also,

$$\mathcal{B}_2 = \{(a, b) : a < b \text{ and both } a \text{ and } b \text{ are rational}\}$$

   is a base for this topology. Both $\mathcal{B}_1$ *and* $\mathcal{B}_2$ *have nice forms, but* $\mathcal{B}_2$ *is itself a countable set whereas* $\mathcal{B}_1$ *is uncountable. So one might imagine that there would be situations where one of these choices would be preferable to the other.*

4. $\mathcal{B} = \{(a, b) : a < b\} \cup \{\{x\} : x \in \mathbb{R} \backslash \mathbb{Q}\}$ *is a base for the Michael line* $\mathbb{M}$*. We could also use just the intervals with rational endpoints in the first set of this union, or just the intervals with irrational endpoints, to describe bases for* $\mathbb{M}$*, depending on what properties we wanted to illustrate. Notice, for example, that* (a, b) *is both open and closed in* $\mathbb{M}$ *when* a *and* b *are irrational. So there is a base for* $\mathbb{M}$ *whose elements are both open and closed. Such spaces are called* **zero-dimensional** *.*

We often use a base to define a topology, and thus we will need to see what properties a collection must have in order to be a base for *some* topology. Before doing that, however, we introduce another idea, the idea of a subbase.

Definition 4.17 *Suppose X is a topological space. A **subbase** for the topology on X, also called a subbase for X, is a collection S of open subsets of X such that the collection of all intersections of finite subcollections of S is a base for X.*

Example 4.18 *Let $S = \{(-\infty, b) : b \in \mathbb{R}\} \cup \{(a, \infty) : a \in \mathbb{R}\}$. Since $(-\infty, b) \cap (a, \infty) = (a, b)$ whenever $a < b$, we see that S is a subbase for the usual topology on $\mathbb{R}$.*

Theorem 4.19 *If X is a set, and S is any collection of subsets of X, then S is a subbase for some topology on X.*

Proof. Let $\mathcal{B} = \{(\cap \mathcal{F}) \cap X : \mathcal{F} \subset S \text{ and} \mathcal{F} \text{ is finite}\}$, and let $\tau = \{\cup \mathcal{G} : \mathcal{G} \subset \mathcal{B}\}$. Notice that $(\cap \mathcal{F}) \cap X = \cap \mathcal{F}$ if $\mathcal{F} \neq \varnothing$. We will show that τ is a topology on X. First we note that $\varnothing \subset \mathcal{B}$, and $\cup \varnothing = \varnothing$. So $\varnothing \in \tau$. To see that $X \in \tau$, notice that $\cap \varnothing = \{x : x \in A \text{ for every } A \in \varnothing\}$. Since there are no sets A with $A \in \varnothing$, every point of X satisfies the condition to be an element of this class. Hence $(\cap \varnothing) \cap X = X$, and $\varnothing$ is certainly finite. So $X \in \mathcal{B}$, and $\{X\} \subset \mathcal{B}$ Thus $X = \cup \{X\} \in \tau$. That τ is closed under arbitrary unions is obvious. It remains only to show that τ is closed under finite intersections. Suppose $U \in \tau$ and $V \in \tau$, and let $x \in U \cap V$. Since U is a union of elements of $\mathcal{B}$, we can choose $B \in \mathcal{B}$ such that $x \in B \subset U$. Choose $\mathcal{F}_1$ a finite subset of S such that $B = \cap \mathcal{F}_1$. Similarly, choose a finite $\mathcal{F}_2 \subset S$ such that $x \in \cap \mathcal{F}_2 \subset V$. Now $\mathcal{F}_1 \cup \mathcal{F}_2$ is a finite subset of S, so $\cap(\mathcal{F}_1 \cup \mathcal{F}_2) \in \mathcal{B}$, and $x \in \cap(\mathcal{F}_1 \cup \mathcal{F}_2) \subset (\cap \mathcal{F}_1) \cap (\cap \mathcal{F}_2) \subset U \cap V$. Hence $U \cap V$ is the union of elements of $\mathcal{B}$, and we have shown that $U \cap V \in \tau$. This completes the proof. ∎

Another word of caution: if there is a topology already present on X, we have *not* shown that any collection of subsets is a subbase for *that* topology. We have shown that any collection of subsets is a subbase for *some* topology on X, namely the one defined in the proof.

Theorem 4.20 *A family $\mathcal{B}$ of subsets of a set X is a base for a topology on X if and only if $\mathcal{B}$ has the following two properties:*

B1 $\cup \mathcal{B} = X$

B2 *If $U \in \mathcal{B}$, $V \in \mathcal{B}$, and $x \in U \cap V$, then $\exists W \in \mathcal{B}$ such that $x \in W \subset U \cap V$.*

Proof. Suppose that $\mathcal{B}$ is a base for a topology on X. Since X must be open in this topology, there exists $\mathcal{G} \subset \mathcal{B}$ with $\cup \mathcal{G} = X$. Now we have $X = \cup \mathcal{G} \subset \cup \mathcal{B} \subset X$, so B1 is true. Now if $U, V \in \mathcal{B}$, then U and V are open sets. So $U \cap V$ is an open set, and if $x \in U \cap V$, then there must be a basic open set W with $x \in W \subset U \cap V$. So B2 is true.

To prove the converse, suppose $\mathcal{B}$ is a collection of subsets of X such that B1 and B2 are satisfied. We let $\tau = \{\cup \mathcal{G} : \mathcal{G} \subset \mathcal{B}\}$ and show that τ is a topology. Since $\varnothing \subset \mathcal{B}$, $\varnothing = \cup \varnothing \in \tau$. By B1, $X = \cup \mathcal{B} \in \tau$. It is clear that τ is closed under arbitrary unions. Suppose $U \in \tau$ and $V \in \tau$, and let $x \in U \cap V$. Choose $\mathcal{U} \subset \mathcal{B}$ and $\mathcal{V} \subset \mathcal{B}$ such that $U = \cup \mathcal{U}$ and $V = \cup \mathcal{V}$. Choose $U_1 \in \mathcal{U}$ and $U_2 \in \mathcal{V}$ such that $x \in U_1$ and $x \in U_2$. Since $\mathcal{B}$ satisfies B2, we may choose $W_x \in \mathcal{B}$ such that $x \in W_x \subset U_1 \cap U_2 \subset U \cap V$. We do this for each point $x \in U \cap V$, and we have $U \cap V = \cup \{W_x : x \in U \cap V\}$. Since $\{W_x : x \in U \cap V\} \subset \mathcal{B}$, we have $U \cap V \in \tau$. ∎

Example 4.21 *The collection $\{[a, b) : a < b\}$ satisfies the conditions of theorem 4.20 and is therefore a base for a topology on $\mathbb{R}$. When we give $\mathbb{R}$ the topology which this base generates, we denote the space by $\mathbb{S}$ and call this the **Sorgenfrey line**.*

This example was created by Robert H. Sorgenfrey who was student of the famous R. L. Moore. We will therefore call this example the *Sorgenfrey line*. This topology is a natural modification of the usual topology and has applications in areas of mathematics outside topology. For this reason, it has been "discovered" on more than just this one occasion, and thus it has appeared with other names. Some authors call this the lower-limit topology on $\mathbb{R}$. It is the topology that makes the cumulative distribution functions, which you may recall from probability, be continuous functions (see definition 5.5).

Definition 4.22 *Suppose X is a topological space and $x \in X$. A **neighborhood base** at x is a collection $\mathcal{B}_x$ of neighborhoods of x such that if U is any neighborhood of x, then $\exists B \in \mathcal{B}_x$ such that $B \subset U$.*

Definition 4.23 *If $\mathcal{B}_x$ is a neighborhood base at x whose elements are open neighborhoods of x, then we call $\mathcal{B}_x$ a **local base** at x.*

Remark 4.24 *We list some examples and make a few remarks about local bases.*

1. *Some books use the phrases "neighborhood base" and "local base" interchangeably.*

2. *If $\mathcal{B}_x$ is a neighborhood base at x, then $\{IntB : B \in \mathcal{B}_x\}$ is a local base at x. Further, any local base is a neighborhood base.*

3. *If X is a metric space and $x \in X$, then $\mathcal{B}_x = \{B(x,\varepsilon) : \varepsilon > 0\}$ is a local base at x. It is also true that $\mathcal{B}_x^1 = \{B(x, \frac{1}{2^n}) : n \in \omega\}$ is also a local base at x.*

4. *In the reals with the usual topology, $\{[x - \frac{1}{n}, x + \frac{1}{n}] : n \in \mathbb{N}\}$ is a neighborhood base at x, but it is not a local base at x.*

5. *If $\mathcal{B}_x$ is a local base for each $x \in X$, then $\mathcal{B} = \bigcup_{x \in X} \mathcal{B}_x$ is a base for X.*

6. *If $\mathcal{B}$ is a base for X, then for each $x \in X$, $\mathcal{B}_x = \{B \in \mathcal{B} : x \in B\}$ is a local base at x.*

Definition 4.25 *If a space X has a countable neighborhood base (or a countable local base) at each of its points, then we call X a **first-countable** space. If a space X has a countable base, then we call X a **second-countable** space.*

The classical terminology for first-countable spaces is that X satisfies the *first axiom of countability*. Similarly, second-countable spaces are sometimes called spaces that satisfy the *second axiom of countability*.

It is clear, from 6 above, that second countable spaces are first countable. From 3, all metric spaces are first countable. There are first-countable spaces that are not second countable, for instance, the Michael line (example 4.4), the Sorgenfrey line (example 4.21), and any uncountable discrete space (example 4.4) are all first-countable spaces that are not second countable. Any uncountable set with the cofinite topology (example 4.4) is not first countable. Statements 7 and 8 of example 4.4 are also examples of non-first-countable spaces.

In the next few results, we explore what is needed for metric spaces to be second countable.

Definition 4.26 *We say a space X is a **separable** space if and only if there is a countable subset of X that is dense in X.*

Example 4.27 *Notice that $\mathbb{Q}$ is dense in both the usual topology and the Sorgenfrey topology on $\mathbb{R}$.*

1. *$\mathbb{R}$ is separable.*

2. *$\mathbb{S}$ is separable.*

3. *The Michael line is not separable since any dense subset of $\mathbb{M}$ must contain all of the irrational points.*

Theorem 4.28 *Every second-countable space is separable.*

Proof. Suppose $\mathcal{B} = \{B_n : n \in \omega\}$ is a countable base for X with $B_n \neq \varnothing$ for each n. Note that if $\varnothing \in \mathcal{B}$, for some base $\mathcal{B}$, then $\mathcal{B} \backslash \{\varnothing\}$ is also a base, so the assumption that the elements are nonempty is justified. For each n, choose $d_n \in B_n$, and let $D = \{d_n : n \in \omega\}$. It is clear that D is countable. We show that D is dense. Suppose $x \in X$. If U is any neighborhood of x, then we can choose $n \in \omega$ such that $x \in B_n \subset U$, thus $d_n \in U$. Hence every neighborhood of x intersects D, and thus $x \in \overline{D}$. Hence $\overline{D} = X$. ∎

Example 4.29 *The converse is not true. The Sorgenfrey line is separable, but $\mathbb{S}$ is not second countable.*

Proof. To see that $\mathbb{S}$ is not second countable, suppose $\mathcal{B} = \{[a_n, b_n) : n \in \omega\}$ is a proposed countable base (we will see shortly that if there is a countable base, there must be one of this form, see theorem 4.36). If $x \notin \{a_n : n \in \omega\}$, then there is no element $B \in \mathcal{B}$ such that $x \in B \subset [x, x+1)$. Thus $\mathcal{B}$ is not a base. ∎

So second countable implies separable, but separable does not imply second countable in general. However, for metric spaces these are equivalent.

Theorem 4.30 *Every separable metric space is second countable.*

Proof. Suppose (X, ρ) is a metric space, and D is a countable dense subset of X. We let $\mathcal{B} = \left\{ B(d, \frac{1}{2^n}) : d \in D, n \in \omega \right\}$ and claim that $\mathcal{B}$ is a base for X. Since $\mathcal{B}$ is clearly countable (it is indexed by the countable set $D \times \omega$), this will complete the proof. Suppose U is open, and $x \in U$. Choose $n \in \omega$ such that $B(x, \frac{1}{2^n}) \subset U$. Since D is dense, choose $d \in D \cap B(x, \frac{1}{2^{n+1}})$. By symmetry, $x \in B(d, \frac{1}{2^{n+1}})$, and if $z \in B(d, \frac{1}{2^{n+1}})$, then $\rho(z, x) \leq \rho(z, d) + \rho(d, x) < \frac{1}{2^{n+1}} + \frac{1}{2^{n+1}} = \frac{1}{2^n}$. Hence $x \in B(d, \frac{1}{2^{n+1}}) \subset B(x, \frac{1}{2^n}) \subset U$. Thus $\mathcal{B}$ is a base. ∎

We turn now to a related property which is the first we have seen of what are called *covering properties*. This property was introduced by Lindelöf's theorem, and it is therefore called the Lindelöf property (see definition 4.33). (E. Lindelöf worked in topology in the very early years of the twentieth century and published this famous theorem in 1903.)

Theorem 4.31 *(Lindelöf's theorem) If X is a second-countable space and $\mathcal{G}$ is a collection of open subsets of X, then there is a countable subcollection $\mathcal{U}$ contained in $\mathcal{G}$ such that $\cup \mathcal{U} = \cup \mathcal{G}$.*

Proof. Suppose $\mathcal{B}$ is a countable base for X, and suppose $\mathcal{G}$ is a collection of open subsets of X. Let $\mathcal{B}^* = \{B \in \mathcal{B} : \exists G \in \mathcal{G} \text{ with } B \subset G\}$. Note that $\mathcal{B}^* \subset \mathcal{B}$, and hence, $\mathcal{B}^*$ is countable. For each $B \in \mathcal{B}^*$, choose one $G_B \in \mathcal{G}$ with $B \subset G_B$. Now the collection $\mathcal{U} = \{G_B : B \in \mathcal{B}^*\}$ is clearly countable. Because $\mathcal{U}$ is a subcollection of $\mathcal{G}$, we have $\cup \mathcal{U} \subset \cup \mathcal{G}$. Suppose $x \in \cup \mathcal{G}$. Choose $G \in \mathcal{G}$ with $x \in G$. Since G is open, there exists $B \in \mathcal{B}$ with $x \in B \subset G$. Thus $B \in \mathcal{B}^*$, and $x \in B \subset G_B$. Hence $x \in \cup \mathcal{U}$. So we have $\cup \mathcal{G} \subset \cup \mathcal{U}$, and the proof is complete. ∎

Definition 4.32 *A collection of subsets of a space X whose union is the space X is called a **cover** of X. If the elements of the cover are open sets, then we call the cover an **open cover**. If the elements of the cover are closed, then we call the cover a **closed cover**, and so on. If $\mathcal{U}$ is a cover of X, and $\mathcal{V}$ is subcollection of $\mathcal{U}$ which is also a cover, then we call $\mathcal{V}$ a **subcover**.*

Definition 4.33 *If a topological space X has the property that every open cover of X has a countable subcover, then we call X a **Lindelöf space**.*

By Lindelöf's theorem, we know that second-countable spaces are Lindelöf. That the converse fails can be seen using the Sorgenfrey line again. It is not trivial to prove that $\mathbb{S}$ is Lindelöf, but it is true nevertheless. As was true for separable, in the setting of metric spaces, we do get the converse.

Theorem 4.34 *Every Lindelöf metric space is second countable.*

Proof. Suppose (X, d) is a Lindelöf metric space. For each $n \in \omega$, let $\mathcal{B}_n = \{B(x, \frac{1}{2^n}) : x \in X\}$. For each n, $\mathcal{B}_n$ is an open cover of X. Choose a countable $\mathcal{U}_n \subset \mathcal{B}_n$ which is also a cover. Let $\mathcal{U} = \bigcup_{n \in \omega} \mathcal{U}_n$ and show $\mathcal{U}$ is the desired countable base. Countability is clear since $\mathcal{U}$ is the union of countably many countable collections.

Suppose U is open, and $x \in U$. Choose $n \in \omega$ with $B(x, \frac{1}{2^n}) \subset U$. Since $\mathcal{U}_{n+1}$ covers X, choose $y \in X$ such that $x \in B(y, \frac{1}{2^{n+1}}) \in \mathcal{U}_{n+1}$. If $z \in B(y, \frac{1}{2^{n+1}})$, then we have $d(x, z) \leq d(x, y) + d(y, z) < \frac{1}{2^{n+1}} + \frac{1}{2^{n+1}} = \frac{1}{2^n}$. Hence $B(y, \frac{1}{2^{n+1}}) \subset B(x, \frac{1}{2^n})$, and we have that $B(y, \frac{1}{2^{n+1}}) \in \mathcal{U}$, and $x \in B(y, \frac{1}{2^{n+1}}) \subset U$. This completes the proof that $\mathcal{U}$ is a base. ∎

Combining the last four theorems, we have the following very nice result.

Theorem 4.35 *Suppose X is a metric space. The following are equivalent:*

1. X is second countable.

2. X *is separable.*

3. X *is Lindelöf.*

To complete our proof that the Sorgenfrey line is not second countable, we need the next result, which says that if a space *has* a countable base then *every* base must contain one. So if we have a base made up of sets that have a particular desirable form and there is a countable base for the space, then we can get a countable base made up of sets with the desirable form.

Theorem 4.36 *If X is second countable, and $\mathcal{B}$ is a base for X, then there is a countable $\mathcal{G} \subset \mathcal{B}$ such that $\mathcal{G}$ is also a base for X.*

Proof. Assume that X is second countable, and let $\{B_n : n \in \omega\}$ be a countable base for X. Suppose $\mathcal{B}$ is a base for X. For each $n \in \omega$, choose $\mathcal{B}_n \subset \mathcal{B}$ such that $B_n = \cup \mathcal{B}_n$. Since each B_n is open and $\mathcal{B}$ is a base, this can be done. By Lindelöf's theorem, for each n, we can choose $\mathcal{U}_n \subset \mathcal{B}_n$ such that $\cup \mathcal{U}_n = \cup \mathcal{B}_n = B_n$. Let $\mathcal{U} = \bigcup_{n \in \omega} \mathcal{U}_n$. Now, $\mathcal{U}$ is a countable subcollection of $\mathcal{B}$. Further, since $\{B_n : n \in \omega\}$ is a base for X, so must be $\mathcal{U}$. ∎

Exercises

1. Prove that the examples in example 4.4 satisfy definition 4.1.

2. Prove that the definition of interior that we gave for metric spaces is equivalent to the definition that we gave in this section for topological spaces. By this we mean that if X is a metric space and $A \subset X$, then let $Int_1 A$ be the interior as defined for metric spaces, let $Int_2 A$ be the interior as defined for the topology generated by the metric, and show $Int_1 A = Int_2 A$.

3. Prove that the definition of closure that we gave for metric spaces is equivalent to the definition that we gave in this section for topological spaces.

4. Prove that any dense subset of the Michael line must contain all the irrationals.

5. Prove that every countable first-countable space is second countable.

6. Is it true that every separable first-countable space is second countable? Prove or find counterexample.

7. Prove that if X has the property that every open cover of X that consists of basic open sets has a countable subcover, then X is Lindelöf.

8. Prove that the Sorgenfrey line is Lindelöf. Hint: Suppose that $\mathcal{U}$ is an open cover of $\mathbb{S}$. For $U \in \mathcal{U}$, let $int_\mathbb{R}(U)$ denote the interior of U in the usual space of real numbers. Let $Y = \cup\{int_\mathbb{R}(U) : U \in \mathcal{U}\}$ and T be the set of all $x \in \mathbb{S}$ such that for each $U \in \mathcal{U}$ with $x \in U$ we have $U \subset [x,\infty)$. Show that T is countable, and for some countable subcollection $\mathcal{U}_0 \subset \mathcal{U}$, we have $Y \subset \cup\mathcal{U}_0$.

9. Prove that the Michael line is not Lindelöf.

10. Consider parts 7 and 8 of example 4.4. Are these spaces separable? Are they Lindelöf? Are they first countable? Are they second countable?

Chapter 5

Basic Constructions: New Spaces From Old

The construction of subspaces, products, and quotients is fundamental in every branch of mathematics. In this chapter, we introduce these basic constructions in the context of topological spaces. In the Part Two of this book we will discuss products and quotients in more depth. In this chapter, we will cover only finite and countable products and continuous functions (precursors of quotients). First, we define subspaces.

Definition 5.1 *Suppose X is a topological space and $Y \subset X$. The **subspace topology** on Y is the collection $\{U \cap Y : U$ is open in $X\}$.*

It is an easy exercise to show that the subspace topology *is* a topology on Y. The subspace topology on Y is also often called the *relative topology* on Y.

Definition 5.2 *When $Y \subset X$ and has the subspace topology, we say Y is a **subspace** of X.*

Example 5.3 *Here are a couple of interesting subspaces.*

1. *$[0,1]$ is a subspace of $\mathbb{R}$. Notice that the set $(\frac{1}{2}, 1]$ is an open set in the subspace $[0,1]$ since $(\frac{1}{2}, 1] = (\frac{1}{2}, 2) \cap [0,1]$. Also note that this is the topology on $[0,1]$ generated by the Euclidean metric restricted to $[0,1]$.*

2. *If we give $\mathbb{R} \backslash \mathbb{Q}$ the subspace topology from $\mathbb{M}$, we have a discrete space.*

3. *If we give $\mathbb{Q}$ the subspace topology from $\mathbb{M}$, we just get the usual Euclidean topology on $\mathbb{Q}$.*

Theorem 5.4 *Suppose X is a topological space, and $A \subset X$ is a subspace of X. The following are true:*

1. *$H \subset A$ is open in A if and only if $\exists G$ open in X with $H = A \cap G$.*

2. *$F \subset A$ is closed in A if and only if $\exists K$ closed in X with $F = A \cap K$.*

3. *If $E \subset A$, then $cl_A E = A \cap cl_X E$.*

4. *If $x \in A$, then V is a neighborhood of x in A if and only if $V = U \cap A$ where U is a neighborhood of x in X.*

5. *If $x \in A$ and $\mathcal{B}_x$ is a neighborhood base at x in X , then $\{B \cap A : B \in \mathcal{B}_x\}$ is a neighborhood base at x in A.*

6. *If $\mathcal{B}$ is a base for X, then $\{B \cap A : B \in \mathcal{B}\}$ is a base for A.*

Proof.   Statement 1 is just the definition of subspace topology. Statement 2 follows immediately from statement 1 by taking complements. To see 3, suppose $\{F_\alpha : \alpha \in \Lambda\}$ is the collection of all closed subsets of A which contain E. For each α, we can choose a closed $K_\alpha \subset X$ such that $F_\alpha = A \cap K_\alpha$. Thus $cl_A E = \bigcap_{\alpha \in \Lambda} F_\alpha = \bigcap_{\alpha \in \Lambda} A \cap K_\alpha \supset \bigcap \{A \cap K : K$ is closed in X and $E \subset K\} = A \cap cl_X E$. Moreover, $A \cap cl_X E$ is a closed subset of A which contains E, and thus $cl_A E \subset A \cap cl_X E$. Hence, $cl_A E = A \cap cl_X E$.

Let us now prove 4. Suppose $x \in A$. If U is a neighborhood of x in X, and $V = A \cap U$, then $A \cap IntU$ is an open set in A, and $A \cap IntU \subset V$. Thus V is a neighborhood of x in A. Conversely, if V is a neighborhood of x in A, then $x \in Int_A V$. Choose open $W \subset X$ such that $A \cap W = Int_A V$. Now since $x \in W$, $V \cup W = U$ is a neighborhood of x in X, and $U \cap A = (V \cup W) \cap A = V$.

To see 5, let $x \in A$ and let $\mathcal{B}_x$ be a neighborhood base at x in X. We know from 4 that $\{B \cap A : B \in \mathcal{B}_x\}$ is a collection of neighborhoods of x in A. Suppose that V is a neighborhood of x in A. Using 4 again, we choose U a neighborhood of x in X such that $V = A \cap U$. There exists $B \in \mathcal{B}_x$ with $B \subset U$, and hence $A \cap B \subset A \cap U = V$. Thus $\{B \cap A : B \in \mathcal{B}_x\}$ is a neighborhood base at x in A.

Finally, to see 6, let $\mathcal{B}$ be a base for X. Suppose that U is open in A. Choose an open set $V \subset X$ with $U = A \cap V$. Choose $\{B_\alpha : \alpha \in \Lambda\} \subset \mathcal{B}$ such that $V = \bigcup_{\alpha \in \Lambda} B_\alpha$. Now $U = A \cap (\bigcup_{\alpha \in \Lambda} B_\alpha) = \bigcup_{\alpha \in \Lambda} (A \cap B_\alpha)$, and U is the union of a subcollection of $\{B \cap A : B \in \mathcal{B}\}$ as desired.   ∎

From 5 and 6 above, it is clear that every subspace of a first-countable space is also first countable, and that every subspace of a second-countable space is also second countable. We call properties for which every subspace inherits the property from the larger space *hereditary properties*. Some of the other properties introduced in chapter 4 are not hereditary. We leave it an exercise to find a Lindelöf space with a non-Lindelöf subspace and to find a separable space with a non-separable subspace. When a space has a property P and all of its subspaces also have property P, then we say the space is hereditarily P. Of particular interest in the development of set-theoretic topology have been the hereditarily Lindelöf spaces and the hereditarily separable spaces. We will see more about these spaces in part two of the book, but suffice it to say here that there is a fascinating duality between hereditarily Lindelöf non-separable spaces (usually called L-spaces) and the hereditarily separable non-Lindelöf spaces (usually called S-spaces). It has been known since the mid-1970's that if the continuum hypothesis is assumed, then one can construct both L-spaces and S-spaces. It was shown in 1982 that there is a model of set theory which does not contain any S-spaces. It is still not known (and it is one of the most interesting problems in set-theoretic topology) whether or not there is a model of set theory that has no L-spaces.

We now return to the concept of a continuous function. As in the metric case, our definition is motivated by the idea (from calculus) that a continuous function is one that maps points that are really close together to points that are really close together. What we are doing is continuing the removal of the algebraic structure (i.e., the absolute values, the differences, the inequalities) that we saw in calculus from the phrase "really close together."

Definition 5.5 *Suppose X and Y are topological spaces, $c \in X$, and $f : X \to Y$. We say f is **continuous at** c if and only if for any neighborhood V of $f(c)$ there exists a neighborhood U of c such that $f(U) \subset V$.*

Definition 5.6 *We say f is a **continuous function** if f is continuous at each $c \in X$.*

The next result is the analogous statement to theorem 3.6 that we remarked on in the metric space section. It generalizes to arbitrary topological spaces.

Theorem 5.7 *Suppose X and Y are topological spaces and $f : X \to Y$. The following are equivalent:*

1. *f is a continuous function.*

2. *for any open set $U \subset Y$, $f^{-1}(U)$ is open in X.*

3. for any closed set $H \subset Y$, $f^{-1}(H)$ is closed in X.

4. for any $A \subset X$, $f(\overline{A}) \subset \overline{f(A)}$.

Proof. We leave the proof to the reader (See theorem 3.6. You will have to work a bit harder to show 4 $\implies$ 1 than in theorem 3.6 because we don't have the characterization of continuity in terms of sequences. ∎

We will see in the exercises that we can replace "open set" with "basic open set," or even "subbasic open set" in 2 above, and still get a characterization of continuity.

We next state a useful result that is an easy consequence of the theorem above.

Theorem 5.8 *If $f : X \to Y$ is continuous and $g : Y \to Z$ is continuous, then the composition $g \circ f : X \to Z$ is also continuous.*

Proof. Left to the reader. ∎

Definition 5.9 *We say a function $f : X \to Y$ is an **open mapping** if and only if $f(U)$ is open in Y for any open $U \subset X$.*

As we mentioned earlier, some authors reserve the word *mapping* for continuous functions, and we have *not* adopted that convention here. In particular, an open mapping is not necessarily continuous.

Remark 5.10 *It will be an exercise to show that f is an open mapping provided $f(B)$ is open for any basic open set B in X.*

Definition 5.11 *A continuous, open, one-to-one, and onto mapping is called a **homeomorphism**. If there is a homeomorphism f mapping X onto Y, then we say X and Y are **homeomorphic**.*

Another way of saying this definition is the following theorem. This is often given as the definition.

Theorem 5.12 *A mapping $f : X \to Y$ is a homeomorphism if and only if f is one-to-one, onto, and both f and f^{-1} are continuous.*

Proof. Left to the reader. ∎

A homeomorphism gives us a one-to-one correspondence between the points of the two spaces, and the same function gives us a one-to-one correspondence between the open sets of the two spaces. So as far as topology is concerned the two spaces are identical.

Example 5.13 *Consider the function $f : (-\frac{\pi}{2}, \frac{\pi}{2}) \to \mathbb{R}$ given by $f(x) = \tan x$. This function is one-to-one and onto, and we recall from calculus that both $\tan$ and Arctan are continuous, even differentiable. Thus we see that $(-\frac{\pi}{2}, \frac{\pi}{2})$ and $\mathbb{R}$ are homeomorphic.*

Example 5.14 *Using a function whose graph is a straight line, we can show that any two open intervals in the real line are homeomorphic.*

Example 5.15 *Since $\mathbb{R}$ is uncountable, and $\mathbb{Q}$ is countable, we see that $\mathbb{R}$ and $\mathbb{Q}$ are not homeomorphic. (Why?)*

Definition 5.16 *A **topological property** is any property so that if X and Y are homeomorphic, then X has the property if and only if Y has the property.*

Definition 5.17 *Suppose $f : X \to Y$, and $A \subset X$. The **restriction** of f to A is the function $f|A : A \to Y$ defined by $f|A(x) = f(x)$ for all $x \in A$. Equivalently, thinking of f as a set of ordered pairs, $f|A = \{(x, y) \in f : x \in A\}$.*

Theorem 5.18 *If X and Y are topological spaces, $A \subset X$, and $f : X \to Y$ is continuous, then $f|A : A \to Y$ is continuous when A has the subspace topology.*

Proof. For any $U \subset Y$, $(f|A)^{-1}(U) = \{x \in A : f(x) \in U\} = A \cap f^{-1}(U)$. Hence, the continuity of f would guarantee the continuity of $f|A$. ∎

It is natural to wonder if the converse is true. Obviously, being continuous at one particular small subspace (like a single point, for instance) is not enough to make a function continuous over a much larger space, but what if the function is continuous on enough subspaces to cover the space?

Example 5.19 *Define $f : \mathbb{R} \to \mathbb{R}$ by $f(x)$ where*

$$f(x) = \begin{cases} 1, & \text{if } x \text{ is rational,} \\ 0, & \text{if } x \text{ is irrational.} \end{cases}$$

We saw earlier that f is not continuous at any point. However, $f|\mathbb{Q}$ and $f|(\mathbb{R} \setminus \mathbb{Q})$ are both constant functions, and both are therefore continuous. Certainly, $\mathbb{R} = \mathbb{Q} \cup \mathbb{R} \setminus \mathbb{Q}$, and so the general result alluded to above is not true.

The consolation prize comes in a couple of forms. We have that the restriction of the function is continuous on enough subspaces to cover the space; but the subspaces must be of a particular type. We need more than that.

Theorem 5.20 *Suppose X and Y are spaces, $f : X \to Y$, and $X = \bigcup_{\alpha \in \Lambda} U_\alpha$ where each U_α is open in X. If $f|U_\alpha$ is continuous for each $\alpha \in \Lambda$, then f is continuous.*

Proof. Assume $f|U_\alpha$ is continuous for each $\alpha \in \Lambda$ and each U_α is open. Suppose V is an open subset of Y. Note that $f^{-1}(V) = f^{-1}(V) \cap (\bigcup_{\alpha \in \Lambda} U_\alpha) = \bigcup_{\alpha \in \Lambda} (f^{-1}(V) \cap U_\alpha) = \bigcup_{\alpha \in \Lambda} (f|U_\alpha)^{-1}(V)$. For each $\alpha \in \Lambda$, $(f|U_\alpha)^{-1}(V)$ is an open subset of U_α, and since U_α itself is open in X, $(f|U_\alpha)^{-1}(V)$ is open in X. Since unions of open sets are open, $f^{-1}(V)$ is open in X. Hence f is continuous. ∎

Theorem 5.21 *Suppose X and Y are spaces, $f : X \to Y$, and $X = A \cup B$ where A, B are closed subsets of X. If $f|A$ and $f|B$ are continuous, then f is continuous.*

Proof. Assume $f|A$ and $f|B$ are continuous and both A and B are closed. We show that inverse images of closed sets are closed. Suppose H is a closed subset of Y. Now $f^{-1}(H) = f^{-1}(H) \cap (A \cup B) = (A \cap f^{-1}(H)) \cup (B \cap f^{-1}(H)) = (f|A)^{-1}(H) \cup (f|B)^{-1}(H)$. Now $(f|A)^{-1}(H)$ is a closed subset of the closed set A, and thus $(f|A)^{-1}(H)$ is closed in X. Similarly, $(f|B)^{-1}(H)$ is a closed subset of X. Since finite unions of closed sets are closed, $f^{-1}(H)$ is closed in X. Hence f is continuous. ∎

Obviously, the theorem for closed sets is much more restrictive than the theorem for open sets. We will see in part two that this can be extended somewhat. However, we cannot get the full analogue of the open subset theorem, since the restriction of any function to a singleton set is continuous on the singleton, and any set is the union of all its singletons. So if the analogue of the open set theorem were true for closed sets, then every function would be continuous on any space where singletons are closed sets, for example, metric spaces. Certainly, this is not the case.

Quotients of topological spaces, in general, will be defined in part two. We will see that a quotient is a special type of continuous function.

We complete this chapter on basic constructions with the introduction of product spaces. As we indicated at the beginning of this chapter, we will consider the topology of only finite products and countably infinite products in part one, but there is no additional complication in defining arbitrary Cartesian products as sets.

The motivating idea for products is careful consideration of the familiar Euclidean spaces $\mathbb{R}^2$ and $\mathbb{R}^3$. These are the familiar sets of ordered pairs and ordered triples of real numbers. What we want to consider is the question, "What is an ordered pair, or an ordered triple?" Well, an ordered pair is an object that has two elements, a first element and a second element, and we can tell which one is first and which one is second. So there must be some function f which can tell these points apart; that is, a function which chooses $f(1)$ to be the first element and $f(2)$ to be the second element. Similarly, an ordered triple should generate a function that tells which element is first $(f(1))$, which element is second $(f(2))$, and which element is third $(f(3))$.

Definition 5.22 *Suppose $\{X_\alpha : \alpha \in A\}$ is a collection of sets. The **Cartesian product** of these sets is denoted by $\prod_{\alpha \in A} X_\alpha$ and is defined by*

$$\prod_{\alpha \in A} X_\alpha = \left\{ x : x \text{ is a function from } A \text{ into } \bigcup_{\alpha \in A} X_\alpha \text{ and } x(\alpha) \in X_\alpha \forall \alpha \right\}.$$

Notation 5.23 *There are several easy observations and notational conventions associated with products that need to be mentioned.*

1. *If $x \in \prod_{\alpha \in A} X_\alpha$, we denote $x(\alpha) = x_\alpha$, and we call this the αth **coordinate** of x.*

2. *For each $\beta \in A$, the set X_β is the called the βth **factor** of the product $\prod_{\alpha \in A} X_\alpha$.*

3. *If $X_n = X$ for all $n \in \omega$, then $\prod_{n \in \omega} X_n$ is the set of all sequences in X.*

4. *We often write $x = (x_\alpha : \alpha \in A)$ for a point $x \in \prod_{\alpha \in A} X_\alpha$, following statement 1 and our conventions for sequences.*

5. *If $A = \{1, 2\}$, we usually write $X_1 \times X_2$ rather than $\prod_{\alpha \in \{1,2\}} X_\alpha$. We also write $x = (x_1, x_2)$ rather than $(x_i : i \in \{1, 2\})$. Of course, these are not really equal. The function $(x_i : i \in \{1, 2\})$ is $\{(1, x_1), (2, x_2)\}$, but no confusion arises from this minor abuse of the notation.*

6. *$\mathbb{R} \times \mathbb{R}$ is $\mathbb{R}^2$.*

7. *$X \times Y$ is the same as $X_1 \times X_2$ where $X = X_1$ and $Y = X_2$.*

8. *We often write $\prod_{k=1}^{n} X_k$ for $\prod_{k \in \{1, 2, \ldots, n\}} X_k$, and $\prod_{n=0}^{\infty} X_n = \prod_{n \in \omega} X_n$.*

9. *If $X_\alpha = X$ for all $\alpha \in A$, then we usually denote $\prod_{\alpha \in A} X_\alpha$ by X^A.*

10. *For each $\beta \in A$, the mapping $\pi_\beta : \prod_{\alpha \in A} X_\alpha \to X_\beta$ defined by $\pi_\beta(x) = x_\beta$ is called the **projection mapping** onto the βth factor, or sometimes the βth **projection**.*

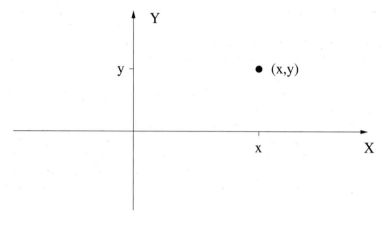

Figure 5.1: $X \times Y$

11. *If the product is nonempty, then the projection mappings are always onto mappings.*

12. *If any factor is empty, then the product is empty.*

We usually envision $X \times Y$ just as you did in calculus. See figure 5.1.

We said in the preface that we would not use any fancy set theory in part one. However, let us take just a moment here to discuss the axiom of choice. The axiom of choice (AC) says that if you have any collection of nonempty sets, then you can choose one point from each of those sets. At first glance, this seems obvious, but think about how you would try to do this if the collection were infinite. You cannot go one set at a time because, even choosing points as fast as you can, by the end of your life you would still have chosen from only finitely many sets. For this reason, AC is not regarded as quite as obvious as the other axioms.

Any element of $\prod_{\alpha \in A} X_\alpha$ is a "choice function" on the collection $\{X_\alpha : \alpha \in A\}$. So, by assuming that $\prod_{\alpha \in A} X_\alpha$ will always have elements if all the factors are nonempty, we are assuming that the axiom of choice is true. Since most people, including this author, accept the axiom of choice as true, this is not much of a leap, but for those inclined to be careful about such things, we note this.

The construction of the topology on finite products is very natural.

Definition 5.24 *Suppose X and Y are topological spaces. The collection $\mathcal{B} = \{U \times V : U$ is open in X and V is open in $Y\}$ is a base for a topology on $X \times Y$. (See Theorem 4.20.) This topology is called the **product topology** . We can*

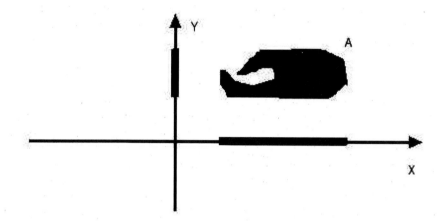

Figure 5.2: Projections $\pi_1(A)$ and $\pi_2(A)$.

extend this definition to finite products as well. A base for the product topology on $\prod_{n=1}^{k} X_n$ is given by $\left\{ \prod_{n=1}^{k} U_n : U_n \text{ is open in } X_n, \ n = 1, 2, ..., k \right\}.$

Theorem 5.25 *The product topology on $\mathbb{R}^2$ is the same topology as that generated by the metrics in Example 2.9. In particular, the product topology is the Euclidean topology on $\mathbb{R}^2$.*

Proof. By theorem 2.16, it is enough to show that the product topology is the same as the topology generated by the metric d_∞. We will do this by showing that the balls in the d_∞ topology are open in the product topology, and that the basic open sets in the product topology are open in the d_∞ topology. This is sufficient to show that these two topologies coincide. (Why?) First, suppose that $x = (x_1, x_2) \in \mathbb{R}^2$ and $\varepsilon > 0$. Notice that $B_{d_\infty}(x, \varepsilon) = (x_1 - \varepsilon, x_1 + \varepsilon) \times (x_2 - \varepsilon, x_2 + \varepsilon)$, which is a basic open set in the product topology on $\mathbb{R}^2$. So the open balls are open in the product topology.

Now suppose $U \times V$ is a basic open set in the product topology, which simply means that both U and V are open in $\mathbb{R}$. Let $x = (x_1, x_2) \in U \times V$. We need to produce a positive number ε for which $B_{d_\infty}(x, \varepsilon) \subset U \times V$. Since U is open and $x_1 \in U$, there is $\delta_1 > 0$ such that $(x_1 - \delta_1, x_1 + \delta_1) \subset U$. Similarly, there is $\delta_2 > 0$

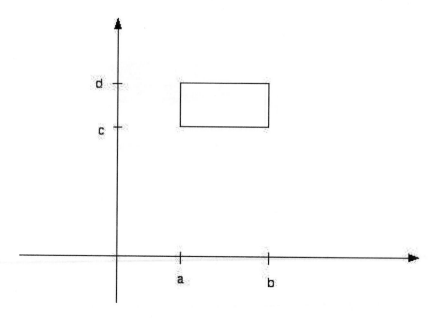

Figure 5.3: $(a, b) \times (c, d)$

such that $(x_2 - \delta_2, x_2 + \delta_2) \subset V$. Let $\varepsilon = \min\{\delta_1, \delta_2\}$. Now

$$
\begin{aligned}
B_{d_\infty}(x, \varepsilon) &= (x_1 - \varepsilon, x_1 + \varepsilon) \times (x_2 - \varepsilon, x_2 + \varepsilon) \\
&\subset (x_1 - \delta_1, x_1 + \delta_1) \times (x_2 - \delta_2, x_2 + \delta_2) \subset U \times V.
\end{aligned}
$$

Since this is true for each $x \in U \times V$, $U \times V$ is open in the metric topology. ∎

Theorem 5.26 *If $X_1 \times X_2$ is a nonempty product space, then the projections π_1 and π_2 are both open, continuous, onto functions.*

Proof. We will verify this for π_2 and leave the π_1 case to the reader. First, to see that π_2 is onto, choose $(a, b) \in X_1 \times X_2$. Now, for any $y \in X_2$, $(a, y) \in X_1 \times X_2$, and $\pi_2((a, y)) = y$. Hence π_2 is onto. If V is an open set in X_2, then $\pi_2^{-1}(V) = X_1 \times V$ is a basic open set in $X_1 \times X_2$. Hence π_2 is continuous. Now if $U \times V$ is any nonempty basic open set in $X_1 \times X_2$, then $\pi_2(U \times V) = V$. This is sufficient to show that $\pi_2(W)$ is open for any open set W in $X_1 \times X_2$. (Why?) ∎

Theorem 5.27 *If $a \in X$, then Y is homeomorphic to the subspace $\{a\} \times Y$ of the product $X \times Y$.*

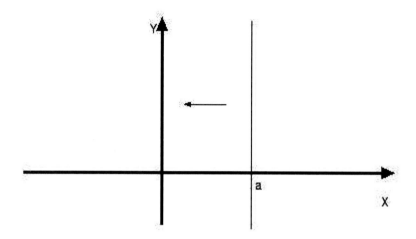

Figure 5.4: $\{a\} \times Y$ is homeomorphic to Y

Proof. $\pi_2|(\{a\} \times Y)$ is a homeomorphism. ∎

So we see that each factor of a nonempty product space is homeomorphic to a subspace of the product. If we have, in addition, that the singletons in each factor are closed sets, then each factor is homeomorphic to a closed subspace of the product.

Before we write the definition of the product topology on an infinite product, let us think for a moment about sequences. If $X_n = X$ for each $n \in \omega$, then the product $\prod_{n=0}^{\infty} X_n$ is just the set of all sequences in X. To make the setting even more familiar, suppose $X_n = \{0, 1, 2, 3, 4, 5, 6, 7, 8, 9\}$, and look at the product $\prod_{n=1}^{\infty} X_n$. Does this look familiar? It is actually just the set of all decimals between 0 and 1. To see this, just write $(x_n : n \in \mathbb{N})$ as $0.x_1 x_2 x_3... = \sum_{n=1}^{\infty} \frac{x_n}{10^n}$. We want our topology to tell us when points are close together, that is, points that are in neighborhoods of each other. If we have two decimals that agree up to the third decimal place, then we would say that they are within 0.001 of each other regardless of what the values for the higher decimal places might be. We will do a similar thing in the product topology. We will define basic open sets by restricting the first finitely many coordinates, after which we will not place any restriction.

Definition 5.28 *The collection*

$$\mathcal{B} = \left\{ \prod_{n=0}^{\infty} U_n : U_n \text{ is open in } X_n \text{ for each } n \in \omega \text{ and } \exists k \text{ such that } U_n = X_n \forall n \geq k \right\}$$

*is a base for a topology on $\prod_{n=0}^{\infty} X_n$. This is called the **product topology**, or the **Tychonoff topology**, on $\prod_{n=0}^{\infty} X_n$.*

Remark 5.29 *It is sometimes helpful to notice that $\prod_{n=0}^{\infty} U_n$, where $U_n = X_n$ for all $n \neq i$, is just the set $\pi_i^{-1}(U_i)$. So the base we have described is the one generated by the subbase $\{\pi_n^{-1}(U_n) : U_n \text{ is open in } X_n \text{ and } n \in \omega\}$. In particular, a basic open set in $\prod_{n=0}^{\infty} X_n$ has form $\bigcap_{n=0}^{k} \pi_n^{-1}(U_n)$ where U_n is open in X_n for each $n \leq k$.*

As we saw above in the finite case (and the proofs are the same), we have the following theorems.

Theorem 5.30 *If $\prod_{n=0}^{\infty} X_n$ is a nonempty product space, then the projections π_n are open, continuous, onto functions for each $n \in \omega$.*

Theorem 5.31 *If, for each $n \in \omega$, $a_n \in X_n$, then for each $k \in \omega$ the factor space X_k is homeomorphic to the subspace A_k where*

$$A_k = \left\{ x \in \prod_{n=0}^{\infty} X_n : \pi_n(x) = a_n \text{ when } n \neq k \right\}.$$

We also remark that we can think of the finite product $X_0 \times X_1$ as a countable product by taking X_n to be a singleton for each $n \geq 2$.

We will have much more to say about products as we progress through our discovery of topological properties, but we will state just one more useful theorem as we close out this chapter on basic constructions.

Theorem 5.32 *Let X be a topological space, let $\prod_{n=0}^{\infty} X_n$ be a product space, and suppose $f : X \to \prod_{n=0}^{\infty} X_n$. The function f is continuous if and only if $\pi_n \circ f$ is continuous for all n.*

Proof. Since each projection is continuous, and the composition of two continuous functions is continuous, the "only if" part follows. Now suppose that $\pi_n \circ f$ is continuous for all n. Let S be a subbasic open set in $\prod_{n=0}^{\infty} X_n$, say $S = \pi_i^{-1}(U_i)$ where U_i is open in X_i. Now $f^{-1}(S) = f^{-1}(\pi_i^{-1}(U_i)) = (\pi_i \circ f)^{-1}(U_i)$ which is open by the continuity of $\pi_i \circ f$. Since this is true for any subbasic open set in $\prod_{n=0}^{\infty} X_n$, f is continuous. ∎

Exercises

1. Prove theorem 5.8.

2. Prove that if $f : X \to Y$, then f is continuous if and only if $f^{-1}(B)$ is open in X for each member B of some base for Y. Prove the same result with "base" replaced by "subbase."

3. Prove that if $f : X \to Y$, then f is open if and only if $f(B)$ is open in Y for each member B of some base for X. What goes wrong if we try to extend this result to "subbase?"

4. Prove theorem 5.12.

5. Prove that if a and b are real numbers with $a < b$, then (a, b) is homeomorphic to $\mathbb{R}$.

6. Since $(0, \frac{1}{4})$ is homeomorphic to $\mathbb{R}$, it is a complete metric space. Consider the function $f : (0, \frac{1}{4}) \to (0, \frac{1}{4})$ defined by $f(x) = x^2$, and show that $|f(x) - f(y)| \le \frac{1}{2}|x - y|$ for each $x, y \in (0, \frac{1}{4})$. Yet, f has no fixed points. Why does this not violate the Banach fixed-point theorem?

7. Suppose D is a dense subset of the space X and U is an open subspace of X. Prove that $D \cap U$ is dense in U. (We assume U has the subspace topology.)

8. Suppose that $A \subset X$ and $B \subset Y$. Prove that $cl_{X \times Y}(A \times B) = \overline{A} \times \overline{B}$ and $Int_{X \times Y}(A \times B) = (IntA) \times (IntB)$.

9. Prove that $\mathbb{S} \times \mathbb{S}$ is separable.

10. Find a subspace of $\mathbb{S} \times \mathbb{S}$ that is not separable. Hence the result in exercise 7 fails if we do not assume the subspace is open. Note that this shows that "separable" is not a hereditary property even though it follows from exercise 7 that "separable" is hereditary to open subspaces.

11. Find a subspace of a Lindelöf space that is not Lindelöf. Note that this shows that "Lindelöf " is not a hereditary property.

12. Prove that every closed subspace of a Lindelöf space is again Lindelöf. So "Lindelöf" is hereditary to closed subspaces.

13. Prove that $\mathbb{S} \times \mathbb{S}$ is not Lindelöf.

14. Suppose F is a closed subspace of X, and $K \subset F$ is closed in F. Prove that K is closed in X.

15. Suppose U is an open subspace of X, and $G \subset U$ is open in U . Prove that G is open in X.

16. Suppose $f : X \rightarrow Y$, and f is continuous and onto Y.

   (a) Prove that if X is separable, then so is Y.

   (b) Prove that if X is Lindelöf, then so is Y.

17. Find a closed subset K of $\mathbb{R} \times \mathbb{R}$ such that $\pi_1(K)$ is not closed in $\mathbb{R}$. So even though the projections are continuous mappings and open mappings, they very well may not be "closed" mappings.

18. Suppose that F_n is a closed subset of X_n for each $n \in \omega$. Prove that $\prod_{n=0}^{\infty} F_n$ is closed in the product space $\prod_{n=0}^{\infty} X_n$. Is the same result true if we replace "closed" with "open?" Why?

19. The Hilbert space l_2 is the set of all square summable sequences of real numbers i.e.,

$$l_2 = \left\{ (x_n : n \in \omega) \in \mathbb{R}^\omega : \sum_{n=0}^{\infty} x_n^2 < \infty \right\}.$$

We define a metric on l_2 by

$$d((x_n : n \in \omega), (y_n : n \in \omega)) = \left[\sum_{n=0}^{\infty} (x_n - y_n)^2 \right]^{\frac{1}{2}}.$$

Show that d is a metric on l_2. How does the metric topology on l_2 compare with the subspace topology from the product $\mathbb{R}^\omega$?

20. Another interesting space from analysis is the space l_∞ of all bounded sequences of real numbers, i.e.,

$$l_\infty = \left\{ (x_n : n \in \omega) \in \mathbb{R}^\omega : \sup \{|x_n| : n \in \omega\} < \infty \right\}.$$

We define a metric on l_∞ by

$$d((x_n : n \in \omega), (y_n : n \in \omega)) = \sup \{|x_n - y_n| : n \in \omega\}.$$

Show that d is a metric on l_∞. How does the metric topology on l_∞ compare with the subspace topology from the product $\mathbb{R}^\omega$?

Chapter 6

Separation Axioms

In this chapter, we introduce the separation axioms. Separation axioms are statements about the richness of a topology. They answer questions like, "Are there enough open sets to tell points apart?" and "Are there enough open sets to tell points from closed sets?" We will see that the higher separation axioms have other interesting characteristics as well. We denote these conditions by T_0, T_1, T_2, T_3, and T_4 in increasing order of richness. The origin of this is the German word *Trennungsaxiome*, which translates as "separation axiom" or "division axiom."

We will develop these separation axioms in this chapter. In the figures below, we hope to help motivate the various axioms. T_0, T_1, and T_2 are axioms that tell how well the open sets (denoted by the oval shaped globs) can separate points from each other. T_3 describes the ability of the open sets to separate points from closed sets, and T_4 is the ability of the open sets to separate disjoint closed sets.

Our first two separation axioms are T_0 and T_1. If a space is T_0, then for any two points of the space, there is a neighborhood of at least one that excludes the other. If the space is T_1, then for any two points of the space, there are neighborhoods of each that exclude the other.

Definition 6.1 *A topological space X is called a T_0-**space** if and only if for any two distinct points of X there is an open subset of X which contains one but not the other.*

Definition 6.2 *A topological space X is called a T_1-**space** (or simply "X is T_1") if and only if for any two distinct points x and y of X there exist open subsets U and V of X such that*

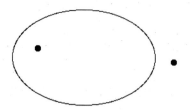

Figure 6.1: T_0

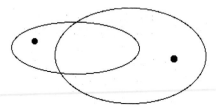

Figure 6.2: T_1

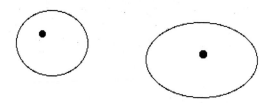

Figure 6.3: T_2

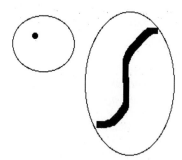

Figure 6.4: T_3

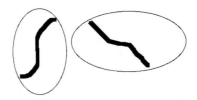

Figure 6.5: T_4

1. $x \in U$ and $y \notin U$, and

2. $y \in V$ and $x \notin V$.

Example 6.3 *Here we list two examples.*

1. *Let X be a set with at least two points, and let $\tau = \{\varnothing, X\}$, the trivial topology. This space is not T_0.*

2. *Let X be the Sierpiński space, that is $X = \{0, 1\}$ and $\tau = \{\varnothing, X, \{0\}\}$. This space is T_0, but it is not T_1.*

Remark 6.4 *Note that $T_1 \implies T_0$.*

We have seen already in this book that it is often convenient to have the singleton sets be closed. This is a characterization of T_1-spaces.

Theorem 6.5 *A topological space X is a T_1-space if and only if for every $x \in X$ the singleton $\{x\}$ is a closed set in X.*

Proof. Suppose X is a T_1-space and $x \in X$. For each $y \neq x$, we choose open $U_y \subset X$ such that $y \in U_y$ and $x \notin U_y$. Now $\bigcup_{y \neq x} U_y$ is open, and $\{x\} = X \backslash (\bigcup_{y \neq x} U_y)$. Thus $\{x\}$ is closed.

Conversely, if $x \neq y$ and both $\{x\}$ and $\{y\}$ are closed, then $U = X \backslash \{y\}$ and $V = X \backslash \{x\}$ play the desired roles in the definition of T_1. ∎

Continuing to build stronger separation, a space is T_2 if and only if for any two points of the space, there are neighborhoods of each which exclude not only the other point, but actually exclude a neighborhood of the other point.

Definition 6.6 *A topological space X is said to be a T_2-space, also called a **Hausdorff space** , if and only if for any two distinct points x and y of X there exist disjoint open subsets U and V of X such that $x \in U$ and $y \in V$.*

Remark 6.7 *Note that $T_2 \implies T_1 \implies T_0$.*

Example 6.8 *We list several examples.*

1. *The Michael line, the Sorgenfrey line, and all metric spaces are Hausdorff.*

2. *Any infinite set with the cofinite topology is a T_1-space which is not T_2.*

In general topology, the weakest separation axiom that is usually studied is the Hausdorff axiom. Indeed, there are books that include this in the definition of the phrase "topological space." If an example is constructed that answers a question, but the example is not at least Hausdorff, most people would regard the question as still being open. The general feeling is that if a space is constructed which is just T_1, and not at least T_2, then the construction is really set theory and not topology. While we will not go so far as to include Hausdorff in the definition, this author certainly shares this point of view.

As was the case in metric spaces, continuous functions with Hausdorff ranges are determined on dense subsets. Compare this next theorem with theorem 3.7.

Theorem 6.9 *If X is a topological space, Y is a Hausdorff space, $f : X \to Y$ is continuous, $g : X \to Y$ is continuous, and $\{x \in X : f(x) = g(x)\}$ is dense in X, then $f = g$.*

Proof. Assume Y is Hausdorff, f and g are continuous, and $\{x \in X : f(x) = g(x)\}$ is dense in X. If $f \neq g$, there is a point $x_0 \in X$ such that $f(x_0) \neq g(x_0)$. Since Y is T_2, we may choose disjoint open sets U and V in Y such that $f(x_0) \in U$ and $g(x_0) \in V$. Since f and g are continuous, $f^{-1}(U) \cap g^{-1}(V)$ is an open set containing x_0. Hence we may choose $z \in f^{-1}(U) \cap g^{-1}(V) \cap \{x \in X : f(x) = g(x)\}$. Now $f(z) = g(z) \in U \cap V = \varnothing$, a contradiction. ∎

We now discuss limits of sequences in topological spaces. Thinking of the metric balls as being basic neighborhoods, the definition is exactly analogous to what we did in metric spaces. Compare this with definition 2.29 and theorem 2.31.

Definition 6.10 *Suppose X is a topological space and $(x_n : n \in \omega)$ is a sequence in X. We say the sequence $(x_n : n \in \omega)$ **converges** to a point $x \in X$ if and only if for every neighborhood U of x there exists $N \in \omega$ such that if $n \geq N$, then $x_n \in U$. We denote this by $x_n \to x$, by $\lim_{n \to \infty} x_n = x$, or by $\lim x_n = x$.*

Definition 6.11 *When $x_n \to x$, we call x the **limit point** of the sequence $(x_n : n \in \omega)$.*

The use of the phrase "the limit point" and the notation in definition 6.10 both seem to imply that a sequence can have only one limit point. For Hausdorff spaces, that is indeed the case, but we will see in the exercises that this is not a property of spaces that are "only T_1."

Theorem 6.12 *(Uniqueness of limits) Suppose X is a Hausdorff space and $(x_n : n \in \omega)$ is a sequence in X. If $x_n \to x$ and $x_n \to y$, then $x = y$.*

Proof. Left to the reader. Compare with theorem 2.32. ∎

These last two theorems give us two of the nicest things about Hausdorff spaces, namely, continuous functions are determined on dense subsets and limits are unique.

We now turn to preservation of Hausdorff spaces under basic constructions.

Theorem 6.13 *(Preservation theorem for T_2 spaces)*

1. *If X is a Hausdorff space and A is a subspace of X, then A is a Hausdorff space.*

2. *If X_n is Hausdorff for each $n \in \omega$, then $\prod_{n \in \omega} X_n$ is also Hausdorff.*

3. *Continuous images of Hausdorff spaces may fail to be Hausdorff, but homeomorphic images of Hausdorff spaces are Hausdorff.*

Proof. For 3, take any Hausdorff space and map it by the identity onto the same set with the trivial topology. We leave the positive result for homeomorphisms as an exercise.

Let us prove 1. Suppose X is a T_2 space, and A is a subspace of X. Let $x, y \in A$ with $x \neq y$. Since X is Hausdorff, we choose disjoint open sets U and V in X with $x \in U$ and $y \in V$. Now $U \cap A$ and $V \cap A$ are disjoint open sets in A which separate x and y. Thus A is Hausdorff.

Finally, suppose X_n is Hausdorff for each $n \in \omega$, and let $x, y \in \prod_{n \in \omega} X_n$ with $x \neq y$. Choose $k \in \omega$ such that $x_k \neq y_k$. Since X_k is T_2, we choose disjoint open $U, V \subset X_k$ with $x_k \in U$ and $y_k \in V$. Now $\pi_k^{-1}(U)$ and $\pi_k^{-1}(V)$ are disjoint open sets in $\prod_{n \in \omega} X_n$ which contain x and y respectively. ∎

From 3 above, we note that Hausdorff is a topological property.

Definition 6.14 *A topological space X is called **regular** if and only if for any closed $F \subset X$ and any point $x \in X \setminus F$ there exist disjoint open subsets U and V of X such that $x \in U$ and $F \subset V$. A regular T_1-space is called a T_3-**space**.*

Remark 6.15 $T_3 \implies T_2 \implies T_1 \implies T_0$.

Proof. Left to the reader. ∎

Example 6.16 *We give a couple of examples to show that the remark really cannot be improved.*

1. *If X is a set with at least two points, and X is given the trivial topology, i.e. $\tau = \{\varnothing, X\}$, then we have a space which is regular (prove this!), but it is not even T_0.*

2. *Let $X = \mathbb{R}$, and let*

$$\mathcal{B} = \{U : U \text{ is Euclidean open in } \mathbb{R}\} \cup \left\{ (-\varepsilon, \varepsilon) \backslash \left\{ \frac{1}{n} : n \in \mathbb{N} \right\} : \varepsilon > 0 \right\}.$$

It is easy to check that $\mathcal{B}$ is a base for a topology on X. In this space, we can still separate points since the Euclidean open sets $(x - \frac{|x-y|}{2}, x + \frac{|x-y|}{2})$ and $(y - \frac{|x-y|}{2}, y + \frac{|x-y|}{2})$ are still open. So this space is T_2. However, $\{\frac{1}{n} : n \in \mathbb{N}\}$ is now a closed set which cannot be separated from the point 0. Thus this space is not regular. Hence $T_2 \nRightarrow T_3$.

Example 6.17 *The Michael line, the Sorgenfrey line, and all metric spaces are regular.*

Proof. Left to the reader. ∎

The following characterization of regularity is often more useful than the definition itself. It says that not only can you squeeze down on a point with neighborhoods, but you can get a neighborhood base consisting of *closed* neighborhoods.

Theorem 6.18 *A space X is regular if and only if whenever $U \subset X$ is open and $x \in U$ there exists an open set $V \subset X$ such that $x \in V \subset \overline{V} \subset U$.*

Proof. Suppose that X is regular, U is an open subset of X, and $x \in U$. Now, $X \backslash U$ is closed and $x \notin X \backslash U$, so we may choose disjoint open V and W such that $x \in V$ and $X \backslash U \subset W$. Now $X \backslash W$ is a closed set, $X \backslash W \subset U$, and since $V \cap W = \varnothing$, $V \subset X \backslash W$. Hence $\overline{V} \subset X \backslash W \subset U$, and we have $x \in V \subset \overline{V} \subset U$.

For the converse, assume that whenever $U \subset X$ is open and $x \in U$, there exists an open set $V \subset X$ such that $x \in V \subset \overline{V} \subset U$. Suppose $F \subset X$ is closed and $x \in X \backslash F$. Since F is closed, $X \backslash F$ is open. Choose open V such that

$x \in V \subset \overline{V} \subset X \backslash F$. Now V and $X \backslash \overline{V}$ provide the separation to witness regularity. ∎

Part 2 of the next theorem is a really nice exercise in using the product topology, and part 1 is easy. So we won't spoil it by hinting at the proof.

Theorem 6.19 *(Preservation theorem for T_3 spaces)*

1. *Every subspace of a regular (or T_3) space is regular (respectively, T_3).*

2. *If X_n is regular (or T_3) for each $n \in \omega$, then $\prod_{n \in \omega} X_n$ is regular (respectively, T_3).*

3. *Continuous images need not preserve regularity, even if the domains and ranges are T_1.*

Proof. We leave 1 and 2 as exercises. For part 3, consider $\mathbb{R}$ with the cofinite topology. This space does not have any disjoint nonempty pairs of open sets. So it fails to be regular, or even Hausdorff. On the other hand, the identity mapping is a continuous function onto this space from $\mathbb{R}$ with the usual topology. ∎

Note that the converse to part 2 is true if the product is nonempty since, in that case, each factor is homeomorphic to a subspace of the product.

The next separation axiom we give is the most interesting to the most topologists. Even though it is just the next natural step in the progression from separating points to separating points from closed sets, it seems not to be a separation axiom at all because it fails to have the type of preservation theorem that the others have. It also seems to have additional properties that are much stronger than mere separation. As topologists have done since the invention of the subject, we will spend a little extra time on this property, *normality*.

Definition 6.20 *A topological space X is called a **normal** space if and only if for any pair H and K of disjoint closed subsets of X there exists a pair U and V of disjoint open subsets of X such that $H \subset U$ and $K \subset V$. A normal T_1-space is a $\mathbf{T_4}$-space .*

Remark 6.21 $T_4 \implies T_3 \implies T_2 \implies T_1 \implies T_0$.

Example 6.22 *We list a few examples that show some of the strange behavior of normal spaces.*

1. *The Sierpiński space, $X = \{0, 1\}$ and $\tau = \{\varnothing, X, \{0\}\}$, is normal and T_0 but not regular. So T_0 is enough to get from regular to T_1 (or T_2), but it is not enough to get from normal to T_1.*

2. *The Sorgenfrey line $\mathbb{S}$ is normal, but $\mathbb{S} \times \mathbb{S}$ is not normal. However $\mathbb{S} \times \mathbb{S}$ can be embedded as a dense subspace of a T_4-space. This shows the following facts:*

   (a) *products of normal spaces need not be normal, even if they are T_2. So normality is not preserved by products.*

   (b) *subspaces of T_4-spaces need not be normal, even though they are T_3. So normality is not preserved by subspaces.*

3. *The product $\mathbb{M} \times (\mathbb{R} \backslash \mathbb{Q})$ of the Michael line and the irrationals is not normal. It is somewhat more difficult to prove, but the product $\mathbb{M} \times \mathbb{R}$ is normal, which provides another example of a nonnormal subspace of a normal space.*

4. *The real line with the cofinite topology is a continuous image of $\mathbb{R}$ (under the identity map). So normality is not preserved by continuous images.*

Proof. To prove statement 2, we use the fact that $\mathbb{S} \times \mathbb{S}$, as well as most of the common examples, can be embedded in some normal Hausdorff space. This is a fact that we will establish in part two of the book. That $\mathbb{S} \times \mathbb{S}$ is not normal is an easy result if you have Jones's lemma, a statement involving some cardinal arithmetic, which we will prove in part two. We can now give a proof that $\mathbb{S} \times \mathbb{S}$ is not normal by using the Baire category theorem. This is a good illustration of using the Baire category theorem, so we will leave it as a (difficult) exercise. Here is a hint. Let $H = \{(x, -x) : x \in \mathbb{Q}\}$ and $K = \{(x, -x) : x \in \mathbb{R} \backslash \mathbb{Q}\}$. Show that H and K are disjoint closed subsets of $\mathbb{S} \times \mathbb{S}$ that cannot be separated by any pair of disjoint open subsets of $\mathbb{S} \times \mathbb{S}$. To do this, note that if V is an open set containing K, then there is a fixed positive natural number n such that $[x, x + \frac{1}{n}) \times [-x, -x + \frac{1}{n}) \subset V$ for a set of x's which is dense in some open interval (this is where you use the Baire category theorem). This doesn't leave room for U.

For statement 3, to show that $\mathbb{M} \times (\mathbb{R} \backslash \mathbb{Q})$ is not normal, show that $\Delta = \{(x, x) : x \in \mathbb{R} \backslash \mathbb{Q}\}$ and $\mathbb{Q} \times (\mathbb{R} \backslash \mathbb{Q})$ are disjoint closed subsets. The proof that they cannot be separated by open sets is another Baire category argument similar in spirit and difficulty to the one for $\mathbb{S} \times \mathbb{S}$. ∎

The following is a very useful characterization of normality. It is exactly analogous to the corresponding result for regularity, and the proof is the same.

Theorem 6.23 *A space X is normal if and only if whenever $F \subset X$ is closed, $U \subset X$ is open, and $F \subset U$, there exists an open set V such that $F \subset V \subset \overline{V} \subset U$.*

Proof. This is just like the proof of theorem 6.18. We leave it for the reader. ∎

Theorem 6.24 *Every metric space is T_4.*

Proof. Suppose that (X, d) is a metric space, and H and K are disjoint closed subsets of X. For each $x \in H$, we know that $x \notin K$ because of the disjointness, and $X \backslash K$ is open. So we choose $\varepsilon_x > 0$ such that $B(x, \varepsilon_x) \subset X \backslash K$, that is, $B(x, \varepsilon_x) \cap K = \varnothing$. Similarly, for each $y \in K$, we choose $\delta_y > 0$ such that $B(y, \delta_y) \cap H = \varnothing$. Let

$$U = \cup \left\{ B(x, \frac{\varepsilon_x}{2}) : x \in H \right\}$$

and let

$$V = \cup \left\{ B(y, \frac{\delta_y}{2}) : y \in K \right\}.$$

Clearly, U and V are open sets, $H \subset U$, and $K \subset V$. Suppose that $z \in U \cap V$. Choose $x \in H$ and $y \in K$ such that $z \in B(x, \varepsilon_x) \cap B(y, \delta_y)$. Now either $\varepsilon_x \leq \delta_y$ or $\delta_y \leq \varepsilon_x$. Assume $\varepsilon_x \leq \delta_y$; the other case is done in exactly the same way. We know that

$$d(x, y) \leq d(x, z) + d(z, y) < \frac{\varepsilon_x}{2} + \frac{\delta_y}{2} \leq \frac{\delta_y}{2} + \frac{\delta_y}{2} = \delta_y.$$

Hence $x \in H \cap B(y, \delta_y)$, which contradicts the choice of δ_y. Hence $U \cap V = \varnothing$, as desired. ∎

Theorem 6.25 *Every T_3 Lindelöf space is T_4.*

Proof. Suppose X is a Lindelöf T_3 space, and H and K are disjoint closed subsets of X. For each $x \in H$, since $H \cap K = \varnothing$, $x \notin K$. Thus by regularity, we can choose an open set G_x such that $x \in G_x \subset \overline{G_x} \subset X \backslash K$. Similarly, for each $y \in K$, we choose open W_y such that $y \in W_y \subset \overline{W_y} \subset X \backslash H$. Since H and K are closed subsets of the Lindelöf space X, H and K are also Lindelöf. So we choose countable subcollections $\{G_n^* : n \in \omega\} \subset \{G_x : x \in H\}$ and $\{W_n^* : n \in \omega\} \subset \{W_y : y \in K\}$ such that $H \subset \bigcup_{n \in \omega} G_n^*$ and $K \subset \bigcup_{n \in \omega} W_n^*$. We construct by induction two new sequences $\{U_n : n \in \omega\}$ and $\{V_n : n \in \omega\}$ as follows. Starting at $n = 0$, we let

$$\begin{aligned} U_0 &= G_0^* \text{ and } V_0 = W_0^* \backslash \overline{U_0}, \\ U_1 &= G_1^* \backslash \overline{V_0} \text{ and } V_1 = W_1^* \backslash (\overline{U_0} \cup \overline{U_1}) \\ U_2 &= G_2^* \backslash (\overline{V_0} \cup \overline{V_1}) \text{ and } V_2 = W_2^* \backslash (\overline{U_0} \cup \overline{U_1} \cup \overline{U_2}) \end{aligned}$$

and continue, for $n > 0$,

$$U_n \;=\; G_n^* \backslash \bigcup_{k<n} \overline{V_k} \text{ and } V_n = W_n^* \backslash \bigcup_{k\leq n} \overline{U_k}.$$

We let $U = \bigcup_{n=0}^{\infty} U_n$ and $V = \bigcup_{n=0}^{\infty} V_n$. Notice that both U and V are unions of open sets, and thus they are open. If $x \in H$, choose $n \in \omega$ such that $x \in G_n^*$, and since $\overline{V_k} \subset \overline{W_k^*} \subset X \backslash H$, $x \notin \bigcup_{k<n} \overline{V_k}$. Hence $H \subset U$. Similarly, $K \subset V$. We need to show that $U \cap V = \varnothing$. Suppose $x \in U \cap V$. We choose $n, m \in \omega$ such that $x \in U_n$ and $x \in V_m$. If $n \leq m$, then the definition of V_m gives us that $x \in V_m \implies x \notin \overline{U_n}$ and in particular, $x \notin U_n$, a contradiction. If $n > m$, then the definition of U_n gives that $x \in U_n \implies x \notin \overline{V_m}$, again a contradiction. Hence no such x can exist, and we have shown that $U \cap V = \varnothing$. Hence, X is normal, and the proof is complete. ∎

This is a very pretty argument, and it is often referred to as "climbing the chimney" because we work back and forth between the two sequences, building in the disjointness while keeping the sets open, much as a mountain climber works back and forth between the two sides when climbing a rock chimney.

From the two theorems above, we see that many spaces are normal, but the product theorem for normality fails badly. Indeed, we see that the Sorgenfrey line is normal, even Lindelöf as proved in the exercises in chapter 5, while the product with itself fails to be normal. In an effort to obtain some kind of product theorem, early topologists looked at products of the type $X \times Y$ where X is normal, and Y is "nice." The definition of "nice" could be a number of things. The success of the project was that many interesting examples were created, such as both $\mathbb{S}$ and $\mathbb{M}$; the failure of the project was that it wasn't possible to prove that $X \times [0, 1]$ was normal whenever X was, and whatever "nice" meant, surely $[0, 1]$ would be nice. An example of a normal space X whose product with $[0, 1]$ is not normal is called a **Dowker space**. It is easy to construct examples of Dowker spaces assuming certain special set-theoretic axioms, but it is extremely difficult to get Dowker spaces within the usual axioms for set theory. In 1970, Mary Ellen Rudin constructed the first Dowker space in ZFC (the usual axioms for set theory). It is a monster! Its cardinality is very large, the minimum cardinality of a base is very large, minimum cardinalities of local bases are very large, and, pretty much, by any measure, it is very large. In 1993, Zoltan Balogh constructed a real Dowker space which was "small" in the sense that it is of size equal to that of the real numbers. At this point in time, these two are the only known constructions of Dowker spaces in ZFC.

The next result is a classic theorem due to Urysohn. One of the charms of this theorem, and its proof, is that it gives us a nuts and bolts construction of a function where none was assumed to exist. Within this proof is what is usually called an inductive construction, or more correctly a recursive construction. We will use this powerful technique many more times in this book.

Theorem 6.26 *(Urysohn's lemma) A space X is normal if and only if for every pair A and B of disjoint closed subsets of X there exists a continuous function $f : X \to [0,1]$ such that $A \subset f^{-1}(0)$ and $B \subset f^{-1}(1)$.*

Before we jump into the proof of Urysohn's lemma, let us say a few words about what we want to do. If you number the fibers according to their image values, you can think of a continuous function from X into $[0,1]$ as determining a "scale" on the space X. We want to build such a scale so that A stays on the 0 line and B stays on the 1 line. We will use the dyadic rationals, the rationals whose denominators are powers of 2, to build this scale, by succesively dividing in the middle using the characterization of normality given in theorem 6.23. In the figure opposite we show the construction for the dyadic rationals with denominator $8 = 2^3$. The entire region to the left and below the line marked as $3/8$ is the set $U_{\frac{3}{8}}$, etc. The solid blob in the lower left is A, and the blob in the upper right is B. We will assign a real number to a point x of the space by finding the supremum of the t for which U_t is "below" x.

Proof. First we prove the "if" part of the statement. Suppose that A and B are disjoint closed subsets of X and $f : X \to [0,1]$ is continuous with $A \subset f^{-1}(0)$ and $B \subset f^{-1}(1)$. We wish to construct disjoint open sets U and V with $A \subset U$ and $B \subset V$. Let $U = f^{-1}([0,\frac{1}{2}))$ and $V = f^{-1}((\frac{1}{2},1])$. Since f is a function, these sets are disjoint, and since f is continuous, they are open. Clearly, $A \subset U$ and $B \subset V$. Hence X is normal.

Now we prove the "only if." Suppose X is normal, and let A and B be disjoint closed subsets of X. Let D be the set of dyadic rationals in $(0,1)$, that is $D = \left\{ \frac{m}{2^n} : n \in \mathbb{N}, m \in \mathbb{N}, m < 2^n \right\}$. We construct a collection $\{U_t : t \in D\}$ of open sets such that if $s < t$, then $A \subset U_s \subset \overline{U_s} \subset U_t \subset \overline{U_t} \subset X \backslash B$. We denote this sequence of inclusions by (1). We do this construction recursively on the exponent of the denominator.

To start with $n = 1$, note that A is closed, $X \backslash B$ is open, and $A \subset X \backslash B$. By theorem 6.23, we choose an open set $U_{\frac{1}{2}}$ such that

$$ A \subset U_{\frac{1}{2}} \subset \overline{U_{\frac{1}{2}}} \subset X \backslash B. $$

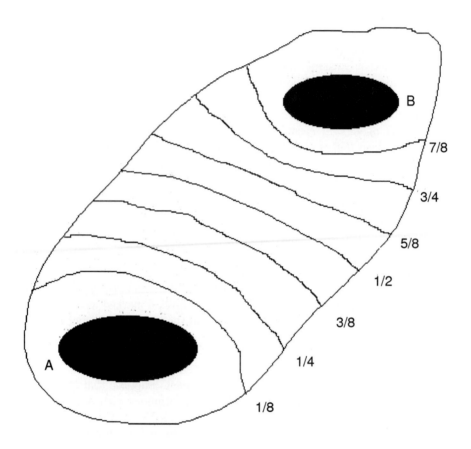

Suppose that $U_{\frac{j}{2^n}}$ has been chosen for $0 < j < 2^n$ such that (1) is true for $s = \frac{i}{2^n}$, $t = \frac{j}{2^n}$, $i < j$. Since A is closed, $U_{\frac{1}{2^n}}$ is open, and $A \subset U_{\frac{1}{2^n}}$, by theorem 6.23, we choose an open set $U_{\frac{1}{2^{n+1}}}$ such that

$$A \subset U_{\frac{1}{2^{n+1}}} \subset \overline{U_{\frac{1}{2^{n+1}}}} \subset U_{\frac{1}{2^n}}.$$

Similarly, by theorem 6.23, for each j with $1 \le j < 2^n - 1$, we choose an open set $U_{\frac{2j+1}{2^{n+1}}}$ such that

$$\overline{U_{\frac{j}{2^n}}} \subset U_{\frac{2j+1}{2^{n+1}}} \subset \overline{U_{\frac{2j+1}{2^{n+1}}}} \subset U_{\frac{j+1}{2^n}}$$

Finally, we again use theorem 6.23 to choose an open set $U_{\frac{2^{n+1}-1}{2^{n+1}}}$ such that

$$\overline{U_{\frac{2^n-1}{2^n}}} \subset U_{\frac{2^{n+1}-1}{2^{n+1}}} \subset \overline{U_{\frac{2^{n+1}-1}{2^{n+1}}}} \subset X\backslash B.$$

This completes the $n+1$ level of the construction. It is easy to see that we have (1) satisfied. Complete the construction of $\{U_t : t \in D\}$ by recursion.

We are now ready to define the function $f : X \to [0,1]$. We define

$$f(x) = \begin{cases} 0, & \text{if } x \in U_t \text{ for all } t \in D, \\ \sup\{t \in D : x \notin U_t\}, & \text{otherwise.} \end{cases}$$

Clearly, $f(x) = 0$ for all $x \in A$, and $f(x) = 1$ for all $x \in B$. It remains only to show that f is continuous. Let $\mathcal{S} = \{[0,a) : a \in (0,1)\} \cup \{(b,1] : b \in (0,1)\}$. Since $\mathcal{S}$ is a subbase for $[0,1]$, it suffices to show that $f^{-1}(S)$ is open in X for each $S \in \mathcal{S}$.

First suppose $a \in (0,1)$. We will show that $f^{-1}([0,a)) = \cup\{U_t : t < a\}$. To see this, suppose $f(x) < a$. Since D is dense in $(0,1)$, there exists $t \in D$ such that $f(x) < t < a$. Now since $f(x) < t$, $x \in U_t$. Hence $f^{-1}([0,a)) \subset \cup\{U_t : t < a\}$. To see the other inclusion, suppose $t < a$, and $x \in U_t$. Now for every $s \in D$ with $s \geq t$ we have $U_t \subset U_s$ so $x \in U_s$ for all $s \geq t$. Thus $f(x) \leq t < a$. Hence $x \in f^{-1}([0,a))$. Thus we have $f^{-1}([0,a)) \supset \cup\{U_t : t < a\}$. So $f^{-1}([0,a)) = \cup\{U_t : t < a\}$, and $f^{-1}([0,a))$ is open.

Now let $b \in (0,1)$. We will show that $f^{-1}((b,1]) = \cup\{X\backslash\overline{U_t} : t > b\}$. Again, we argue pointwise. Suppose $f(x) > b$. Choose t_1 and t_2 in D such that $b < t_1 < t_2 < f(x)$. Thus $x \notin U_{t_2} \supset \overline{U_{t_1}}$, and $x \in X\backslash\overline{U_{t_1}}$ with $t_1 > b$. So we have $f^{-1}((b,1]) \subset \cup\{X\backslash\overline{U_t} : t > b\}$. Now suppose $x \notin \overline{U_{t_0}}$ for some $t_0 > b$. This means $x \notin U_{t_0}$ and thus $f(x) = \sup\{t : x \notin U_t\} \geq t_0 > b$. Thus $f^{-1}((b,1]) \supset \cup\{X\backslash\overline{U_t} : t > b\}$. Hence we have $f^{-1}((b,1]) = \cup\{X\backslash\overline{U_t} : t > b\}$, which is the union of open sets. Hence f is continuous, so the proof is complete. $\blacksquare$

This powerful theorem will now be used to prove the extension theorem for normal spaces. This is a recurring theme in mathematics. We have a nice function defined on a subspace, and we want to extend the function to the entire space while preserving the nice property.

Theorem 6.27 *(Tietze extension theorem) A space X is normal if and only if whenever $A \subset X$ is closed and $f : A \to [a,b]$ is continuous, there is a continuous $\widehat{f} : X \to [a,b]$ such that $f(x) = \widehat{f}(x)$ for all $x \in A$, that is, $\widehat{f}$ is an extension of f.*

Proof. First we prove the "if" part of the statement. Suppose that H and K are disjoint closed subsets of X, and assume that X has the extension property. We define $f : H \cup K \to [0,1]$ by

$$f(x) = \left\{ \begin{array}{ll} 0, & \text{if } x \in H, \\ 1, & \text{if } x \in K. \end{array} \right.$$

Since H and K are disjoint, f is well defined, and since H and K are closed and $f|H$ and $f|K$ are continuous, f is continuous. Now $\widehat{f} : X \to [0,1]$ is continuous, $H \subset \widehat{f}^{-1}(0)$, and $K \subset \widehat{f}^{-1}(1)$, i.e., $\widehat{f}$ is an *Urysohn function*. So by Urysohn's lemma, X is normal.

Now we prove the "only if" portion. Suppose that X is normal, and suppose that $A \subset X$ is closed. We wish to prove that the extension property holds. So we suppose $f : A \to [-1,1]$ is continuous and show that f has a continuous extension to X. By translation and scaling, we can move any closed bounded interval homeomorphically onto any other closed bounded interval. So this will complete the proof.

Before proceeding with the construction, we also note that by composing an Urysohn function with an appropriately chosen translation and scaling, we can obtain from normality that for any two real numbers $r_1 < r_2$ and for any two disjoint closed subsets H and K of X we have a continuous function $\phi : X \to [r_1, r_2]$ such that $H \subset \phi^{-1}(r_1)$ and $K \subset \phi^{-1}(r_2)$.

We construct by induction two sequences $(f_n : n \in \omega)$ and $(g_n : n \in \omega)$ of functions and two sequences $(A_n : n \in \omega)$ and $(B_n : n \in \omega)$ of closed subsets of X. Let

$$\begin{aligned} f_0 &= f \\ A_0 &= f_0^{-1}\left(\left[-1, -\frac{1}{3}\right]\right) \\ B_0 &= f_0^{-1}\left(\left[\frac{1}{3}, 1\right]\right). \end{aligned}$$

Note that A_0 and B_0 are the "upper third" and "lower third" of A as determined by the scale given by f. By the continuity of f, A_0 and B_0 are disjoint closed subsets of A, and since A is closed, we see that A_0 and B_0 are disjoint closed subsets of X. Using Urysohn's lemma, choose a continuous function

$$g_0 : X \to \left[-\frac{1}{3}, \frac{1}{3}\right]$$

such that

$$g_0(A_0) \subset \left\{ -\frac{1}{3} \right\}$$

and

$$g_0(B_0) \subset \left\{ \frac{1}{3} \right\}.$$

Now let

$$f_1 = f_0 - g_0$$

and note that

$$|f_1(x)| \leq \frac{2}{3} \text{ for all } x \in A.$$

Now, let

$$
\begin{aligned}
A_1 &= f_1^{-1}([-\tfrac{2}{3}, (-\tfrac{1}{3})(\tfrac{2}{3})]) = \{x : f_1(x) \leq (-\tfrac{1}{3})(\tfrac{2}{3})\} \\
B_1 &= f_1^{-1}([(\tfrac{1}{3})(\tfrac{2}{3}), \tfrac{2}{3}]) = \{x : f_1(x) \geq (\tfrac{1}{3})(\tfrac{2}{3})\},
\end{aligned}
$$

Since f_0 and g_0, and, consequently, f_1, are continuous on A, the sets A_1 and B_1 are disjoint closed sets in A. Hence, A_1 and B_1 are disjoint closed subsets of X. By Urysohn's lemma, choose a continuous

$$g_1 : X \to [(-\tfrac{1}{3})(\tfrac{2}{3}), (\tfrac{1}{3})(\tfrac{2}{3})]$$

such that

$$g_1(A_1) \subset \{(-\tfrac{1}{3})(\tfrac{2}{3})\} \text{ and } g_1(B_1) \subset \{(\tfrac{1}{3})(\tfrac{2}{3})\}.$$

Now let

$$f_2 = f_1 - g_1 = f_0 - (g_0 + g_1),$$

and note that

$$|f_2(x)| \leq \left(\frac{2}{3} \right)^2.$$

Continue this construction recursively to construct sequences $(f_n : n \in \omega)$ and $(g_n : n \in \omega)$ of functions and $(A_n : n \in \omega)$ and $(B_n : n \in \omega)$ of closed sets such that for each $n \in \omega$,

$$f_n : A \to [-(\tfrac{2}{3})^n, (\tfrac{2}{3})^n]$$

and

$$g_n : X \to [(-\tfrac{1}{3})(\tfrac{2}{3})^n, (\tfrac{1}{3})(\tfrac{2}{3})^n]$$

are both continuous and $f_n = f_0 - (g_0 + g_1 + ... + g_{n-1})$ on A, and g_n is a Urysohn function mapping $A_n = \{x : f_n(x) \leq -\frac{1}{3}(\frac{2}{3})^n\}$ to $\{-\frac{1}{3}(\frac{2}{3})^n\}$ and $B_n = \{x : f_n(x) \geq \frac{1}{3}(\frac{2}{3})^n\}$ to $\{\frac{1}{3}(\frac{2}{3})^n\}$.

For each $x \in X$, consider the series $\sum_{n=0}^{\infty} g_n(x)$. Since $|g_n(x)| \leq (\frac{1}{3})(\frac{2}{3})^n$ $\forall x \in X$, by the Weierstrass M-test (the proof of which has nothing to do with the domain actually being contained in the reals), the series converges uniformly to a function $\widehat{f} : X \to [-1, 1]$, which is then continuous by the uniform convergence (that proof also has nothing to do with the domain being contained in the reals). Note that $|\widehat{f}(x)| = |\sum_{n=0}^{\infty} g_n(x)| \leq \sum_{n=0}^{\infty} |g_n(x)| \leq \sum_{n=0}^{\infty}(\frac{1}{3})(\frac{2}{3})^n = 1$. So the range is contained in $[-1, 1]$. To see that $\widehat{f}$ extends f, suppose $x \in A$, then $|f(x) - \widehat{f}(x)| = \lim_{n\to\infty} |f(x) - \sum_{k=0}^{n-1} g_k(x)| = \lim_{n\to\infty} |f_n(x)| \leq \lim_{n\to\infty} (\frac{2}{3})^n = 0$. ∎

Exercises

1. Prove theorem 6.12.

2. Prove that $\mathbb{M}$, $\mathbb{S}$, and all metric spaces are T_2.

3. Suppose X is an infinite set with the cofinite topology. Prove that X is T_1, but not T_2.

4. In the same example as exercise 3, suppose $(x_n : n \in \omega)$ is a sequence in X with $x_n \neq x_m$ whenever $n \neq m$. What is "the" limit of this sequence?

5. Prove that every finite T_1-space is discrete. (This is why you don't see very many finite examples in topology books.)

6. Prove that if X is any topological space, $A \subset X$, and there is a sequence $(x_n : n \in \omega)$ in A with $x_n \to x$ for some point $x \in X$, then $x \in \overline{A}$. Find an example to show that the converse to this fails. That is, find an example of a space X, a subspace $A \subset X$, and a point $x \in \overline{A}$ such that no sequence in A converges to x.

7. Prove that if X is a first countable space, $A \subset X$, and a point $x \in \overline{A}$, then there is a sequence in A which converges to x.

8. Prove theorem 6.19.

9. A space X is called *completely regular* if and only if for any closed $A \subset X$ and any point $x \in X \backslash A$ there is a continuous function $f : X \to [0, 1]$ such that $f(x) = 0$ and $A \subset f^{-1}(1)$. Prove that any completely regular space is

regular. Note: A completely regular T_1-space is usually called a Tychonoff space, but in the T-subscript tradition, such spaces are often called $T_{3.5}$ or $T_{3\frac{1}{2}}$-spaces. This probably started as a joke, but it has become fairly standard. Some people have also suggested that such spaces be called T_π-spaces since they are "closer" to regular than they are to normal.

10. Recall from example 4.16 that a space is called *zero-dimesional* if each of its points has a neighborhood base consisting of sets which are both open and closed. Prove that every zero-dimensional space is completely regular.

11. Prove that both $\mathbb{M}$, the Michael line, and $\mathbb{S}$, the Sorgenfrey line, are zero-dimensional.

12. Prove that $\mathbb{M}$ is normal.

13. Complete the proof that $\mathbb{M} \times (\mathbb{R}\backslash\mathbb{Q})$ is not normal.

14. From results that we have established, it is clear now that $\mathbb{S}$ is normal. (Why?) Complete the proof that $\mathbb{S} \times \mathbb{S}$ is not normal.

15. Prove theorem 6.23.

Project: the Urysohn Metrization Theorem

The following sequence of results leads to the Urysohn metrization theorem.

1. Suppose that (X_n, d_n) is a metric space for each $n \in \omega$, and each d_n is bounded by 1. Let $X = \prod_{n=0}^\infty X_n$, and define d on X by $d(x,y) = \sum_{n=0}^\infty \frac{d_n(x_n,y_n)}{2^n}$. Prove that d is a metric on X which generates the product topology on X.

2. Suppose that (X_n, d_n) is a separable metric space for each $n \in \omega$. Prove that $\prod_{n=0}^\infty X_n$ is also a separable metric space.

3. Suppose that X is a separable metric space. Prove that every subspace of X is also a separable metric space.

4. Suppose that X is a normal, second countable, T_1-space and $\mathcal{B}$ is a countable base for X. Let $\mathcal{A} = \{(U,V) : U, V \in \mathcal{B}$ and $\overline{U} \subset V\}$.

   (a) For each $(U,V) \in \mathcal{A}$, where $V \neq X$, prove that there exists a continuous function $f_{UV} : X \to [0,1]$ such that $f_{UV}(\overline{U}) = \{0\}$ and $f_{UV}(X\backslash V) = \{1\}$.

(b) Let $\mathcal{F} = \{f_{UV} : (U, V) \in \mathcal{A}\}$. Prove that $\mathcal{F}$ is countable.

(c) For each $f \in \mathcal{F}$, let $I_f = [0, 1]$. Define $e : X \to \prod_{f \in \mathcal{F}} I_f$ by $\pi_f(e(x)) = f(x)$. The mapping e is called the evaluation mapping. Notice that since $\mathcal{F}$ is countable, we have the topology defined on this product. Prove that e is a homeomorphism of X onto $e(X)$.

5. (Urysohn metrization theorem) Suppose that X is a T_1-space. Prove that X is regular and second countable if and only if X is a separable metrizable space.

Thus you have shown that every regular second countable space is metrizable. So every countable first countable T_3-space must be metrizable. For example, it would follow that $\mathbb{Q}$ with the Sorgenfrey topology must be metrizable, even though it would take a little thought to figure out what this metric would be!

Notice that something you have done in this project is show that the product space $[0, 1]^\omega$ is a "universal space" for the class of separable metric spaces. That is, in the proof above, you showed that every separable metric space is homeomorphic to a subspace of a countable product of copies of the closed interval $[0, 1]$. A space X is said to be a **universal space** for property P provided that every space which satisfies property P is homeomorphic to a subspace of X. The existence of universal spaces of various sorts has been an interesting area of research for many years.

Another major area of research in set-theoretic topology, which you have begun in this project, is the metrization problem. That is, what conditions are needed on a topological space in order to show that the topology is generated by a metric? Of course, the definitions provide a characterization, but what we are looking for here is a list of properties of the open sets themselves, not external structures like a distance function.

As we said above, you have shown that if a T_3-space has a countable base, then there is a metric which generates the topology. So we now have a simply stated sufficient condition for metrizability that talks only about the space and the open sets. Is it possible to give a simple condition on the topology, or on a base for the topology, which is both necessary and sufficient? This is a question to which we will return in part two.

Chapter 7

Compact Spaces

Other than metric spaces, the most important class of topological spaces is the class of compact spaces. This is the class of spaces where most of analysis works well: continuous real-valued functions are uniformly continuous and therefore Riemann integrable, continuous real-valued functions have maxima and minima, sequences always have cluster points, and Cauchy sequences always converge. Many of the most interesting topological spaces are infinite, but intuitively at least, what makes the compact spaces nice is that in many ways they *act like* finite sets.

Definition 7.1 *A topological space X is **compact** if and only if every open cover of X has a finite subcover.*

Example 7.2 *We list a few easy examples.*

1. *Any finite space is compact.*

2. $\mathbb{R}$ *is not compact.*

3. *Any compact space is Lindelöf.*

4. *If $(x_n : n \in \omega)$ is a sequence in a topological space X, and $x_n \to x$ in X, then the subspace $A = \{x\} \cup \{x_n : n \in \omega\}$ is compact.*

5. *Any set with the cofinite topology is compact.*

Proof. We prove 4 and leave the rest to the reader. If $\mathcal{U}$ is an open cover of A, then we choose $V \in \mathcal{U}$ such that $x \in V$. We can find an open set W in X such that $V = W \cap A$. Since $x_n \to x$, choose $N \in \omega$ such that if $n > N$, then

$x_n \in W$. In particular, if $n > N$, then $x_n \in V$. For $n \leq N$, we can choose $U_n \in \mathcal{U}$ with $x_n \in U_n$ since $\mathcal{U}$ covers A. Now $\{V\} \cup \{U_n : 0 \leq n \leq N\}$ is the desired finite subcover. ∎

In the example above of the convergent sequence, the subspace consisting of just the points of the sequence without the limit point would not be compact, if the space X were, for instance, the reals. So we cannot hope to prove a result that says that subspaces will inherit this property. We have the following consolation prize.

Theorem 7.3 *Any closed subspace of a compact space is compact.*

Proof. Suppose X is compact and K is a closed subspace of X. Let $\{G_\alpha : \alpha \in \Lambda\}$ be an open cover of the space K. For each $\alpha \in \Lambda$, choose U_α open in X such that $G_\alpha = K \cap U_\alpha$. Now, $\{X \backslash K\} \cup \{U_\alpha : \alpha \in \Lambda\}$ is an open collection in X, and it clearly covers X. So by the compactness of X, choose $\alpha_1, \alpha_2, ... \alpha_n$ in Λ such that $X = (X \backslash K) \cup \bigcup_{k=1}^{n} U_{\alpha_k}$. Hence we see that $K = \bigcup_{k=1}^{n} (K \cap U_{\alpha_k}) = \bigcup_{k=1}^{n} G_{\alpha_k}$, and thus $\{G_{\alpha_k} : k = 1, 2, ..., n\}$ is the desired finite subcover. ∎

For compact spaces which are also Hausdorff, the converse holds.

Theorem 7.4 *Any compact subset of a Hausdorff space is closed.*

Proof. Suppose X is T_2 and K is a compact subspace of X. Suppose that $x \in X \backslash K$. For each $y \in K$, we choose a pair of disjoint open sets U_y and V_y such that $x \in U_y$ and $y \in V_y$. Since $\{V_y : y \in K\}$ is a collection of open sets in X which covers K, and K is compact, we choose $y_1, y_2, ..., y_n$ in K such that $K \subset \bigcup_{i=1}^{n} V_{y_i}$. Let $U = \bigcap_{i=1}^{n} U_{y_i}$. Note that U is open, and $x \in U$. Further, $U \cap \bigcup_{i=1}^{n} V_{y_i} = \varnothing$, so $U \subset X \backslash K$. Thus $X \backslash K$ is an open set. ∎

Theorem 7.5 *Any continuous image of a compact space is also compact.*

Proof. Suppose X is compact, and $f : X \rightarrow Y$ is a continuous mapping onto Y. Let $\mathcal{U}$ be an open cover of Y. By continuity, $\{f^{-1}(U) : U \in \mathcal{U}\}$ is an open cover of X. Choose $U_1, U_2, ... U_n$ such that $X = \bigcup_{k=1}^{n} f^{-1}(U_k)$. Now $Y = f(X) = f(\bigcup_{k=1}^{n} f^{-1}(U_k)) = \bigcup_{k=1}^{n} f(f^{-1}(U_k)) = \bigcup_{k=1}^{n} U_k$ since f is onto Y. Hence $\{U_1, U_2, ..., U_n\}$ is the desired finite subcover of $\mathcal{U}$. ∎

From the last two results, we get the following interesting corollary. We leave the proof as an exercise.

Theorem 7.6 *If X is compact, Y is T_2, and $f : X \to Y$ is one-to-one, onto and continuous, then f is a homeomorphism.*

The next result is a key result in analysis. It also gives us many more examples that we know are compact.

Theorem 7.7 *(Heine-Borel theorem) A subset $K \subset \mathbb{R}$ is compact if and only if K is closed and bounded.*

Proof. The "only if" part is easy. Since $\mathbb{R}$ is Hausdorff, a compact subset K must be closed, and the collection $\{(-n, n) : n \in \mathbb{N}\}$ must have a finite subcollection which covers K, so K must be bounded.

Now we prove the "if" part. Suppose K is a closed bounded subset of $\mathbb{R}$. Since K is bounded, we choose $a < b$ such that $K \subset [a, b]$. Since K is closed, by theorem 7.3, we will be finished if we can prove $[a, b]$ is compact.

Suppose $\mathcal{U}$ is an open cover of $[a, b]$. If $[a, b]$ is covered by some finite subcollection of $\mathcal{U}$, we have nothing to do. Suppose that is not the case. Let

$$B = \{x \in [a, b] : [a, x] \text{ is not covered by any finite subcollection of } \mathcal{U}\}.$$

Since $b \in B$, $B \neq \varnothing$ and B is clearly bounded. Using the fact that every nonempty bounded subset of $\mathbb{R}$ has a greatest lower bound, we let $m = \inf B$. Note that $m > a$ since there is some $U \in \mathcal{U}$ with $a \in U$, and thus $\exists \varepsilon > 0$ with $[a, a + \varepsilon) \subset U$. So $[a, a + \frac{\varepsilon}{2}]$ is covered by a singleton subcollection of $\mathcal{U}$. Hence we have $a < a + \frac{\varepsilon}{2} \leq m$.

Now, since $\mathcal{U}$ covers $[a, b]$, we choose $U_0 \in \mathcal{U}$ such that $m \in U_0$. Since U_0 is open, choose an open interval (x, y), $a < x < m$, with $m \in (x, y) \cap [a, b] \subset U_0$. Since $x < m$, there must be a point $z \in (x, m)$ and $z \notin B$. So we choose a finite subcollection $\mathcal{U}_0 \subset \mathcal{U}$ with $[a, z] \subset \cup \mathcal{U}_0$. If $m < b$, then choose $w \in (m, y) \cap [a, b]$. We now have that $\mathcal{U}_0 \cup \{U_0\}$ is a finite subcollection of $\mathcal{U}$ which covers $[a, w]$, and thus $\inf B \geq w > m = \inf B$, a contradiction. Thus $m = b$, and $\mathcal{U}_0 \cup \{U_0\}$ is a finite subcollection of $\mathcal{U}$ which covers $[a, b]$, and the proof is complete. ∎

We said that the Heine-Borel theorem is a key result in analysis. Perhaps we should explain just a bit what we mean. This is the key observation in proving that continuous functions on closed bounded intervals are integrable. It is also the key to proving that continuous functions on closed bounded intervals have maximum and minimum values. While independently interesting, that result is the key to proving Rolle's lemma and the mean-value theorem, and virtually every result in calculus after the middle of the first semester is an application of the mean-value theorem, including of course, the fundamental theorem of calculus.

Theorem 7.8 *Every compact T_2-space is normal.*

Proof. Suppose that X is a compact Hausdorff space. We first show that X is regular. Suppose $F \subset X$ is closed and $x \in X \backslash F$. For each $y \in F$, we choose disjoint open sets U_y and V_y such that $x \in U_y$ and $y \in V_y$. Now the collection $\{V_y : y \in F\}$ is a collection of open sets which covers F, and, by theorem 7.3, F is compact. So we choose $y_1, y_2, ..., y_n$ in F such that $F \subset \bigcup_{k=1}^n V_{y_k}$. Let $U = \bigcap_{k=1}^n U_{y_k}$ and $V = \bigcup_{k=1}^n V_{y_k}$, and we see that U and V are disjoint open sets containing x and F, respectively. Hence X is regular.

Now we can repeat, in essence, the argument above to get from regular to normal (and this is a good exercise), or we can simply cite theorem 6.25 to conclude that X is normal. ∎

Our next theorem is one whose importance can hardly be overstated. We will state the theorem in full generality here even though we have only defined the product topology for countably many factors at this point. To give a proof for countable products now is possible, though really quite messy. We will defer the proof until Part Two when we will have the machinery, namely ultrafilters, to make the proof easy. The proof for two factors is a good exercise.

Theorem 7.9 *(Tychonoff theorem) A nonempty product space is compact if and only if each factor space is compact.*

We turn now to several properties similar to compactness. Each has been proposed as the definition of compactness in some setting. We show that they are all equivalent for metric spaces. In particular, for most of analysis, any of these can be taken as the definition.

Definition 7.10 *A space X has the **Bolzano-Weierstrass property** if and only if every infinite subset of X has a cluster point in X.*

The Bolzano-Weierstrass theorem, from analysis, is the statement that every closed bounded subset of $\mathbb{R}$ has the Bolzano-Weierstrass property.

Definition 7.11 *A space X is called **sequentially compact** if and only if every sequence in X has a convergent subsequence.*

Definition 7.12 *A space X is called **countably compact** if and only if every countable open cover of X has a finite subcover.*

It is clear that a space X is compact if and only if X is countably compact and Lindelöf. In particular, since $\mathbb{R}$ is hereditarily Lindelöf, a subset of the real line will be compact if and only if it is countably compact.

Theorem 7.13 *Suppose X is a T_1-space. The space X is countably compact if and only if X has the Bolzano-Weierstrass property.*

Proof. First, suppose that X is countably compact and let A be an infinite subset of X with no cluster points in X. Choose $B = \{x_n : n \in \omega\} \subset A$ indexed so that $x_i \neq x_j$ if $i \neq j$. Note that since B also has no cluster points, B is a closed set in X. For each $n \in \omega$, x_n is not a cluster point of B, so we may choose an open set U_n with $B \cap U_n = \{x_n\}$. Now $\{U_n : n \in \omega\} \cup \{X \backslash B\}$ is a countable open cover of X with no finite subcover, a contradiction.

Now suppose that X has the Bolzano-Weierstrass property, and suppose that $\{U_n : n \in \omega\}$ is an open cover of X that has no finite subcover. We use recursion to contruct an infinite set with no cluster points. Choose $x_0 \in X$ and $k_0 \in \omega$ such that k_0 is the first $m \in \omega$ with $x_0 \in U_m$. Suppose that $n \in \omega$, and x_m, k_m have been chosen for all $m \leq n$ so that $x_m \in U_{k_m}$ and if $j < k_m$, then $x_m \notin U_j$. Since $\{U_n : n \in \omega\}$ has no finite subcover, $\{U_j : j \leq k_n\}$ does not cover X. Choose $x_{n+1} \in X \backslash (\bigcup_{j=0}^{k_n} U_j)$, and choose $k_{n+1} \in \omega$ such that k_{n+1} is the first $m \in \omega$ such that $x_{n+1} \in U_m$. Continue recursively to construct the set $A = \{x_n : n \in \omega\}$. It is clear from the construction that A is an infinite set.

We now show that A has no cluster points. Suppose that $x \in X$. Choose $m \in \omega$ such that $x \in U_m$. Note that in our construction, $k_n \geq n$ for each n. So we can choose N to be the first natural number such that $m \leq k_N$. Notice that if $n > N$, then $k_n > k_N$, so $x_n \notin U_m$, and $\{x_j : j \leq N$ and $x \neq x_j\}$ is a finite set and thus is closed. Let $U = U_m \backslash \{x_j : j \leq N$ and $x \neq x_j\}$, and note that U is an open set containing x. Further, $U \cap A \subset \{x\}$, and thus x is not a cluster point of A. Since this is true for every $x \in X$, A has no cluster points. This is a contradiction, and so no such cover can exist. ∎

Corollary 7.14 *Every sequentially compact T_1-space is countably compact.*

Proof. It is easy to see that sequentially compact spaces must satisfy the Bolzano-Weierstrass property since any infinite set A contains the range of a one-to-one sequence, and the limit of any convergent subsequence of that sequence would be a cluster point of A. ∎

Theorem 7.15 *If X is a first countable T_1 space, then X is countably compact if and only if X is sequentially compact.*

Proof. By the preceding corollary, we only need prove that if X is a countably compact, first-countable, T_1 space, then X is sequentially compact. So we suppose that X has those properties, and we suppose that $(x_n : n \in \omega)$ is a sequence in X. If $\{x_n : n \in \omega\}$ is a finite set, then $(x_n : n \in \omega)$ has a constant subsequence, and that would certainly be a convergent subsequence. So we assume that $A = \{x_n : n \in \omega\}$ is infinite. By theorem 7.13, we know that $\{x_n : n \in \omega\}$ has a cluster point, call it x^*. By the first countability, we choose a countable open neighborhood base $\{B_n : n \in \omega\}$ at x^*.

We now construct a subsequence that converges to x^*. Let $k_0 = 0$, and suppose that for $n \in \omega$ we have chosen $k_0, k_1, ..., k_n$ in an increasing way. Let $U = (\bigcap_{m \leq k_n} B_m) \backslash \{x_j : j \leq k_n \text{ and } x_j \neq x^*\}$. Now U is an open neighborhood of x^* , and x^* is a cluster point of A. So we choose a point $x_m \in A \cap (U \backslash \{x^*\})$. Let $k_{n+1} = m$, and note that $k_{n+1} \neq j$ for any $j \leq k_n$, and thus $k_{n+1} > k_n$. By recursion, we have the sequence $(k_n : n \in \omega)$. Note that this sequence is increasing. Hence $(x_{k_n} : n \in \omega)$ is a subsequence of $(x_n : n \in \omega)$.

Suppose that V is an open neighborhood of x^*. Choose $N \in \omega$ such that $B_N \subset V$. Now by the construction, if $n \geq N$, then $N \leq k_n$, and hence $x_{k_{n+1}} \in B_N \subset V$. Thus $x_{k_n} \to x^*$. Hence every sequence in X has a convergent subsequence. ∎

We will see in part two of this book an example of a non-first-countable space which is countably compact (actually even compact) but is not sequentially compact. So this theorem cannot be extended to all spaces.

We are now ready to pull these results together to obtain the result that for metric spaces, all these notions of compactness are equivalent.

Theorem 7.16 *Suppose that X is a metric space. The following are equivalent:*

1. *X is compact.*

2. *X is countably compact.*

3. *X has the Bolzano-Weierstrass property.*

4. *X is sequentially compact.*

Proof. Since metric spaces are first countable and T_1, we have $2 \iff 3 \iff 4$ It is obvious that compact spaces are countably compact. So it remains only to show that countably compact metric spaces are compact. We will do this by first showing that such a space is separable, from which it follows that the space is Lindel öf, and the proof will then be complete.

Suppose that (X, d) is a countably compact metric space. Fix $n \in \omega$. Let

$$\mathbb{P}_n = \left\{ A \subset X : \text{if } x, y \in A \text{ and } x \neq y \text{ , then } d(x,y) \geq \frac{1}{2^n} \right\}.$$

We order $\mathbb{P}_n$ by subset inclusion, and note that each singleton set in X is an element of $\mathbb{P}_n$. Thus $\mathbb{P}_n$ is a nonempty partially ordered set. If $\mathcal{C}$ is a linearly ordered subset of $\mathbb{P}_n$, then for distinct points x and y of $\cup \mathcal{C}$, we may choose $C_1 \in \mathcal{C}$ and $C_2 \in \mathcal{C}$ such that $x \in C_1$ and $y \in C_2$, but by the linear ordering of $\mathcal{C}$, either $C_1 \subset C_2$ or $C_2 \subset C_1$ so $d(x,y) \geq \frac{1}{2^n}$. Hence every linearly ordered subset of $\mathbb{P}_n$ has an upper bound in $\mathbb{P}_n$. By Zorn's lemma, $\mathbb{P}_n$ has a maximal element, which we call D_n. For any point $x \in X$, $B(x, \frac{1}{2^{n+1}}) \cap D_n$ can be at most one point by the triangle inequality. Hence D_n can have no cluster points. By the Bolzano-Weierstrass property, D_n must be finite. Do this construction for each $n \in \omega$, and let $D = \bigcup_{n \in \omega} D_n$. Now, clearly, D is a countable set.

We claim that D is dense in X. Suppose it is not, then we can choose $x \in X \backslash \overline{D}$. Choose $\varepsilon > 0$ such that $B(x, \varepsilon) \cap D = \varnothing$, and choose $n \in \omega$ such that $\frac{1}{2^n} \leq \varepsilon$. Now $d(x,y) \geq \frac{1}{2^n}$ for every $y \in D$, and thus $\{x\} \cup D_n \in \mathbb{P}_n$. This contradicts the maximality of D_n, so no such x can exist. Thus X is a separable metric space, and so X is Lindelöf. Clearly, any countably compact, Lindelöf space is compact, and the proof is complete. ∎

We close this chapter with three theorems that are especially useful in analysis.

Theorem 7.17 *Every compact metric space is complete.*

Proof. Suppose X is a compact metric space and $(x_n : n \in \omega)$ is a Cauchy sequence in X. Since X is sequentially compact, $(x_n : n \in \omega)$ has a convergent subsequence. It is a routine exercise (assigned in chapter 2) to show that if a Cauchy sequence has a convergent subsequence, then it must converge. ∎

We recall from chapter 2 that completeness is a property of the metric, and not generally of the topology. Here that is not the case. We have shown above that any metric that is compatible with a compact topology must be a complete metric. The converse to that is not true of course, since $\mathbb{R}$ is complete with the Euclidean metric, but not compact.

Definition 7.18 *Suppose that $\mathcal{G}$ is an open cover of a metric space (X, d). A real number δ is called a **Lebesgue number** for $\mathcal{G}$ if and only if whenever $A \subset X$ and the diameter of A is less than δ, then there exists $G \in \mathcal{G}$ such that $A \subset G$.*

Theorem 7.19 *In a compact metric space, every open cover has a Lebesgue number.*

Proof. Suppose (X, d) is a compact metric space and $\mathcal{G}$ is an open cover of X. For each $x \in X$, choose $G_x \in \mathcal{G}$ such that $x \in G_x$. For each $x \in X$, since G_x is open, choose $\varepsilon_x > 0$ such that $B(x, \varepsilon_x) \subset G_x$. Now $\left\{ B(x, \frac{\varepsilon_x}{2}) : x \in X \right\}$ is an open cover of X. By compactness, we can choose a finite subcover, say $\left\{ B(x_i, \frac{\varepsilon_{x_i}}{2}) : 1 \leq i \leq n \right\}$. Let $\delta = \min \left\{ \frac{\varepsilon_{x_i}}{2} : 1 \leq i \leq n \right\}$. Suppose $A \subset X$ and $diam(A) < \delta$. If $A = \varnothing$, then choose any $G \in \mathcal{G}$, and $A \subset G$. Suppose $A \neq \varnothing$; choose $x_0 \in A$. Since $\left\{ B(x_i, \frac{\varepsilon_{x_i}}{2}) : 1 \leq i \leq n \right\}$ is a cover, choose i with $x_0 \in B(x_i, \frac{1}{2}\varepsilon_{x_i})$. We will show that $A \subset G_{x_i}$. Suppose that $x \in A$. Since $diam(A) < \delta$, $d(x, x_0) < \delta \leq \frac{1}{2}\varepsilon_{x_i}$. Thus $d(x, x_i) \leq d(x, x_0) + d(x_0, x_i) < \delta + \frac{1}{2}\varepsilon_{x_i} \leq \varepsilon_{x_i}$. Thus $x \in B(x_i, \varepsilon_{x_i}) \subset G_{x_i}$. Hence $A \subset G_{x_i}$ and δ is a Lebesgue number for $\mathcal{G}$. ∎

Theorem 7.20 *If (X, d) and (Y, ρ) are metric spaces, X is compact, and $f : X \to Y$ is continuous, then f is uniformly continuous.*

Proof. Suppose $\varepsilon > 0$ is given. The collection $\left\{ f^{-1}(B(y, \frac{\varepsilon}{2})) : y \in Y \right\}$ is an open cover of X. Let δ be the Lebesgue number for this cover. If $d(x, z) < \delta$, then the diameter of the set $\{x, z\}$ is less than δ. Thus there is a $y \in Y$ such that $\{x, z\} \subset f^{-1}(B(y, \frac{\varepsilon}{2}))$. Thus we have that $\rho(f(x), f(z)) \leq \rho(f(x), y) + \rho(y, f(z)) < \frac{\varepsilon}{2} + \frac{\varepsilon}{2} = \varepsilon$, as desired. ∎

From this result and from the Heine-Borel theorem, we have that any continuous real-valued function on a closed interval $[a, b]$ is uniformly continuous. This is exactly what you need to prove that such functions are Riemann integrable.

Recall from calculus how we define the Riemann integral in terms the convergence of Riemann sums. Suppose that f is continuous on $[a, b]$. For any $\varepsilon > 0$, if we choose the uniform δ given by the Lebesgue number of the open cover $\left\{ f^{-1}(B(y, \frac{\varepsilon}{2})) : y \in Y \right\}$, and if we take a partition P of $[a, b]$ whose norm is smaller than δ, then for any two Riemann sums whose partitions refine P, the difference between the values of the Riemann sums will be less than $\varepsilon(b - a)$. This is enough to bring about convergence.

Exercises

1. Prove that any set with the cofinite topology is a compact space.

2. Prove that if X is a T_2-space and every subspace of X is compact, then X is discrete.

3. We say a collection $\mathcal{A}$ has the *finite intersection property* provided that whenever $\mathcal{B}$ is a finite subcollection of $\mathcal{A}$, then $\cap \mathcal{B} \neq \varnothing$. Prove that a space X is compact if and only if whenever $\mathcal{F}$ is a collection of closed subsets of X with the finite intersection property, $\cap \mathcal{F} \neq \varnothing$.

4. Suppose that X is a T_2 space and $(K_n : n \in \omega)$ is a decreasing sequence of nonempty compact subsets of X. Show that $\bigcap_{n \in \omega} K_n \neq \varnothing$. Note that this result gives you the nested-interval theorem from analysis, which says that $\bigcap_{n=0}^{\infty} [a_n, b_n] \neq \varnothing$ whenever $([a_n, b_n] : n \in \omega)$ is a nested sequence of closed bounded intervals.

5. Suppose that X is compact, and $f : X \to \mathbb{R}$ is continuous. Prove that there exist points x^* and x_* in X such that $f(x_*) \leq f(x) \leq f(x^*)$ for every $x \in X$. Note that this result gives you the "extreme-value theorem" from calculus, which says that every continuous real-valued function on a closed bounded interval has a maximum value and a minimum value.

6. Suppose that X is a compact T_2 space, U is an open subset of X, and $x \in U$. Prove that there exists a compact subset $K \subset X$ with $x \in Int(K)$ and $K \subset U$.

7. (*Tube lemma*) Suppose that $\mathcal{U}$ is a collection of open sets in $X \times Y$, where Y is compact, and $\mathcal{U}$ covers $\{x\} \times Y$, where $x \in X$. Show that there exists a finite subcollection $\mathcal{U}_1 \subset \mathcal{U}$ and an open set $V \subset X$ with $x \in V$ such that $\mathcal{U}_1$ covers $V \times Y$.

8. Prove that if X and Y are compact spaces, then $X \times Y$ is also compact. Do this directly from the definition, and don't cite the Tychonoff theorem. Hint: use the tube lemma.

9. Prove that if X is compact, Y is T_2, and $f : X \to Y$ is one-to-one, onto, and continuous, then f is a homeomorphism. This is theorem 7.6, so you should prove it directly and not cite that result.

10. Suppose (X, τ) is a compact Hausdorff space. Show that if σ is a topology on X with τ a proper subset of σ, then (X, σ) cannot be compact. Show that if σ is a topology on X with σ a proper subset of τ, then (X, σ) cannot be Hausdorff.

11. Let $C(X)$ denote the set of all continuous real-valued functions on the space X. Prove that if X is compact, then the distance function d on $C(X)$ given by $d(f, g) = \sup \{|f(x) - g(x)| : x \in X\}$ is a metric on $C(X)$.

Project: the Countable Ordinals

We present by this sequence of theorems and exercises the very useful example ω_1, the space of countable ordinals. We assume that the reader has some familiarity with the idea of partially ordered, linearly (or totally) ordered, and well-ordered sets.

Here is a very quick reminder. A **partial order** is a relation on a set that is reflexive, transitive, and antisymmetric. That is, a relation $\leq$ on a set P is a partial order if and only if the following are true for all $x, y, z \in P$:

1. $x \leq x$ (reflexive).

2. if $x \leq y$ and $y \leq z$, then $x \leq z$ (transitive).

3. if $x \leq y$ and $y \leq x$, then $x = y$ (antisymmetric).

A natural prototype for partial orders is subset inclusion.

A **total order** , or **linear order** , is a partial order in which any two elements are comparable. That is, a linear order on P is a partial order on P which also satisfies the following: for any $x, y \in P$, either $x \leq y$ or $y \leq x$.

A **well order** is a partial order so that any nonempty set has a first element. Note that a well order is always a linear order.

Theorem 7.21 *There is an uncountable well-ordered set Ω with largest element ω_1 such that if $\alpha \in \Omega$ and $\alpha \neq \omega_1$, then $\{\beta \in \Omega : \beta < \alpha\}$ is countable.*

Proof. Suppose A is any uncountable set, and $\leq$ is a well ordering of A. Let $U = \{x \in A : \{y \in A : y < x\}$ is uncountable$\}$. If $U \neq \varnothing$, then let ω_1 be the first element of U, and let $\Omega = \{x \in A : x \leq \omega_1\}$, so Ω has the desired properties. If $U = \varnothing$, then choose a point that is not an element of A (there are some, $\{A\}$ for instance) and call that point ω_1. Now let $\Omega = A \cup \{\omega_1\}$ and extend the order from A to Ω by defining $x < \omega_1$ for all $x \in A$. So Ω has the properties that we claim. ∎

We have used the fact that any set can be well ordered, and this is equivalent to the axiom of choice. This is not strictly necessary at this point, but as most topologists do, I believe the axiom of choice to be true and do not shrink from using it when convenient.

You notice that we have constructed Ω using any uncountable set and any well ordering on it that satisfies the conditions set forth in the theorem. We are interested only in the order structure of the set, and the actual underlying set doesn't matter. In the first part of the project, we outline the "usual" way of describing the ordinals. This uses a very special set and a very special order, but this set will be order isomorphic to any other way of describing Ω. The reader may wish to just use the theorem above and move directly to the next section: the space of countable ordinals.

Ordinals

Definition 7.22 *A set α is called an **ordinal** provided the following are true*

1. *if $x \in y \in \alpha$, then $x \in \alpha$ (i.e. α is $\in$-transitive), and*

2. *if $x \in y \in z \in \alpha$, then $x \in z$ (i.e. each element of α is $\in$-transitive).*

The class of all ordinals is usually denoted by Ord or sometimes On.

Problem 7.23 *Prove the following:*

1. *$\varnothing$ is an ordinal.*

2. *If $\alpha \in Ord$, then $\alpha \cup \{\alpha\} \in Ord$.*

3. *If A is a subset of Ord, then $\cup A \in Ord$.*

4. *If A is a nonempty subset of Ord, then $\cap A \in Ord$.*

Theorem 7.24 *Ord is well ordered by "$\in$ or $=$", that is $\alpha < \beta$ if and only if $\alpha \in \beta$.*

Proof. That "$\in$ or $=$" is a partial order is easy. Trichotomy is difficult, but it follows from the axiom of regularity (see any standard book on set theory, Monk, for instance). That any nonempty subset has a first element follows from 4. ∎

Definition 7.25 *If $\alpha \in Ord$, then we define the **successor** $\alpha + 1 = \alpha \cup \{\alpha\}$.*

Problem 7.26 *Assume that α and β are ordinals, and prove the following:*

1. *$\alpha \in \beta$ if and only if $\alpha \subset \beta$ and $\alpha \neq \beta$.*

2. *If $\alpha \in \beta$, then either $\alpha + 1 = \beta$ or $\alpha + 1 \in \beta$.*

3. *$\alpha = \cup(\alpha + 1)$.*

4. *Either $\alpha = \cup\alpha$ or $\alpha = (\cup\alpha) + 1$.*

5. $\alpha = \{\beta \in Ord : \beta < \alpha\}$.

Example 7.27 *Here are some ordinals*

define $0 = \varnothing$

$1 = 0 + 1 = \varnothing \cup \{\varnothing\} = \{0\}$

$2 = 1 + 1 = 1 \cup \{1\} = \{0, 1\}$

$3 = 2 + 1 = 2 \cup \{2\} = \{0, 1, 2\}$, *and so on, to*

$n + 1 = n \cup \{n\} = \{0, 1, 2, ..., n - 1, n\}$.

In this way, by recursion we construct the set of natural numbers, which we denote by ω, or sometimes ω_0.

We note (prove this!) that ω is an ordinal. It is called the first infinite ordinal, and we also point out (prove this too!) that $\omega = \cup \omega$. Continuing in this way, the following is a partial listing of in order of the elements of Ω:

$0, 1, 2, ..., \omega, \omega + 1, \omega + 2, ..., \omega + \omega (= \omega \cdot 2), \omega \cdot 2 + 1, \omega \cdot 2 + 2, ..., \omega \cdot 3, ..., \omega \cdot \omega, ...$

The first uncountable set on this list is ω_1.

Definition 7.28 *If $\alpha \in Ord$, and there is $\beta \in Ord$ with $\alpha = \beta + 1$, then we say α is a **successor ordinal**. Otherwise, α is a **limit ordinal**.*

Remark 7.29 *α is a limit ordinal if and only if $\alpha = \cup \alpha$.*

The space of countable ordinals

We now present the topological space of countable ordinals. The main thing that we need from the first part of the project is that an ordinal is equal to the set of all smaller ordinals. Following the usage in analysis, for a subset $A \subset Ord$, sup A is the least β such that $\alpha \leq \beta$ for all $\alpha \in A$.

Theorem 7.30 *If A is a countable subset of ω_1, then sup $A < \omega_1$.*

Proof. Let A be a countable subset of ω_1. For each $\alpha \in A$, the set $\{\beta : \beta < \alpha\}$ is countable, hence $\cup A$ is countable. Now $\omega_1 \setminus \cup A \neq \varnothing$ since ω_1 is uncountable. Let γ be the first element of $\omega_1 \setminus \cup A$. Now it is easy to check (and you should!) that sup $A \leq \gamma < \omega_1$. ∎

Theorem 7.31 *The set ω_1 has no infinite decreasing sequences.*

Proof. This is a property of any well-ordered set. If $(\alpha_n : n \in \omega)$ were decreasing, which one would be the first element in the well order? ∎

Proposition 7.32 *The collection of all sets having one of the forms $[0, \alpha)$ or $(\alpha, \omega_1]$ is a subbase for a topology on the set $\Omega = \omega_1 + 1$.*

Definition 7.33 *The topology generated by the subbase in proposition 7.32 is called the **order topology**.*

The order topology will be the *usual topology* on $\omega_1 + 1$, and the subspace topology which is inherited will be the *usual topology* on ω_1.

Remark 7.34 *Here are a few easy, but interesting facts. (You should check these!)*

1. If $\alpha < \beta$, then $[0, \beta + 1) \cap (\alpha, \omega_1] = (\alpha, \beta + 1) = (\alpha, \beta]$ is a basic open set.

2. If $\alpha = \beta + 1$, then $(\beta, \beta + 2) = (\beta, \alpha] = \{\alpha\}$. So every successor is an isolated point.

3. If α is a limit ordinal, then $\{(\beta, \alpha] : \beta < \alpha\}$ is a neighborhood base at α.

Now to complete the project, do the following exercises.

Project exercise 1. ω_1 is first countable, but $\omega_1 + 1$ is not.

Project exercise 2. In the space $\omega_1 + 1$, the point ω_1 is in the closure of the subset ω_1, but no sequence converges to it.

Project exercise 3. $\omega_1 + 1$ is regular and T_1, and thus ω_1 has these properties as well.

Project exercise 4. ω_1 is not separable, not Lindelöf, and not second countable.

Project exercise 5. $\omega_1 + 1$ is not separable and not second countable.

Project exercise 6. $\omega_1 + 1$ is Lindelöf, in fact compact, and thus normal.

Project exercise 7. If A and B are disjoint closed subsets of ω_1, then at least one of them must be bounded.

Project exercise 8. Use exercise 7 to show that ω_1 is normal.

Project exercise 9. ω_1 is countably compact.

Chapter 8

Locally Compact Spaces

We now turn to locally compact spaces. These are spaces that may fail to be compact, but each point in the space has a neighborhood that is compact. So the space *seems* compact at least "locally."

Definition 8.1 *A space X is called **locally compact** if and only if each point of X has a compact neighborhood in X.*

Example 8.2 *We list a few examples.*

1. *Any compact space is locally compact.*

2. *$\mathbb{R}$ is locally compact.*

3. *Any discrete space is locally compact.*

4. *The Michael line and the Sorgenfrey line are NOT locally compact.*

Proof. We prove only 4 The others are left to the reader. For the Michael line, suppose q is rational, and $V \subset \mathbb{M}$ is open with $\overline{V}$ being compact and $q \in V$. Choose irrationals a and b with $q \in (a, b) \subset [a, b] \subset V$. Since $[a, b]$ is closed in $\mathbb{R}$, it is also closed in $\mathbb{M}$, and $[a, b] \subset \overline{V}$. Thus $[a, b]$ must be compact in the Michael topology. List the rationals in $[a, b]$ as $\{q_n : n \in \mathbb{N}\}$. For each $n \in \mathbb{N}$, let $U_n = (q_n - \frac{b-a}{2^{n+2}}, q_n + \frac{b-a}{2^{n+2}})$, and note that the length of U_n is $\frac{b-a}{2^{n+1}}$. Now the union of all these sets would have total length covered of at most $\sum_{n=1}^{\infty} \frac{b-a}{2^{n+1}} = \frac{b-a}{2}$. Hence $[a, b] \backslash \bigcup_{n=1}^{\infty} U_n$ must be an infinite set of irrationals, and thus $\{U_n : n \in \mathbb{N}\} \cup \{\{x\} : x \in [a, b] \backslash \bigcup_{n=1}^{\infty} U_n\}$ is an open cover of $[a, b]$ in the Michael topology with no finite subcover. This contradiction shows that actually none of the rational points has a compact neighborhood.

For the Sorgenfrey line, we show that no point has a compact neighborhood. Suppose $x \in \mathbb{S}$, and suppose V is open in the Sorgenfrey topology, $x \in V$, and $\overline{V}$ is compact. Choose $b > x$ such that $[x, b) \subset V$. Since $[x, b)$ is also closed in the Sorgenfrey topology, we know that $[x, b)$ is compact. For each $n \in \omega$, let $x_n = b - \frac{b-x}{2^n}$. Note that $x_0 = x$, and $[x_n, x_{n+1}) \cap [x_{n+1}, x_{n+2}) = \varnothing$. Hence $\{[x_n, x_{n+1}) : n \in \omega\}$ is an open cover of $[x, b)$ with no finite subcover, a contradiction. ∎

One of the nice properties of locally compact spaces is that any locally compact Hausdorff space can be embedded into a compact Hausdorff space. The simplest construction of such a compact Hausdorff space is called the Alexandroff one-point compactification.

Theorem 8.3 *If X is a locally compact T_2 space, then there is a space X^* containing X such that the following are true:*

1. *X^* is a compact T_2-space.*

2. *$X^*\backslash X$ is one point.*

3. *If X is not compact, then X is dense in X^*.*

Proof. Choose a point not in X, and call it ∞. Let $X^* = X \cup \{\infty\}$. We define $\tau = \{U \subset X^* : U$ is an open subset of X, or $X^*\backslash U$ is a compact subset of $X\}$.

We claim that τ is a topology on X^*. Since $X^*\backslash X^* = \varnothing$, which is a compact subset of X, $X^* \in \tau$, and since $\varnothing$ is an open subset of X, $\varnothing \in \tau$. Suppose U and V are elements of τ. If one of U and V is a subset of X, then $U \cap V = U \cap V \cap X = (U \cap X) \cap (V \cap X)$, and since X is T_2, both $U \cap X$ and $V \cap X$ are open in X. If neither is a subset of X, then $X^*\backslash(U \cap V) = (X^*\backslash U) \cup (X^*\backslash V)$, which is the union of two compact subsets of X, and thus is compact. In either case, we have shown that $U \cap V \in \tau$. Finally suppose that $\{U_\alpha : \alpha \in I\} \subset \tau$. If $\infty \in U_\beta$ for some $\beta \in I$, then $X^*\backslash \bigcup_{\alpha \in I} U_\alpha = \bigcap_{\alpha \in I}(X^*\backslash U_\alpha) = (X^*\backslash U_\beta) \cap \bigcap_{\alpha \neq \beta}(X\backslash U_\alpha)$. This is a closed subset of the compact space $X^*\backslash U_\beta$, and hence it is compact. If $\infty \notin U_\alpha$ for all $\alpha \in I$, then $\bigcup_{\alpha \in I} U_\alpha$ is an open subset of X. In either case, $\bigcup_{\alpha \in I} U_\alpha \in \tau$. Thus τ is a topology on X^*.

It is obvious that 2 and 3 are satisfied, so it remains only to show that X^* is a compact T_2-space. To see that it is Hausdorff, suppose that $x, y \in X^*$. If $\{x, y\} \subset X$, then using the fact that X is T_2, we can choose disjoint open subsets of X, say U and V, such that $x \in U$ and $y \in V$. Since U and V are open in X, they are also open in X^*. If one of x and y is ∞, say $x = \infty$, then since X is locally compact, we can choose a compact neighborhood K of y in X. Now let $U = X^* \backslash K$ and $V = Int_X(K)$. So we have that U and V are disjoint open sets in X^*, and $\infty \in U$, $y \in V$. Hence X^* is T_2.

Now to see that X^* is compact, suppose that $\mathcal{U}$ is an open cover of X^*. Choose $U_0 \in \mathcal{U}$ with $\infty \in U_0$. We know that $X^* \backslash U_0$ is a compact subset of X , so choose some finite $\mathcal{V} \subset \mathcal{U}$ such that $X \subset \cup \mathcal{V}$. Thus $\{U_0\} \cup \mathcal{V}$ is the finite subcover we seek, and we see that X^* is compact. ∎

Theorem 8.4 *If X is a locally compact Hausdorff space, then X is an open subset of X^*, and if X is not compact itself, then X is dense in X^*.*

Proof. We leave this as an exercise. ∎

In general, for a space to be a *compactification* of X, we will want X to be embedded as a dense subset. For locally compact X, X is dense in X^* only when X is not compact itself. Nevertheless, we will follow the custom of calling X^* the one point compactification of X.

Definition 8.5 *The space X^* constructed in the proof of theorem 8.3 is called the* **Alexandroff one-point compactification** *(or just the one-point compactfication) of X.*

Thus we see that any noncompact, locally compact T_2 space can be embedded as an open dense subset of a compact Hausdorff space. This is a general theme that we will revisit in part two. We will want to see which spaces can be embedded as dense subsets of compact Hausdorff spaces. We will find that this is something that is true of all completely regular T_2 spaces. A research topic that has been very useful in set-theoretic topology is the study of what kind of subset we get when we do this embedding. For the locally compact spaces, we have just seen that we get an open set. People have also studied the class of spaces which can be embedded as G_δ-sets (the intersection of countably many open sets). These are called Čech complete spaces . It turns out that a metric space is Čech complete if and only if it has a compatible complete metric. Other classes of so-called "generalized metric spaces" have been defined in terms of more general subsets of the compact Hausdorff space into which our space can be embedded.

Example 8.6 *Here are some examples that you may have seen in analysis.*

1. *The one-point compactification of $\mathbb{R}$ is homeomorphic to a circle.*

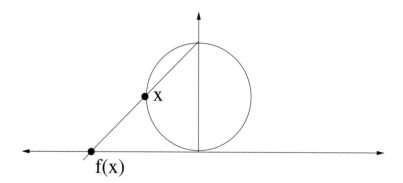

Figure 8.1: Homeomorphism of the circle onto $\mathbb{R}^*$

2. *The one-point compactification of the complex plane is the Riemann sphere. Of course, this is also the one-point compactification of $\mathbb{R}^2$.*

3. *In general, the one-point compactification of $\mathbb{R}^n$ is the surface of the unit ball in $\mathbb{R}^{n+1}$.*

Proof. We indicate how the homeomorphism from the set C, where

$$ C = \left\{ (x, y) : x^2 + (y - \frac{1}{2})^2 = (\frac{1}{2})^2 \right\} $$

onto $\mathbb{R}^*$ is defined. The details to show that this is a homeomorphism are left to the reader. We map $f((0,1)) = \infty$, where $\mathbb{R}^* = \mathbb{R} \cup \{\infty\}$. For all other x, we define f as in the picture. We define $f(x)$ to be the point on the horizontal axis where the line through x from the point $(0,1)$ crosses the horizontal axis. The construction is similar in higher dimensions. ■

Corollary 8.7 *Every locally compact T_2-space is completely regular.*

Proof. Let X be a locally compact T_2-space. The space X^* is compact and T_2, and thus X^* is normal. By Urysohn's lemma, X^* must be completely regular. It is easy to see that every subspace of a completely regular space is also completely regular. Hence X is completely regular. ■

The next theorem is a characterization that could be given as the definition of locally compact space, and it sometimes is.

Theorem 8.8 *If X is a Hausdorff space, then X is locally compact if and only if each point of X has a neighborhood base consisting of compact sets.*

Proof. The "if" part is obvious. Conversely, suppose X is locally compact. Suppose $x \in X$, and K is a compact neighborhood of x. Let $\mathcal{B}_x$ be any neighborhood base at x. Define $\mathcal{C}_x = \{\overline{B} \cap K : B \in \mathcal{B}_x\}$. Clearly, $\mathcal{C}_x$ is a collection of neighborhoods of x. Suppose that U is open, and $x \in U$. Since X is a locally compact T_2 space, we know that X is regular, so we choose $B \in \mathcal{B}_x$ with $x \in B \subset \overline{B} \subset U$. Thus $\overline{B} \cap K \subset U$. Hence $\mathcal{C}_x$ is a neighborhood base at x consisting of compact sets. ∎

The locally compact spaces have the same "Baire property" as the completely metrizable spaces. We will finish this chapter by proving this result.

Theorem 8.9 *(Baire category theorem) Any locally compact Hausdorff space is second category in itself.*

Proof. Suppose that X is a locally compact Hausdorff space. We need to show that X cannot be written as $\bigcup_{n \in \omega} F_n$ where each F_n is nowhere dense in X. Suppose, to the contrary, that we do have X written in this way. We will do an inductive construction to show that this is not possible. Since X is open, $\overline{F_0} \neq X$. So choose $x_0 \in X \backslash \overline{F_0}$ and an open neighborhood U_0 of x_0 with $U_0 \cap F_0 = \varnothing$. Let H_0 be an open neighborhood of x_0 with $x_0 \in H_0 \subset \overline{H_0} \subset U_0$, and $\overline{H_0}$ is compact. Now since $Int\overline{F_1} = \varnothing$, choose $x_1 \in H_0 \backslash \overline{F_1}$. Also choose U_1 an open neighborhood of x_1 such that $U_1 \cap F_1 = \varnothing$, and choose an open neighborhood H_1 of x_1 with $H_1 \subset \overline{H_1} \subset U_1 \cap H_0$. Note that $\overline{H_1}$ is compact since $\overline{H_1} \subset \overline{H_0}$. We continue inductively. For the $n+1$ step, choose $x_{n+1} \in H_n \backslash \overline{F_{n+1}}$, U_{n+1} and open neighborhood of x_{n+1} such that $U_{n+1} \cap F_{n+1} = \varnothing$. Let H_{n+1} be an open neighborhood of x_{n+1} with $H_{n+1} \subset \overline{H_{n+1}} \subset U_{n+1} \cap H_n$. By induction we construct the decreasing sequence $(H_n : n \in \omega)$ of open sets with $\overline{H_n} \cap F_n = \varnothing$ for all n. We note that by the compactness, there is a point $x \in \bigcap_{n \in \omega} \overline{H_n}$. For each n, we have $x \notin F_n$, and thus $x \in X \backslash \bigcup_{n \in \omega} F_n$, a contradiction. ∎

As we noted in the metric case, we could restate the Baire category theorem in terms of open dense subsets.

Corollary 8.10 *If X is a locally compact Hausdorff space and $(U_n : n \in \omega)$ is a sequence of open dense subsets of X, then $\bigcap_{n \in \omega} U_n \neq \varnothing$, in fact, $\bigcap_{n \in \omega} U_n$ is dense.*

Of course, compact Hausdorff spaces are locally compact, and so we have these results for compact spaces as well.

Corollary 8.11 *(Baire category theorem) Any compact Hausdorff space is second category in itself.*

Corollary 8.12 *If X is a compact Hausdorff space, and $(U_n : n \in \omega)$ is a sequence of open dense subsets of X, then $\bigcap_{n \in \omega} U_n \neq \varnothing$, and in fact $\bigcap_{n \in \omega} U_n$ is dense.*

Exercises

1. Prove that the examples in example 8.2 have the properties claimed.

2. Prove that if X is not compact, then X is dense in X^*.

3. Prove that ω_1 is locally compact. (This exercise requires that you have completed the project in chapter 7.)

4. Find the one-point compactification of ω_1. (This exercise requires that you have completed the project in chapter 7.)

5. Prove that if both X and Y are locally compact, then so is $X \times Y$.

6. Prove that any closed subspace of a locally compact space is locally compact.

7. Prove that any open subspace of a locally compact T_2-space is locally compact.

Chapter 9

Connected Spaces

Of all the topological properties that arise in calculus, compactness and connectedness are the most important. They are the properties behind the max-min theorem (a statement about compactness) and the intermediate-value theorem (a statement about connectedness). In the last two chapters, we discussed compactness. We turn now to the idea of connectedness.

For a space to be connected what our intuition demands is that the space be all in one piece; that is, it is not in two or more pieces.

Definition 9.1 *Let X be a topological space. We say X is **disconnected** if and only if there exist two disjoint nonempty open subsets H and K of X such that $X = H \cup K$. In this case, we call $\{H, K\}$ a disconnection of X, or we sometimes say $\{H, K\}$ disconnects X.*

Definition 9.2 *We say X is **connected** if and only if X is not disconnected.*

Theorem 9.3 *A space X is disconnected if and only if there exist two disjoint nonempty closed subsets A and B of X such that $X = A \cup B$.*

Proof. Suppose that there exist two disjoint nonempty closed subsets A and B of X such that $X = A \cup B$. Since $X \backslash A = B$ and $X \backslash B = A$, these sets are also open, thus $\{A, B\}$ is a disconnection of X. Conversely, if $\{H, K\}$ disconnects X, then $H = X \backslash K$ and $K = X \backslash H$ are disjoint closed sets with $H \cup K = X$. ∎

Another useful characterization of connectedness is that $\varnothing$ and X are the only subsets that are simultaneously open and closed.

Theorem 9.4 *A space X is connected if and only if the only subsets of X that are both open and closed are $\varnothing$ and X.*

Proof. If $\{H, K\}$ is a disconnection of X, then H is both open and closed, and $H \notin \{\varnothing, X\}$. Conversely, if A is both open and closed, then $\{A, X \backslash A\}$ is a disconnection of X unless $A \in \{\varnothing, X\}$. ∎

Our next result provides us with many useful examples. In particular, it tells us exactly which subsets of $\mathbb{R}$ are connected. First we recall what is meant by "interval."

Definition 9.5 *A subset $I \subset \mathbb{R}$ is an **interval** iff whenver $x, y \in I$ and $x < z < y$, we have $z \in I$.*

Theorem 9.6 *A subspace X of the real line with the usual topology is connected if and only if X is an interval.*

Proof. First suppose that X is not an interval. There must exist x, y, z with $x < y < z$, $x \in X$, $z \in X$, and $y \notin X$. Now $\{X \cap (-\infty, y), X \cap (y, \infty)\}$ is a disconnection of X. To see the converse, suppose that X is an interval, and hoping to produce a contradiction, suppose that $\{H, K\}$ is a disconnection of X. Since both H and K are nonempty, choose $x \in H$ and $z \in K$, and, without loss of generality, we assume that $x < z$. Let $y = \sup \{t < z : t \in H\}$. Note that $y \in [x, z] \subset X$ since X is an interval, and thus $y \in X$. For every $\varepsilon > 0$, by the definition of sup, we know that $H \cap (y - \varepsilon, y] \neq \varnothing$. Hence $y \in \overline{H} = H$. On the other hand, for every $\varepsilon > 0$, $[y, y + \varepsilon) \cap K \neq \varnothing$, and thus $y \in \overline{K} = K$. Thus $y \in H \cap K$, a contradiction. Hence X is connected. ∎

Corollary 9.7 *The following are true.*

1. $\mathbb{R}$ *is connected.*

2. *Subsets of connected spaces need not be connected.*

Proof. Left to the reader. ∎

Our next result gives us the familiar intermediate-value theorem from calculus.

Theorem 9.8 *Every continuous image of a connected space is also connected.*

Proof. Suppose that X and Y are spaces and $f : X \to Y$ is continuous and onto Y. If $\{H, K\}$ were a disconnection of Y, then $\{f^{-1}(H), f^{-1}(K)\}$ would be a disconnection of X. ∎

Corollary 9.9 *(intermediate-value theorem) If f is a continuous real-valued function on $[a, b]$ and m lies between $f(a)$ and $f(b)$, then there is a point $c \in [a, b]$ with $f(c) = m$.*

Proof. By theorems 9.6 and 9.8, $f([a, b])$ must be connected. So by theorem 9.6, $f([a, b])$ must be an interval. Hence $m \in f([a, b])$, as desired. ∎

Theorem 9.10 *A space X is disconnected if and only if there is a continuous function from X onto the two point discrete space $\{0, 1\}$.*

Proof. Suppose that $f : X \to \{0, 1\}$ is continuous and onto. Now

$$\{f^{-1}(\{0\}), f^{-1}(\{1\})\} \text{ is a disconnection of } X.$$

For the converse, suppose that $\{H, K\}$ disconnects X. Define $f : X \to \{0, 1\}$ by

$$f(x) = \begin{cases} 0, & \text{if } x \in H, \\ 1, & \text{if } x \in K. \end{cases}$$

Since $H \cap K = \varnothing$, f is well defined. Both $f|H$ and $f|K$ are constant functions, and thus they are continuous. Since $H \cup K = X$, by theorem 5.20, f is continuous. ∎

It is a bit of a stretch to call the next result a theorem, but it needs to be mentioned since it dispels one of the favorite mistakes by students.

Example 9.11 *Even if A and B are connected subspaces of the space X, $A \cap B$ may not be connected.*

Proof. In the picture opposite, both the line and the parabola are obviously connected subspaces of the plane, but the intersection is a two-point discrete space. ∎

We turn now to a condition that will allow us to tell when subsets of a space will be connected in the subspace topology.

Definition 9.12 *Sets H and K are **mutually separated** in a space X if and only if $H \cap \overline{K} = \overline{H} \cap K = \varnothing$.*

Theorem 9.13 *A subspace E of a space X is connected if and only if there is no pair of nonempty mutually separated sets H and K in X such that $E = H \cup K$.*

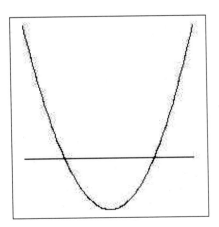

Proof. First, suppose $\{H, K\}$ is a disconnection of E. Both H and K are nonempty closed subsets of E. Thus,

$$
\begin{aligned}
H \cap cl_X(K) &= (H \cap cl_X(K)) \cap E = H \cap (cl_X(K) \cap E) \\
&= H \cap cl_E(K) = H \cap K = \varnothing.
\end{aligned}
$$

Similarly, $K \cap cl_X(H) = \varnothing$. Thus H and K are mutually separated, and $E = H \cup K$.

Conversely, suppose that $E = H \cup K$ where H and K are mutually separated. This means that

$$
\begin{aligned}
cl_E(H) &= E \cap cl_X(H) = (H \cup K) \cap cl_X(H) \\
&= (H \cap cl_X(H)) \cup (K \cap cl_X(H)) = H \cup \varnothing = H.
\end{aligned}
$$

Thus H is closed in E. Similarly, K is closed in E. Thus $\{H, K\}$ is a disconnection of E. ∎

Corollary 9.14 *If H and K are mutually separated in the space X and E is a connected subset of $H \cup K$, then either $E \subset H$ or $E \subset K$.*

Proof. In such a case, $E \cap H$ and $E \cap K$ are also mutually separated, and so we must have that at least one of these is empty. Since $E = (E \cap H) \cup (E \cap K)$, we then have either $E = E \cap H$, or $E = E \cap K$, and the result follows. ∎

Theorem 9.15 *Suppose X is a topological space.*

1. *If $X = \bigcup_{\alpha \in I} X_\alpha$, where each X_α is connected and $\bigcap_{\alpha \in I} X_\alpha \neq \varnothing$, then X is connected.*

2. *If each pair $\{x, y\} \subset X$ is contained in a connected set $E_{xy} \subset X$, then X is connected.*

3. *If $X = \bigcup_{n=1}^{\infty} X_n$, where each X_n is connected and $X_n \cap X_{n+1} \neq \varnothing$ for each n, then X is connected.*

Proof. To see 1, assume $X = \bigcup_{\alpha \in I} X\alpha$ where each $X\alpha$ is connected and $\bigcap_{\alpha \in I} X\alpha \neq \varnothing$. Suppose that H and K are mutually separated in X. For each $\alpha \in I$, by corollary 9.14, either $X_\alpha \subset H$ or $X_\alpha \subset K$. Choose $\beta \in I$. Suppose that $X_\beta \subset H$ (the argument would be analogous in the other case). For any $\alpha \in I$, $X_\alpha \cap X_\beta \neq \varnothing$, so $X_\alpha \cap H \neq \varnothing$, and thus $X_\alpha \subset H$. Hence $X = \bigcup_{\alpha \in I} X_\alpha \subset H$, that is $K = \varnothing$, and thus X is connected.

For 2, assume each pair $\{x, y\} \subset X$ is contained in a connected set $E_{xy} \subset X$. Choose a point $x_0 \in X$. By part 1, $X = \bigcup_{y \in X} E_{x_0 y}$ is connected.

To see 3, assume $X = \bigcup_{n=1}^{\infty} X_n$, where each X_n is connected and $X_n \cap X_{n+1} \neq \varnothing$ for each n. Note that we can use part 1 and induction to show that for each n, $A_n = \bigcup_{k=1}^{n} X_k$ is connected. Now using part 1 again, since $\bigcap_{n=1}^{\infty} A_n = X_1 \neq \varnothing$, we have that $X = \bigcup_{n=1}^{\infty} X_n = \bigcup_{n=1}^{\infty} A_n$ is connected. ∎

The previous result showed that a space is connected if it can be covered in a nice way by subspaces which are connected. The next result is along those same lines, but in this case we show that if there is one large connected subspace, then the space is connected.

Lemma 9.16 *If E is a connected subspace of X, and $E \subset A \subset \overline{E}$, then A is connected.*

Proof. Suppose E is connected, $E \subset A \subset \overline{E}$, and $\{H, K\}$ is a disconnection of A. Since $cl_A(E) = A$, and both H and K are open in A, we have that $H \cap E \neq \varnothing$ and $K \cap E \neq \varnothing$. Thus $\{H \cap E, K \cap E\}$ is a disconnection of E, a contradiction. ∎

Theorem 9.17 *If a space X has a dense connected subspace, then X is connected.*

Proof. This is immediate from lemma 9.16. ∎

We now turn to discussion of products of connected spaces. Here the behavior is very nice.

Theorem 9.18 *A nonempty product space $\prod_{n \in \omega} X_n$ is connected if and only if each factor space X_n is connected.*

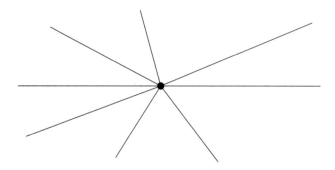

Figure 9.1: One point common to all subspaces—subspaces are connected.

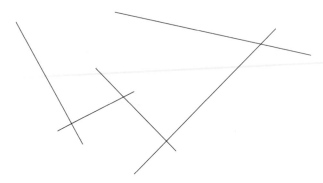

Figure 9.2: Each subspace intersects the next—subspaces are connected.

Proof. If $\prod_{n \in \omega} X_n \neq \varnothing$, then each projection mapping is continuous and onto, so each factor is connected if the product is, by theorem 9.8.

Now we suppose that X_n is connected for all $n \in \omega$. Choose a point $a \in \prod_{n \in \omega} X_n$. Let E be the set of all points in $\prod_{n \in \omega} X_n$ that lie in some connected subset of $\prod_{n \in \omega} X_n$ which contains a. By theorem 9.17, we will be finished if we can show that E is dense in $\prod_{n \in \omega} X_n$. Suppose that $U = \bigcap_{i=0}^{k} \pi_i^{-1}(U_i)$ is a nonempty basic open set in $\prod_{n \in \omega} X_n$. For each i, choose $b_i \in U_i$. We define $E_0, E_1, \dots E_k$ as follows:

$$E_0 = \left\{ c \in \prod_{n \in \omega} X_n : c_0 \text{ is arbitrary and } c_n = a_n \text{ if } n \neq 0 \right\},$$

$$E_1 = \left\{ c \in \prod_{n \in \omega} X_n : c_0 = b_0, c_1 \text{ is arbitrary, and } c_n = a_n \text{ otherwise} \right\},$$

$$E_1 = \left\{ c \in \prod_{n \in \omega} X_n : c_0 = b_0, c_1 = b_1, c_2 \text{ is arbitrary and } c_n = a_n \text{ otherwise} \right\}$$

$$E_k = \left\{ \begin{array}{c} c \in \prod_{n \in \omega} X_n : c_i = b_i \text{ for } i = 0, 1, 2, ..., k-1, \; c_k \text{ is arbitrary,} \\ \text{and } c_n = a_n \text{ otherwise} \end{array} \right\}.$$

For each i, E_i is homeomorphic to X_i, and hence E_i is connected for $i = 0, 1, 2, \ldots, k$. For each i, $E_i \cap E_{i+1} \neq \varnothing$, and thus $F = \bigcup_{i=0}^{k} E_i$ is connected. Since $a \in E_0 \subset F$, we have that $F \subset E$. Since $E_k \cap U \neq \varnothing$, we know that $F \cap U \neq \varnothing$, and thus $E \cap U \neq \varnothing$. Hence E intersects every nonempty basic open set, and thus E is dense. ■

Figure 9.3 is a picture of this proof for $\mathbb{R}^2$. To show that the set E is dense, we need to show that every basic open set in the product contains a point which is in a connected set which contains a. We take the horizontal line containing a and follow it to a point vertically aligned with the basic open set, and then we follow the vertical line to find an intersection. The union of the horizontal line and the vertical line is connected, and we have the point we seek. The same thing is done, in general; we follow the copies of the factor, changing one coordinate at a time, until we reach a point of the basic open set.

Certainly, singleton sets must be connected. So in any given space, a particular point will be contained in *some* connected subset. The largest connected set containing a point is called the *component* of the point.

Definition 9.19 *Suppose x is an element of the space X. We define C_x as $C_x = \bigcup \{C \subset X : x \in C, C \text{ is connected}\}$. The set C_x is called the **component** of x.*

Theorem 9.20 *Suppose X is a topological space. The following are true.*

1. *For each $x \in X$, C_x is connected.*

2. *For each $x \in X$, C_x is closed.*

3. *If $x, y \in X$, then either $C_x \cap C_y = \varnothing$ or $C_x = C_y$.*

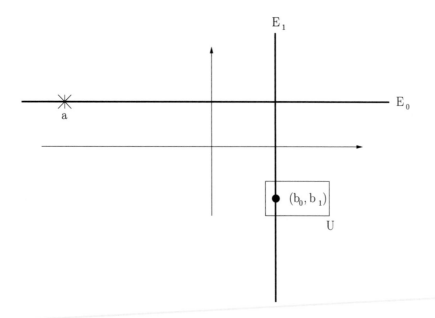

Figure 9.3: The product space of connected factor spaces is also connected.

Proof. Let $\mathcal{C}_x = \{C \subset X : x \in C \text{ and } C \text{ is connected}\}$. Since $x \in \cap \mathcal{C}_x$ and $\cap \mathcal{C}_x \neq \varnothing$ as well as $\cup \mathcal{C}_x = C_x$, be theorem 9.15,C_x is connected. Since C_x is connected, so is $\overline{C_x}$. Hence $\overline{C_x} \in \mathcal{C}_x$, and thus $\overline{C_x} \subset \cup \mathcal{C}_x = C_x$. Thus C_x is closed. Finally, if $C_x \cap C_y \neq \varnothing$, then $C_x \cup C_y$ is connected, and so $C_x \cup C_y \subset C_x$ and $C_x \cup C_y \subset C_y$. Hence $C_x = C_y$. ∎

Example 9.21 *Here are some examples of components.*

1. *If X is connected, then $C_x = X$ for all $x \in X$. In particular, in $\mathbb{R}$, $\mathbb{R}^n$, any interval, or a circle, components are the entire space.*

2. *If $X = (0, 1) \cup (5, 9) \subset \mathbb{R}$, then $C_7 = (5, 9)$.*

3. *If X is discrete space, then for each $x \in X$, $C_x = \{x\}$.*

4. *If X is a T_1 zero-dimensional space (see Example 4.16), then for each $x \in X$, $C_x = \{x\}$. In particular, in $\mathbb{Q}$, $\mathbb{R}\backslash\mathbb{Q}$, $\mathbb{S}$, and $\mathbb{M}$, components are singletons.*

Proof. We leave these to the reader. ∎

We close this chapter with one more property of connected spaces. This property is useful in building "paths" between points. We first define a "simple chain."

Definition 9.22 *A **simple chain** connecting two points a and b of a space X is a finite sequence $U_1, U_2, ..., U_n$ of open sets in X such that $a \in U_1$ only, $b \in U_n$ only, and $U_i \cap U_j \neq \varnothing$ if and only if $|i - j| \leq 1$.*

Theorem 9.23 *If X is connected and $\mathcal{U}$ is an open cover of X, then any two points of X can be connected by a simple chain consisting of elements of $\mathcal{U}$.*

Figure 9.4: Simple chain.

Proof. Assume X is connected and $\mathcal{U}$ is an open cover of X. Suppose $a \in X$. Let Z be the set of all points of X that can be connected to a by a simple chain consisting of elements of $\mathcal{U}$. Clearly, Z is nonempty since $a \in Z$, and Z is open since each point of the last "link" is in Z for any simple chain starting at X.

We will show that Z is also closed. Suppose that $z \in \overline{Z}$. Choose $U \in \mathcal{U}$ with $z \in U$. Since $U \cap Z \neq \varnothing$, we choose $x \in U \cap Z$. Choose a simple chain $U_1, U_2, ..., U_n$ of elements of $\mathcal{U}$ that connects a and x.

If $z \in U_k$ for some k, then let $j = \min \{k : z \in U_k\}$, and $U_1, U_2, ..., U_j$ will be a simple chain connecting a and z consisting of elements of $\mathcal{U}$, so that $z \in Z$.

Otherwise, we know that $U \cap U_n \neq \varnothing$, and we let $i = \min \{k : U \cap U_k \neq \varnothing\}$. Now $U_1, U_2, ..., U_i, U$ is a simple chain connecting a and z consisting of elements of $\mathcal{U}$. Again, this gives us $z \in Z$. Hence Z is nonempty and both open and closed, and thus $X = Z$. ∎

Exercises

1. Prove corollary 9.7.

2. Prove that if X is a T_1 zero-dimensional space and $x \in X$, then $C_x = \{x\}$.

3. Find an example of a space X and a subspace $E \subset X$ such that E is disconnected but there do not exist disjoints sets U and V which are open in X with $E \subset U \cup V$, $E \cap U \neq \varnothing$, and $E \cap V \neq \varnothing$. This shows that the obvious strengthening of theorem 9.13 is not possible.

4. Prove that if X is a connected, completely regular T_1 space, then X is uncountable.

5. Prove that if $f : [0,1] \to [0,1]$ is continuous, then f has a fixed point, i.e., a point x where $f(x) = x$.

6. Prove that any polynomial of odd degree has at least one real root.

7. Suppose that X is connected and $f : X \to Y$ is continuous. Prove that $G = \{(x, f(x)) : x \in X\}$ is a connected subspace of $X \times Y$. Think about a converse to this result. Suppose $f : X \to Y$, and $G = \{(x, f(x)) : x \in X\}$ is a connected subspace of $X \times Y$. Must X be connected? Why? Must f be continuous? Why?

Project: the Cantor Set

We will now work through the construction and some of the properties of the Cantor set, also known as the Cantor ternary set and the Cantor discontinuum.

Project exercise 1. Show that any point in $[0,1]$ can be represented as $\sum_{n=1}^{\infty} \frac{a_n}{3^n}$, where for each n, $a_n \in \{0, 1, 2\}$.

Project exercise 2. We denote $x = \sum_{n=1}^{\infty} \frac{a_n}{3^n}$ as $x = 0.a_1 a_2 a_3....$ Show that this "ternary" representation is unique except when $x = \frac{q}{3^n}$, and in this case there are exactly two representations.

The easiest way to describe the Cantor set for determining its combinatorial properties is the following.

Definition The **Cantor set** $\mathcal{C}$ is the set of all elements of $[0,1]$ that have a ternary representation consisting entirely of 0's and 2's.

The easiest way to see the geometric and topological properties of the Cantor set is the following.

Project exercise 3. Let $F_0 = [0,1]$ and $U_1 = (\frac{1}{3}, \frac{2}{3})$. Let $F_1 = F_0 \backslash U_1$. So we have removed the open middle third of F_0 to obtain F_1. Now let $U_2 = (\frac{1}{9}, \frac{2}{9}) \cup (\frac{7}{9}, \frac{8}{9})$, and let $F_2 = F_1 \backslash U_2$. Continue in this way, letting U_{n+1} be the union of the open intervals forming the middle thirds of the 2^n closed intervals making up F_n and $F_{n+1} = F_n \backslash U_{n+1}$. Show that the Cantor set is the set $\bigcap_{n=1}^{\infty} F_n$.

0			1

0	1/3	2/3	1

0	1/3	2/3	1

0	1/3	2/3	1

●

●

●

Project exercise 4. Show that the total length of the intervals that we have removed in the construction of $\mathcal{C}$ is 1. (So that, even though we have not defined measure on $\mathbb{R}$, it seems like $\mathcal{C}$ should have measure 0. When you get Lebesgue measure defined in succeeding courses, you will see that this is indeed the case.)

Project exercise 5. Giving $\mathcal{C}$ the subspace topology from the reals, show that $\mathcal{C}$ is a compact metric space.

Project exercise 6. Show that $|\mathcal{C}| = |2^{\mathbb{N}}|$, and in fact $\mathcal{C}$ is homeomorphic to the product space $2^{\mathbb{N}}$, thinking of $2 = \{0,1\}$ as the two point discrete space. Hint: consider $f(0.a_1 a_2 a_3...) = (\frac{1}{2}a_n : n \in \mathbb{N})$.

In particular, $\mathcal{C}$ is uncountable, and actually $|\mathcal{C}| = |\mathbb{R}|$. So that even though $\mathcal{C}$ is small in a measure sense, it is quite large in a cardinality sense.

Project exercise 7. Show that if $x \in \mathcal{C}$, then the component of x in the space $\mathcal{C}$ is $C_x = \{x\}$. Spaces that have the property that the components are singletons are called *totally disconnected*. In particular, you have now shown that $\mathcal{C}$ is totally disconnected.

Project exercise 8. Show that if $A \subset \mathbb{R}$, and A is uncountable, then there exists $x \in \mathbb{R}$ such that both $A \cap (-\infty, x)$ and $A \cap (x, \infty)$ are uncountable.

Project exercise 9. Show that if $A \subset \mathbb{R}$, and A is uncountable and closed, then A contains a copy of $\mathcal{C}$.

Project exercise 10. Show that any uncountable closed subset of $\mathbb{R}$ has the same cardinality as $\mathbb{R}$.

This last result is a weak form of the continuum hypothesis. The *continuum hypothesis* is the statement that any uncountable subset of $\mathbb{R}$ has the same cardinality as $\mathbb{R}$. This was one of the most famous unsolved problems in topology and set theory of the first half of the twentieth century. The question of the truth of the continuum hypotheiss has now been shown to be an undecidable question. That is, assuming it to be true causes no contradictions, and assuming it to be false also causes no contradictions.

Chapter 10

Other Types of Connectivity

In the last chapter, we explored the idea of a connected space. The idea is that the space should be, in some sense, all in one piece. The sense we developed in the last chapter was that the space could not be separated into disjoint open sets. In this chapter, we will describe another way of getting at this same idea. This time our notion of being all in one piece is that we can find a continuous path from any point to any other point. We will also develop "local" versions of these properties analogous to the notion of local compactness.

Definition 10.1 *A space X is called **pathwise connected** if and only if for any two points x and y in X, there is a continuous function $f : [0,1] \to X$ such that $f(0) = x$ and $f(1) = y$. We call such a function (as well as its range) a **path** from x to y.*

There is a similar idea called *arcwise connected* in which the path connecting any two points can be chosen so that its range is a homeomorphic copy of $[0,1]$. For completeness sake, we will write this definition below, but we will not really do anything with it. The reason for our omission is a remarkable result that these notions are equivalent for T_2-spaces.

Definition 10.2 *A space X is called **arcwise connected** if and only if for any two points x and y in X, there is a continuous function $f : [0,1] \to X$ such that $f(0) = x$, $f(1) = y$, and f is a homeomorphism onto its range.*

If X is T_2, then by the compactness of $[0,1]$, all that is needed to make a path f into a homeomorphism is that f is one-to-one.

119

Theorem 10.3 *Every pathwise connected space is connected.*

Proof. Suppose X is pathwise connected. If $\{H, K\}$ disconnects X, then choose $x \in H$, $y \in K$, and a path $f : [0, 1] \to X$ from x to y. Now $\{f^{-1}(H), f^{-1}(K)\}$ disconnect $[0, 1]$, a contradiction. Thus X is connected. ■

The converse is not true, as is shown by the following famous example. This example is often called the topologist's sine curve.

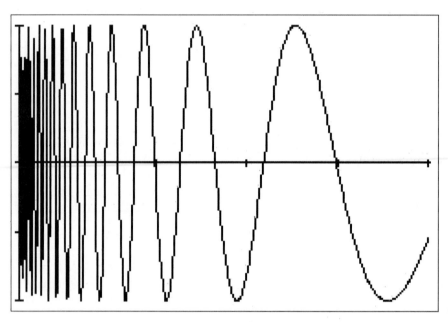

Example 10.4 *(Topologist's sine curve) Let $X = \{(x, \sin \frac{1}{x}) : x > 0\} \cup (\{0\} \times [-1, 1])$, and give X the subspace topology from $\mathbb{R}^2$. We see that X is connected, but not pathwise connected.*

Proof. We leave the details to the reader. Since the sine graph is connected and dense in X, we know that X is connected. Show that there cannot be a path from any point on the sine graph to any point on $\{0\} \times [-1, 1]$. ■

To get from connectedness to pathwise connectedness, we need an additional property of being able to connect in small neighborhoods by paths.

Definition 10.5 *A space X is called **locally pathwise connected** if and only if each point of X has a neighborhood base consisting of pathwise connected subsets.*

Before we proceed to the proof that connected, locally pathwise connected spaces are pathwise connected, we need the notion of *adding paths*.

Definition 10.6 *Suppose that* $f : [0,1] \to X$ *and* $g : [0,1] \to X$ *are paths such that* $f(1) = g(0)$. *We define the **sum** $h = f * g$ to be the function from* $[0,1]$ *into* X *defined by*

$$h(t) = \begin{cases} f(2t), & \text{if } 0 \le t \le \frac{1}{2}, \\ g(2t - 1), & \text{if } \frac{1}{2} \le t \le 1. \end{cases}$$

We leave it to the reader to verify that $f * g$ is continuous. Note that what this does is follow the f path to its last point, which is the first point of g, and then follow the g path. So we are laying the two paths end to end.

Theorem 10.7 *If a space X is connected and locally pathwise connected, then X is pathwise connected.*

Proof. Assume X is connected and locally pathwise connected. Suppose $a \in X$. We let H be the set of all points in X that can be joined to a by a path in X. Since $a \in H$, we know that $H \ne \varnothing$. We will show that H is both open and closed so that $H = X$, and hence X is pathwise connected.

Suppose $b \in H$. Let U be a pathwise connected neighborhood of b. For each point $x \in U$, there is a path connecting b to x, and there is a path connecting a to b since $b \in H$. Adding these two paths, we get a path from a to x. Hence $U \subset H$. Thus H is open.

To see that H is closed. Suppose $x \in \overline{H}$. Let U be a pathwise connected neighborhood of x. Since $U \cap H \ne \varnothing$, choose a point $b \in U \cap H$. There is a path from a to b, and there is a path from b to x. Adding these two paths, we get a path from a to x. Hence $x \in H$. Thus H is closed. ∎

Since line segments in $\mathbb{R}^n$ are clearly paths, balls are pathwise connected. Thus we get the following result which is important in both real and complex analysis. By this result, any open connected domain will contain a path from each point to each other point. Thus we will be able to construct line integrals.

Corollary 10.8 *Every open connected subset of $\mathbb{R}^n$ is pathwise connected.*

Local connectedness is an unusual property in that it neither implies connectedness, nor is it implied by connectedness. Of course, local properties rarely imply global properties, but global properties often imply local ones. That is not the case here.

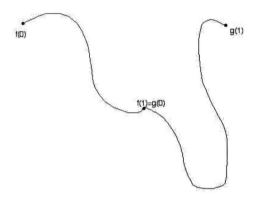

Figure 10.1: $f * g$

Definition 10.9 *A space X is called **locally connected** if and only if every point of X has a neighborhood base consisting of connected sets.*

Example 10.10 *We list two examples that show that there is no implication between connectedness and local connectedness.*

1. *Let $X = (0,1) \cup (2,3)$ with subspace topology from $\mathbb{R}$. The space X is locally connected but not connected.*

2. *Let $A = \{0\} \cup \{\frac{1}{n} : n \in \mathbb{N}\}$, and let $Y = ([0,1] \times A) \cup (\{0\} \times [0,1])$ viewed as a subspace of $\mathbb{R}^2$. The space Y is connected, even pathwise connected, but it is not locally connected.*

Proof. We will leave this as an exercise. ∎

To belabor the obvious just slightly, we note that locally pathwise connected spaces are locally connected. So the open connected subsets of Euclidean spaces provide many examples of locally connected spaces. Below we give a useful characterization of locally connected spaces.

Theorem 10.11 *A space X is locally connected if and only if for any open $U \subset X$ and $x \in U$, the component of the point x in the space U is open.*

Proof. Suppose that X is locally connected, $U \subset X$ is open, and $x \in U$. Let C_x be the component of x in the space U. For $y \in C_x$, we know that $C_y = C_x$. Moreover, there is an open connected set V in X with $y \in V \subset U$. Thus $y \in V \subset C_y = C_x$. Hence C_x is open.

For the converse, suppose that every component of every open subspace is open, and let $x \in X$. Suppose that U is open and $x \in U$. Let $C(x, U)$ be the component of x in the space U. Since $C(x, U)$ is open in U and U is open in X, $C(x, U)$ is open in X. Thus $\{C(x, U) : x \in U \text{ and } U \text{ is open in } X\}$ is a neighborhood base at x consisting of connected sets. Hence X is locally connected. ∎

Corollary 10.12 *If X is locally connected, then the components of X are both open and closed.*

Exercises

1. Complete the proof that the topologist's sine curve, example 10.4, is connected but not pathwise connected.

2. Show that the points on the y-axis of the topologist's sine curve do not have neighborhood bases consisting of connected sets. So this example is neither locally pathwise connected nor locally connected.

3. Show that part 2 of example 10.10 is pathwise connected but not locally connected.

4. Prove that any compact locally connected space has only a finite number of components.

5. Prove that any continuous image of a pathwise connected space is also pathwise connected.

6. Prove that any open continuous image of a locally connected space is also locally connected.

7. Prove that the product of two pathwise connected spaces is pathwise connnected. Is this true for infinite products?

8. Prove that the product of two locally connected spaces is locally connected. Is this true for infinite products?

Chapter 11

Continua

This chapter introduces continuum theory. Continuum theory is the area of topology where Robert Lee Moore made many of the main contributions to the mathematics. So perhaps we should take a few lines here to talk a little about this remarkable man. R. L. Moore is arguably the most successful trainer of mathematicians in the history of North America. The number of his students, and their students, and their students, etc., is well over 500 by this writing. Most of the topologists, at least the general, set-theoretic, or geometric topologists, as well as the continuum theorists, can trace their academic heritage to R. L. Moore, and almost all know in exactly what way. For example, I am not really a direct descendant of Moore, but rather sort of a great nephew. My advisor, Howard Wicke, had a long and productive collaboration with John Worrell, who was a student of Moore. Also my "academic grandfather," E. W. Chittendon, was a student with Moore at the University of Chicago where they both worked with E. H. Moore (no relation to R. L., incidentally). My first publication was coauthored with G. M. Reed and M. L. Wage. Reed was a student of Ben Fitzpatrick, who was a student of R. L. Moore, and Wage was a student of Mary Ellen Rudin, who was a student of R. L. Moore. That paper dealt with weakly uniform bases, a concept introduced by R. W. Heath, who was a student of F. B. Jones, who was a student of R. L. Moore. (To correct one minor historical inaccuracy: Ben Fitzpatrick was a graduate student at the University of Texas working with Moore when a open problem was mentioned in a differential equations course that he was taking. He solved the problem, and that was his Ph.D. dissertation. So technically, Fitzpatrick received his degree in differential equations, but his research, and that of his students, has all been in topology, and Ben considered himself to be a Moore student.)

The "Moore school" branched out to include set-theoretic topology, led by

Mary Ellen Rudin at Wisconsin; geometric topology, which includes many people but certainly R. D. Anderson at Louisiana State and Edwin Moise at Michigan were two of the principal players; and continuum theory, Moore's own area but certainly extended and expanded by R. H. Bing at Wisconsin and Texas. Moore's students from the second half of the decade of the 1940's, Richard D. Anderson, R. H. Bing, Edwin E. Moise, and Mary Ellen Rudin, were a pretty special group. Bing served a term as president of the American Mathematical Society; Anderson, Moise, and Rudin served terms as vice-president of the American Mathematical Society; and Anderson, Bing, and Moise served terms as president of the Mathematical Association of America. They have collectively dominated their areas of topology for over fifty years.

The accomplishments of Moore's students should not diminish the appreciation of his own. We will see a couple of his results later in this chapter, but this is really the tip of a very large iceberg. He had several papers in the Proceedings of the National Academy of Sciences, which is probably the most prestigious journal in the United States. Moore had 50 Ph. D. students, and almost all of them went on to do great things. I have mentioned only a few, and there are certainly others who deserve to be mentioned, but let us return to the task at hand.

We will not do anything more than introduce the reader to continuum theory in this book. This is a very large area of topology, and to do it in more than a very introductory way would require an entire course. We will also mention several results without proof in this chapter.

Definition 11.1 *A compact connected T_2-space is called a* **continuum**.

Many people require a space to be metrizable before calling it a continuum. This group would probably include most of the people working in continuum theory today. However, as we will see in the student project at the end of this chapter, there are some interesting nonmetric continua.

Example 11.2 *Here are some continua.*

1. $[0,1]$.

2. $S^1 =$ *the unit circle.*

3. $S^1 \times S^1 =$ *the torus.*

Proof. We leave it to the reader to prove that these are continua. ∎

Theorem 11.3 *Any product of continua is a continuum.*

Proof. Since any product of connected spaces is connected, any product of Hausdorff spaces is Hausdorff, and any product of compact spaces is compact, the result follows. ∎

We will say that a collection $\{K_\alpha : \alpha \in I\}$ of subsets of a space is *directed* by inclusion provided that if $\alpha, \beta \in I$, there is $\gamma \in I$ such that $K_\gamma \subset K_\alpha \cap K_\beta$. This is a somewhat more general notion than being linearly ordered by inclusion, since in that case either $\gamma = \alpha$ or $\gamma = \beta$ would suffice. In particular, decreasing sequences of sets are directed. Also, note that a neighborhood base at a point is directed.

Theorem 11.4 *If* $\mathcal{K} = \{K_\alpha : \alpha \in I\}$ *is a nonempty directed collection of continua in a space X, then $\bigcap_{\alpha \in I} K_\alpha$ is a continuum.*

Proof. Fix $\alpha_0 \in I$. Since K_{α_0} is a continuum and $\bigcap_{\alpha \in I} K_\alpha \subset K_{\alpha_0}$, we could as well assume that $X = K_{\alpha_0}$. So we shall assume that X is a compact T_2-space. Since X is T_2, each K_α is closed in X, and thus $\bigcap_{\alpha \in I} K_\alpha$ is closed in X and therefore compact. We only need to show that $\bigcap_{\alpha \in I} K_\alpha$ is connected.

Suppose, to the contrary, that $\{H, K\}$ is a disconnection of $\bigcap_{\alpha \in I} K_\alpha$. Since $\bigcap_{\alpha \in I} K_\alpha$ is closed in X, we know that H and K are disjoint closed subsets of X, and X is normal. Thus, we may choose disjoint open sets U and V in X such that $H \subset U$ and $K \subset V$. For each $\alpha \in I$, $K_\alpha \backslash (U \cup V) \neq \varnothing$, since otherwise $\{U, V\}$ would disconnect K_α. Let $\mathcal{F} = \{K_\alpha \backslash (U \cup V) : \alpha \in I\}$. Note that $\mathcal{F}$ is a nonempty collection of nonempty closed sets, and, since $\mathcal{K}$ is directed, $\mathcal{F}$ has the finite intersection property. So by compactness of X, $\cap \mathcal{F} \neq \varnothing$. Since

$$\cap\mathcal{F} = \cap\{K_\alpha \backslash (U \cup V) : \alpha \in I\} = \left(\bigcap_{\alpha \in I} K_\alpha\right)\backslash(U \cup V) \subset \left(\bigcap_{\alpha \in I} K_\alpha\right)\backslash(H \cup K) = \varnothing,$$

we have a contradiction, and no such disconnection can exist. Thus $\bigcap_{\alpha \in I} K_\alpha$ is connected, and the proof is complete. ∎

Corollary 11.5 *The intersection of any decreasing sequence of continua is a continuum.*

When we talk about a *subcontinuum* of a continuum K, we mean a subset of K which is a continuum in the subspace topology. This will be used in the next definition.

Definition 11.6 *A continuum K contained in a space X is said to be **irreducible about a subset** A of X provided that $A \subset K$ and no proper subcontinuum of K contains A. If $A = \{a, b\}$, then we say K is **irreducible between** a **and** b.*

Theorem 11.7 *If K is a continuum, and A is a nonempty subset of K, then there is a subcontinuum $C \subset K$ that is irreducible about A.*

Proof. Suppose K is a continuum and A is a nonempty subset of K. Let $\mathcal{K}$ be the set of all subcontinua of K which contain the set A. We order $\mathcal{K}$ by reverse inclusion, i.e. $K_1 \leq K_2$ if and only if $K_2 \subset K_1$. By theorem 11.4, the intersection of any linearly ordered subcollection of $\mathcal{K}$ is again a member of $\mathcal{K}$. This shows that every chain in $\mathcal{K}$ has an upper bound in $\mathcal{K}$. Hence by Zorn's lemma, $\mathcal{K}$ has a maximal element, call it C. By the maximality, C must be irreducible about A. ∎

A major research area in the development of continuum theory is the classification problem; that is, the problem of determining when two continua are homeomorphic. One of the most important ideas in the classification of continua is the idea of a cut point.

Definition 11.8 *Let X be a connected T_1-space. A **cut point** of X is a point $p \in X$ such that $X \backslash \{p\}$ is disconnected. If p is not a cut point, then we call p a **noncut point**. A **cutting** of X is a triple $\{p, U, V\}$ where p is a cut point and $\{U, V\}$ is a disconnection of $X \backslash \{p\}$.*

Example 11.9 *In the interval $[0, 1]$, the points 0 and 1 are noncut points, but all other points are cut points.*

Example 11.10 *In the circle, every point is a noncut point.*

Theorem 11.11 *If K is a continuum and $\{p, U, V\}$ is a cutting of K, then both $U \cup \{p\}$ and $V \cup \{p\}$ are continua.*

Proof. Suppose K is a continuum and $\{p, U, V\}$ is a cutting of K. We define a function $f : K \to U \cup \{p\}$ as follows:

$$f(x) = \begin{cases} x, & \text{if } x \in U \cup \{p\}, \\ p, & \text{if } x \in V. \end{cases}$$

Note that $f|(U \cup \{p\})$ is the identity map, and $f|(V \cup \{p\})$ is a constant. So both are continuous. Being complements of open sets, both $U \cup \{p\}$ and $V \cup \{p\}$ are closed sets, and $K = (U \cup \{p\}) \cup (V \cup \{p\})$. Hence f is continuous. Since $U \cup \{p\}$ is the continuous image of a continuum, $U \cup \{p\}$ is compact and connected. Thus $U \cup \{p\}$ is a continuum. The argument is similar for $V \cup \{p\}$. ∎

We have seen that there is a continuum so that every point is a noncut point (the circle), but the other extreme is not possible. Every continuum must have some noncut points.

Theorem 11.12 *If K is a continuum and $\{p, U, V\}$ is a cutting of K, then each of U and V must contain a noncut point.*

Proof. Suppose that each point $x \in U$ is a cut point, say $\{x, U_x, V_x\}$ is a cutting of K. If both U_x and V_x intersect $V \cup \{p\}$, then they would disconnect it since $U_x \cup V_x = K \setminus \{x\}$ and $x \in U$. Hence one of U_x and V_x must be contained in $U = K \setminus (V \cup \{p\})$. Let us assume that the notation is adjusted so that $U_x \subset U$ for all $x \in U$. Since $U_x \cup \{x\} \subset U$, we have that $V \subset V_x$, for each $x \in U$. Further if $x \in U$ and $y \in U_x$, we will show that $U_y \cup \{y\} \subset U_x$.

To see this, note that U_y and V_y are mutually separated in K, and $V_x \cup \{x\} \subset U_y \cup V_y$, and thus either $V_x \cup \{x\} \subset U_y$ or $V_x \cup \{x\} \subset V_y$. However, $V_x \cup \{x\} \subset U_y$ is impossible since $U_y \cap V_y = \varnothing$ and $V \subset V_x \cap V_y$. Hence $V_x \cup \{x\} \subset V_y$, and thus $U_x = K \setminus (V_x \cup \{x\}) \supset K \setminus V_y = U_y \cup \{y\}$.

Let
$$\mathcal{U} = \{U_x \cup \{x\} : x \in U\},$$

and let
$$\mathbb{P} = \{\mathcal{C} : \mathcal{C} \subset \mathcal{U} \text{ and } \mathcal{C} \text{ is linearly ordered by subset inclusion}\}.$$

Now for each $x \in U$, $\{U_x \cup \{x\}\} \in \mathbb{P}$; so $\mathbb{P} \neq \varnothing$. We order $\mathbb{P}$ by subset inclusion.

Suppose $\mathbb{K}$ is a linearly ordered subset of $\mathbb{P}$. Let us list $\mathbb{K} = \{\mathcal{C}_\alpha : \alpha \in I\}$. Consider $\bigcup_{\alpha \in I} \mathcal{C}_\alpha$. Clearly $\bigcup_{\alpha \in I} \mathcal{C}_\alpha \subset \mathcal{U}$ since $\mathcal{C}_\alpha \subset \mathcal{U}$ for each α. Now suppose $U_{x_1} \cup \{x_1\} \in \bigcup_{\alpha \in I} \mathcal{C}_\alpha$ and $U_{x_2} \cup \{x_2\} \in \bigcup_{\alpha \in I} \mathcal{C}_\alpha$. Choose α_1 and α_2 such that $U_{x_1} \cup \{x_1\} \in \mathcal{C}_{\alpha_1}$ and $U_{x_2} \cup \{x_2\} \in \mathcal{C}_{\alpha_2}$. Since $\mathbb{K}$ is linearly ordered, either $\mathcal{C}_{\alpha_1} \subset \mathcal{C}_{\alpha_2}$ or $\mathcal{C}_{\alpha_2} \subset \mathcal{C}_{\alpha_1}$, and each of these is linearly ordered. Thus either $U_{x_1} \cup \{x_1\} \subset U_{x_2} \cup \{x_2\}$ or $U_{x_2} \cup \{x_2\} \subset U_{x_1} \cup \{x_1\}$. Hence $\bigcup_{\alpha \in I} \mathcal{C}_\alpha$ is linearly ordered, and so $\bigcup_{\alpha \in I} \mathcal{C}_\alpha \in \mathbb{P}$.

By Zorn's lemma, $\mathbb{P}$ has a maximal element, say $\mathcal{C}$. We list $\mathcal{C} = \{U_x \cup \{x\} : x \in U^*\}$ for some $U^* \subset U$. By theorem 11.4, $\cap \mathcal{C}$ is a nonempty continuum and $\cap \mathcal{C} \subset U$. Let $q \in \cap \mathcal{C}$. Since $U_q \neq \varnothing$, we let $r \in U_q$, and, as before, $U_r \cup \{r\} \subset U_q$ and $U_q \cup \{q\} \subset U_x \cup \{x\}$ for all $x \in U^*$. Hence $\mathcal{C} \cup \{U_r \cup \{r\}\}$ is linearly ordered, so, by the maximality of $\mathcal{C}$, we must have $r \in U^*$. This gives us that $q \in U_r \cup \{r\} \subset U_q$, a contradiction. Hence U cannot consist entirely of cut points.

A similar argument shows that V must contain noncut points as well. ∎

The next result is really a corollary to the above, but both of these results are interesting enough that we will call them both theorems.

Theorem 11.13 *Every continuum with more than one point has at least two non-cut points.*

Proof. Suppose K is such a continuum. If K has no cut points, then each of the points is a noncut point, and there are more than one of them. If K has a cut point, then by the previous result it has at least two noncut points. ∎

We have seen already that a closed bounded interval has exactly two noncut points. A natural question is whether there are any other continua with exactly two noncut points. For metrizable continua, the answer is no.

We close this chapter with two results, which are due to R. L. Moore. We leave the proofs for a more advanced continuum-theory course.

Theorem 11.14 *If K is a metrizable continuum with exactly two noncut points, then K is homeomorphic to $[0,1]$.*

Theorem 11.15 *If K is a metrizable continuum such that for any two distinct points $\{a,b\} \subset K$, $K\backslash\{a,b\}$ is disconnected, then K is homeomorphic to a circle.*

Exercises

1. Prove that the spaces in example 11.2 are continua.

2. Suppose that $f : X \to Y$ is a homeomorphism of X onto Y. Show that for each $x \in X$, $f|(X\backslash\{x\})$ is a homeomorphism of $X\backslash\{x\}$ onto $Y\backslash\{f(x)\}$.

3. Prove that $\mathbb{R}$ and $[0,\infty)$ are not homeomorphic.

4. Prove that $[0,1]$ and a circle are not homeomorphic.

5. Prove that $\mathbb{R}$ and $\mathbb{R}^n$, for $n > 1$, are not homeomorphic.

6. Prove that any locally connected compact Hausdorff space is the union of a finite number of continua.

Project: the Lexicographic Square, a Nonmetric Continuum

Here we will develop another interesting and important example. It is easy to produce non-metric continua by taking uncountable products, as defined in part two. This one is nice in that it is elementary, and you can draw pictures.

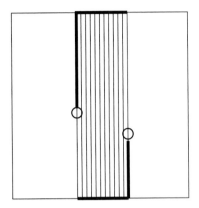

Figure 11.1: $((a, b), (c, d))$ with $a < c$

Definition. The **lexicographic square** is the set $X = [0, 1] \times [0, 1]$ with the dictionary, or lexicographic, order. That is $(a, b) < (c, d)$ if and only if either $a < b$, or $a = b$ and $c < d$. This is a linear order on X, and the example we seek is X with the order topology.

We follow the usual customs for intervals, so that

$$[(a, b), (c, d)) = \{(x, y) \in X : (a, b) \leq (x, y) < (c, d)\}.$$

As usual, the square brackets indicate that the endpoint is included, and the parentheses indicate that the endpoint is not included. A subbase for the order topology on X is the collection of all sets of form $[(0, 0), (a, b))$ or of form $((a, b), (1, 1)]$. In figure 11.1 we draw a picture of the basic open set $((a, b), (c, d))$ where $a < c$.

Project exercise 1. Draw pictures of basic open sets of form $[(0, 0), (0, b))$, $((1, b), (1, 1)]$, and $((a, b), (a, d))$.

Project exercise 2. Show that any basic open set about a point on the "top edge," that is, a point of form $(a, 1)$, where $a < 1$, must intersect the "bottom edge."

Project exercise 3. Show that X is a T_2-space.

Project exercise 4. Show that X is a compact space.

Project exercise 5. Show that X is a connected space.

From exercises 3 through 5, we see that X is a continuum.

Project exercise 6. Show that X has an uncountable collection of pairwise disjoint open sets.

Project exercise 7. Show that X is not separable.

Project exercise 8. Show that X is not a metrizable space.

Chapter 12

Homotopy

In this final chapter of part one, we introduce homotopy and the fundamental group. This is the beginning of the very broad subject of algebraic topology. Here we hope only to give a very brief introduction to the general theme that algebraic objects can be used to determine characteristics of topological spaces. In this chapter, as in the previous, some results will be stated, but their proofs are deferred until a later course.

Definition 12.1 *Suppose f and g are continuous functions from a space X into a space Y. We say f **is homotopic to** g, denoted $f \simeq g$, if and only if there is a continuous function $H : X \times [0,1] \to Y$ such that $H(x,0) = f(x)$ and $H(x,1) = g(x)$ for all $x \in X$. The map H is called a **homotopy** from f to g, and for clarity we often write $H : f \simeq g$.*

If we set $f_t(x) = H(x,t)$ for $x \in X$ and $t \in [0,1]$, we can think of the homotopy H as a family $\{f_t : t \in [0,1]\}$ of maps from X into Y that vary continuously with t. In this way, one can think of a homotopy as a continuous deformation which transforms $f = f_0$ into $g = f_1$.

Example 12.2 *Let f be the identity function on $\mathbb{R}$, i.e. $f(x) = x, \forall x$, and let g be the constant 0 function on $\mathbb{R}$. We see that $H : f \simeq g$ where $H(x,t) = (1-t)x$ for $x \in \mathbb{R}$ and $t \in [0,1]$.*

In this example, the homotopy is just rotating the graph of f about the origin into the graph of g.

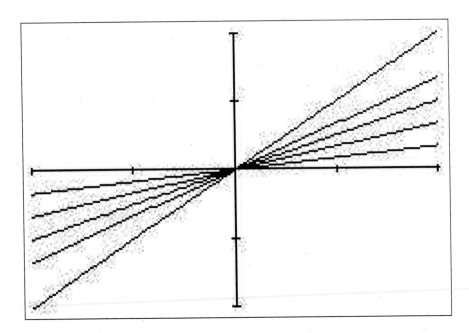

Example 12.3 *Let X be any space, and let Y be a convex subset of $\mathbb{R}^n$. For any two continuous functions f and g mapping X into Y, we have $H : f \simeq g$, where $H(x,t) = tg(x) + (1-t)f(x)$.*

Example 12.4 *The importance of the convexity in the previous example is made clear by looking at the circle and disk in the plane. Let*

$$X = S^1 = \{(x,y) \in \mathbb{R}^2 : x^2 + y^2 = 1\}$$

and let

$$Y = \{(x,y) \in \mathbb{R}^2 : x^2 + y^2 \leq 1\}.$$

Since Y includes the interior and X does not, we see that Y is convex and X is not convex. Consider the following pairs of functions. First $f : X \rightarrow Y$ and $g : X \rightarrow Y$ defined by $f(x) = x$ and $g(x) = (1,0)$. As in the previous example, $f \simeq g$.

Now look at the same functions viewed as mapping into the circle, $\widehat{f} : X \rightarrow X$ and $\widehat{g} : X \rightarrow X$ defined by $\widehat{f}(x) = x$ and $\widehat{g}(x) = (1,0)$. Intuitively, because we no longer have the interior of the circle to drag the graph across, these functions are no longer homotopic.

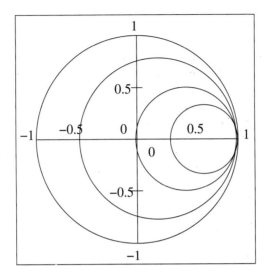

The result that the functions $\widehat{f}$ and $\widehat{g}$ are not homotopic, while it makes sense intuitively, is actually quite a significant result (see theorem 12.18). We will not do a careful proof in this book, but we will leave it to a more advanced algebraic topology course.

Notation 12.5 $C(X,Y)$ *is the set of all continuous functions from X into Y.*

Theorem 12.6 *The relation $\simeq$ is an equivalence relation on $C(X,Y)$.*

Proof. We need to verify reflexivity, symmetry, and transitivity. Suppose $f \in C(X,Y)$. Define $H : X \times [0,1] \to Y$ by $H(x,t) = f(x)$ for all $x \in X$ and $t \in [0,1]$. Now it is easy to see that $H : f \simeq f$. Thus $\simeq$ is reflexive. If $H : f \simeq g$, then we define $\widehat{H}(x,t) = H(x,1-t)$. It is easy to show that $\widehat{H} : g \simeq f$, and thus $\simeq$ is symmetric. Finally, suppose that $H_1 : f \simeq g$ and $H_2 : g \simeq h$. We define

$$H(x,t) = \begin{cases} H_1(x,2t), & \text{if } 0 \leq t \leq \frac{1}{2}, \\ H_2(x,2t-1), & \text{if } \frac{1}{2} \leq t \leq 1. \end{cases}$$

Now we have $H : f \simeq h$.

The only thing left is to show the continuity of the various homotopy functions. The only one that is a little tricky is the transitivity. To do this one, consider the restrictions to the closed subsets $X \times [0,\frac{1}{2}]$ and $X \times [\frac{1}{2},1]$. The details are left to the reader. ∎

Definition 12.7 *The equivalence classes in $C(X,Y)$ under the equivalence rela-tion $\simeq$ are called the **homotopy classes** of $C(X,Y)$.*

Theorem 12.8 *Compositions of homotopic maps are homotopic.*

Proof. Suppose $f_1 \simeq g_1$ where $f_1, g_1 \in C(X,Y)$, and $f_2 \simeq g_2$ where $f_2, g_2 \in C(Y,Z)$. We wish to show that $f_2 \circ f_1 \simeq g_2 \circ g_1$. Notice that these two functions are elements of $C(X,Z)$. Choose $H_1 : f_1 \simeq g_1$ and $H_2 : f_2 \simeq g_2$. Now $f_2 \circ H_1 : f_2 \circ f_1 \simeq f_2 \circ g_1$. We will therefore be finished if we can show $f_2 \circ g_1 \simeq g_2 \circ g_1$. To do this we define $H : X \times [0,1] \rightarrow Z$ by $H(x,t) = H_2(g_1(x),t)$. Now H is the composition of two continuous functions, namely $(x,t) \longmapsto (g_1(x),t)$ and H_2. Thus H is continuous. Hence $H : f_2 \circ g_1 \simeq g_2 \circ g_1$, and, by transitivity, the proof is complete. ∎

Definition 12.9 *A space X is called **contractible** if and only if the identity map $id_X : X \rightarrow X$ is homotopic to a constant map $c : X \rightarrow X$, $c(x) = x_0 \ \forall x$.*

Theorem 12.10 *A space X is contractible if and only if for any space Y, any two continuous maps $f, g : Y \rightarrow X$ are homotopic.*

Proof. The "if" part is easy; simply take $Y = X$, $f = id_X$, and g a constant map. Now suppose X is contractible, and $f : Y \rightarrow X$, $g : Y \rightarrow X$ are both continuous. Let id_X be the identity on X, and choose a constant mapping $c : X \rightarrow X$ such that $id_X \simeq c$. By theorem 12.8, $f = id_X \circ f \simeq c \circ f$ and $g = id_X \circ g \simeq c \circ g$. However, $c \circ f = c \circ g$, and thus by transitivity, $f \simeq g$. ∎

Definition 12.11 *Two spaces X and Y are **homotopy equivalent** if and only if there exist continuous functions $f : X \rightarrow Y$ and $g : Y \rightarrow X$ such that $f \circ g \simeq id_Y$ and $g \circ f \simeq id_X$. The maps f and g are called **homotopy equivalences**, and g is called the **homotopy inverse** of f.*

Remark 12.12 *We leave the following remarks as exercises.*

1. *Homotopy equivalence is an equivalence relation on the class of all topological spaces.*

2. *Homeomorphic spaces are homotopy equivalent, but not necessarily conversely.*

Theorem 12.13 *A space X is contractible if and only if X is homotopy equivalent to a one-point space.*

Proof. Suppose X is contractible, say $id_X \simeq c$, where $c(x) = x_0$ for all $x \in X$ and x_0 is a point of X. Let $Y = \{x_0\}$, and let $j = \{(x_0, x_0)\}$. Note that $j : Y \to X$ is the constant function with value x_0. Now, $j : Y \to X$ is clearly continuous. Moreover, $j \circ c = c \simeq id_X$ and $c \circ j = j = id_Y$. Thus c and j are homotopy equivalences.

For the converse, suppose X is homotopy equivalent to a one-point space. Say $f : X \to Y$ is a homotopy equivalence, where Y is a one-point space, and $g : Y \to X$ is the homotopy inverse of f. The map $g \circ f$ is a constant mapping on X which is homotopic to id_X. Hence X is contractible. ∎

We now turn to the definition of the fundamental group. This is a group in the algebraic sense, whose underlying set is the set of equivalence classes of closed paths, or loops, and the group operation is the sum of paths that we used in chapter 10. We will follow the custom of denoting the group operation as product in this setting since it will not generally be commutative, and the $+$ is usually reserved for commutative operations. So, the operation will be the product of two equivalence classes of loops, which is the equivalence class of the sum of the loops.

Definition 12.14 *Let X be a topological space, and $x_0 \in X$. A **loop based at** x_0 **is a path** $f : [0, 1] \to X$ which is continuous and satisfies $f(0) = f(1) = x_0$. Two loops f and g based at x_0 are called **loop homotopic** provided that there is homotopy $H : [0, 1] \times [0, 1] \to X$, which is continuous, $H(t, 0) = f(t)\forall t$, $H(t, 1) = g(t)\forall t$, and $H(0, s) = H(1, s) = x_0\forall s$. In this case, we write $f \simeq_{x_0} g$, and we say that H witnesses that $f \simeq_{x_0} g$.*

So if two loops with the same base point are loop homotopic, then we can continuously deform one loop into the other by a family of loops indexed by the unit interval.

Lemma 12.15 *The relation $\simeq_{x_0}$ is an equivalence relation on the set of all loops in X based at x_0.*

Proof. We leave this as an exercise. ∎

We will follow the custom that $[f]$ denotes the equivalence class of f. The underlying set for the fundamental group will be the set of all equivalence classes of loops based at a fixed point.

Definition 12.16 *The **fundamental group** of a space X at the base point x_0 is the set of equivalence classes of loops under the relation $\simeq_{x_0}$. The group operation*

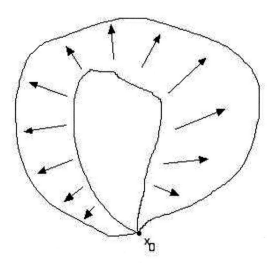

x_0

Figure 12.1: Loop homotopy.

is defined by $[f] \cdot [g] = [f * g]$ for any two loops f and g based at x_0. We denote the fundamental group by $\pi_1(X, x_0)$.

This is often called the first homotopy group, which would seem to imply that there is a second homotopy group. Indeed, that is the case. In fact, there are groups $\pi_n(X, x_0)$ for each $n \in \mathbb{N}$, but we will leave those for a more advanced course. Before we plunge into the proof that this is a group, let us recall the definition of this operation. Essentially, the product of two equivalence classes of loops compresses time so that the first loop is traced in time $[0, \frac{1}{2}]$ and the second is traced in $[\frac{1}{2}, 1]$. More precisely, if $f : [0, 1] \to X$ and $g : [0, 1] \to X$ are loops based at x_0, then $[f] \cdot [g] = [h]$ where h is the loop at x_0 defined by

$$h(t) = \begin{cases} f(2t), & \text{if } t \in [0, \frac{1}{2}], \\ g(1 - 2(1 - t)), & \text{if } t \in [\frac{1}{2}, 1]. \end{cases}$$

Theorem 12.17 *The fundamental group $\pi_1(X, x_0)$ is a group under the operation in definition 12.16.*

Proof. We leave it to the reader to check that the operation is well defined; that is, that the operation is independent of the representative of the equivalence class.

It is clear that this is a binary operation on $\pi_1(X, x_0)$, since the product of two loops is clearly a loop.

To prove associativity, suppose $[f], [g], [h] \in \pi_1(X, x_0)$. We need to show that $([f] \cdot [g]) \cdot [h] = [f] \cdot ([g] \cdot [h])$. We choose f, g, h as the representatives of the equivalence classes, and we must show that $(f * g) * h \simeq_{x_0} f * (g * h)$. As we look at these loops, we see that $(f * g) * h$ traces the f loop in $[0, \frac{1}{4}]$, the g loop in $[\frac{1}{4}, \frac{1}{2}]$, and the h loop in $[\frac{1}{2}, 1]$, while $f * (g * h)$ traces the f loop in $[0, \frac{1}{2}]$, the g loop in $[\frac{1}{2}, \frac{3}{4}]$, and the h loop in $[\frac{3}{4}, 1]$. So what we need is a homotopy which slows the first loop down by half, keeps the second loop at the same pace but translates it over, and speeds the third loop up by a factor of 2. Define $H : [0, 1] \times [0, 1] \to X$ by

$$H(t, s) = \begin{cases} f(\frac{4t}{1+s}), & \text{if } 0 \le t \le \frac{s+1}{4}, \\ g(4t - 1 - s), & \text{if } \frac{s+1}{4} \le t \le \frac{s+2}{4}, \\ h(1 - \frac{4(1-t)}{2-s}), & \text{if } \frac{s+2}{4} \le t \le 1. \end{cases}$$

It is easy to see that H is continuous since the three loops are. We need to check that $H(0, s) = H(1, s) = x_0 \forall s \in [0, 1]$. This is just a simple calculation, $H(0, s) = f(0)$ and $H(1, s) = h(1)$, and we have this part. Finally note that

$$H(t, 0) = \begin{cases} f(4t), & \text{if } 0 \le t \le \frac{1}{4}, \\ g(4t - 1), & \text{if } \frac{1}{4} \le t \le \frac{2}{4}, \\ h(1 - 2(1 - t)), & \text{if } \frac{2}{4} \le t \le 1, \end{cases}$$

$$= ((f * g) * h)(t),$$

for each $t \in [0, 1]$, and

$$H(t, 1) = \begin{cases} f(2t), & \text{if } 0 \le t \le \frac{2}{4}, \\ g(4t - 2), & \text{if } \frac{2}{4} \le t \le \frac{3}{4}, \\ h(1 - 4(1 - t)), & \text{if } \frac{3}{4} \le t \le 1, \end{cases}$$

$$= (f * (g * h))(t),$$

for each $t \in [0, 1]$. Hence we have that $(f * g) * h \simeq_{x_0} f * (g * h)$, i.e. $([f] \cdot [g]) \cdot [h] = [f] \cdot ([g] \cdot [h])$.

The existence of an identity is easy; the equivalence class of the constant loop is such an element. For the existence of inverses, consider a loop f, and let $\widehat{f}$ be the tracing of the loop backwards, that is $\widehat{f}(t) = f(1 - t)$ for each $t \in [0, 1]$. Now we need only show that $[f]^{-1} = [\widehat{f}]$, and we leave this as an exercise. ∎

Theorem 12.18 *The fundamental group of the circle is $\mathbb{Z}$.*

We will leave a careful proof of this to a more advanced course, but the intuition is fairly easy. The only kind of loops we can have must follow the circle all the way around and end at the starting point. So what we have to do is count the number of times the loop goes around the circle and the direction, eliminating any pairs of complete circles in opposite directions. So a loop will correspond to counting the number of the times around, i.e. a natural number, with a direction, i.e. plus or minus.

Example 12.19 *If X is a contractible space, then for any $x_0 \in X$, $\pi_1(X, x_0) = \{0\}$.*

Proof. Suppose X is contractible. Any loop is homotopic to the constant loop $c(x) = x_0 \forall x.$ ∎

We now turn to a discussion of the effects of continuous mappings on fundamental groups.

Theorem 12.20 *Suppose $\varphi : X \to Y$ is continuous. If f is a loop in X based at x_0, then $\varphi \circ f$ is a loop in Y based at $\varphi(x_0) = y_0$. Moreover, if $f \simeq_{x_0} g$, then $\varphi \circ f \simeq_{y_0} \varphi \circ g$.*

Proof. Suppose that f is a loop in X. It is clear that $\varphi \circ f$ is continuous, and $\varphi(f(0)) = \varphi(f(1)) = \varphi(x_0) = y_0$. Now suppose f and g are loops based at x_0. Suppose H is a loop homotopy witnessing that $f \simeq_{x_0} g$. We see that $\varphi \circ H$ is a homotopy witnessing that $\varphi \circ f \simeq_{y_0} \varphi \circ g$. Verifying that $\varphi \circ H$ has the desired properties is a matter of calculation which we leave to the reader. ∎

In particular, under the conditions of theorem 12.20, if $g \in [f]$, then $[\varphi \circ f] = [\varphi \circ g]$. Hence the mapping $\varphi_* : \pi_1(X, x_0) \to \pi_1(Y, \varphi(x_0))$ defined by $\varphi_*([f]) = [\varphi \circ f]$ is a well-defined function. Even more is true.

Theorem 12.21 *Suppose $\varphi : X \to Y$ is continuous, $x_0 \in X$, and $y_0 = \varphi(x_0)$. If $\varphi_* : \pi_1(X, x_0) \to \pi_1(Y, y_0)$ is defined by $\varphi_*([f]) = [\varphi \circ f]$, then φ_* is a group homomorphism.*

Proof. To see that φ_* is a homomorphism, we need only point out that $\varphi \circ (f * g) = (\varphi \circ f) * (\varphi \circ g)$ for any two loops f and g based at x_0. This is just a matter of writing down the definitions. Hence

$$\varphi_*([f] \cdot [g]) = \varphi_*([f * g]) = [\varphi \circ (f * g)] = [(\varphi \circ f) * (\varphi \circ g)] = [\varphi \circ f] \cdot [\varphi \circ g] = \varphi_*([f]) \cdot \varphi_*([g]).$$

∎

Definition 12.22 *If $\varphi : X \to Y$ is continuous, then the homomorphism $\varphi_* : \pi_1(X, x_0) \to \pi_1(Y, \varphi(x_0))$ defined in theorem 12.21 is called the **induced homomorphism**, or the homomorphism induced by φ.*

Theorem 12.23 *Suppose $\varphi : X \to Y$ is continuous, $\psi : Y \to Z$ is continuous, $x_0 \in X$, $y_0 = \varphi(x_0)$, and $z_0 = \psi(\varphi(x_0)) = \psi(y_0)$. The group homomorphism $\psi_* \circ \varphi_* : \pi_1(X, x_0) \to \pi_1(Z, z_0)$ is equal to $(\psi \circ \varphi)_*$.*

Proof. Again, we just do the calculation. If $[f] \in \pi_1(X, x_0)$, then $\psi_* \circ \varphi_*([f]) = \psi_*(\varphi_*([f])) = \psi_*([\varphi \circ f]) = [\psi \circ (\varphi \circ f)] = [(\psi \circ \varphi) \circ f] = (\psi \circ \varphi)_*([f])$. ∎

Theorem 12.24 *If X and Y are homeomorphic, then their fundamental groups, based at corresponding points, are isomorphic.*

Proof. Combining the results we have just established gives us this result once we have made the observation that for the identity map $id_X : X \to X$, $(id_X)_*$ is the identity operator on the group $\pi_1(X, x_0)$. In fact, for any loop f based at x_0 we have that $f = id_X \circ f$, and thus $(id_X)_*([f]) = [id_X \circ f] = [f]$. ∎

We have often referred to "the fundamental group" of a space as if the base point doesn't matter, when referring to the circle for instance. In general, this would be a serious mistake. The next result tells us that the base point doesn't matter if the space is pathwise connected.

Before stating this theorem, let us introduce one more bit of notation. If $p : [0, 1] \to X$ is a path from x_0 to x_1, then $p^- : [0, 1] \to X$ defined by $p^-(t) = p(1-t)$ is the path from x_1 to x_0 which traces p backwards.

Theorem 12.25 *If $p : [0, 1] \to X$ is a path from x_0 to x_1, and we define $\widehat{p} : \pi_1(X, x_0) \to \pi_1(X, x_1)$ by $\widehat{p}([f]) = [p^- * f * p]$, then $\widehat{p}$ is a group isomorphism.*

Proof. We compute

$$\begin{aligned} \widehat{p}([f]) \cdot \widehat{p}([g]) &= [p^- * f * p] \cdot [p^- * g * p] \\ &= [(p^- * f * p) * (p^- * g * p)] \\ &= [p^- * (f * g) * p] = \widehat{p}[f * g]) \end{aligned}$$

This shows that $\widehat{p}$ is a group homomorphism. To see that $\widehat{p}$ has an inverse function, consider the mapping $q : \pi_1(X, x_1) \to \pi_1(X, x_0)$ defined by $q([f]) = [p * f * p^-]$. It is just a calculation to see that $q(\widehat{p}([f])) = [f]$ for all $[f] \in \pi_1(X, x_0)$ and that $\widehat{p}(q([g])) = [g]$ for all $g \in \pi_1(X, x_1)$. We leave these calculations to the reader. ∎

We are again exposing only the tip of a very large iceberg in this brief discussion, but we are in a position to use the construction as we have done it so far to prove a big result in general topology, the Brouwer fixed-point theorem.

Definition 12.26 *A subset A of a topological space X is called a* **retract** *of X provided that there is a continuous $r : X \to A$ such that $r(x) = x$ for all $x \in A$. The mapping r is called a* **retraction**.

Lemma 12.27 *If A is a retract of X, then for any $x_0 \in A$, the homomorphism $j_* : \pi_1(A, x_0) \to \pi_1(X, x_0)$, induced by the inclusion mapping $j : A \to X$, is one-to-one.*

Proof. Suppose that $r : X \to A$ is a retraction. From theorem 12.23, it follows that $r_* \circ j_* = (r \circ j)_*$ which is the identity map on $\pi_1(A, x_0)$. Thus j_* must be one-to-one. ∎

For the remainder of this chapter, we will denote the unit disk in the plane by $\mathbb{D}$. Thus $\mathbb{D} = \{(x, y) \in \mathbb{R}^2 : x^2 + y^2 \le 1\}$.

Lemma 12.28 *The circle S^1 is not a retract of the disk $\mathbb{D}$.*

Proof. Since the disk $\mathbb{D}$ is convex, then for any point $x_0 \in S^1$, we know that $\pi_1(\mathbb{D}, x_0) = \{0\}$. Since $\pi_1(S^1, x_0)$ is infinite, the existence of a retraction of $\mathbb{D}$ onto S^1 would contradict lemma 12.27. ∎

Theorem 12.29 *(Brouwer fixed-point theorem) If $f : \mathbb{D} \to \mathbb{D}$ is continuous, then there is a point $x \in \mathbb{D}$ with $f(x) = x$, that is, f has a fixed point.*

Proof. Suppose $f : \mathbb{D} \to \mathbb{D}$ had no fixed points. For each $x \in \mathbb{D}$, we consider the straight line starting at $f(x)$ and passing through x. Let $r(x)$ be the point, distinct from $f(x)$, where this line intersects the boundary of the disk. From the continuity of f, it is easy to argue the continuity of r, and hence r is a retraction. This is a contradiction, and establishes the result. ∎

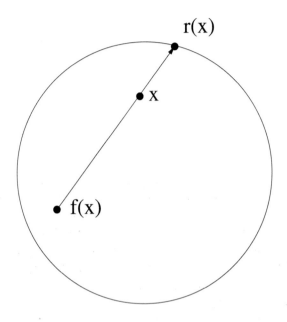

Exercises

1. Complete the details of the proof of theorem 12 that $\simeq$ is an equivalence relation.

2. Prove the results in remark 12.12.

3. Prove that theorem 12.24 can be extended to the statement that if X and Y are homotopy equivalent, then their fundamental groups are isomorphic.

4. Prove lemma 12.15.

5. Complete the proof of theorem 12.17 by showing that the identity and inverses are as indicated.

6. Complete the proof of theorem 12.20.

Part II

Chapter 13

A Little More Set Theory

In this chapter we introduce a little more set theory than we used in part one. We will simply refer the interested reader to any standard introductory set theory book for a development of the axioms of Zermelo and Fraenkel (for instance see Judith Roitman's *Introduction to Modern Set Theory*). These are the eight generally accepted statements on which mathematics has its foundations.

Axiom 13.1 *(Extensionality) Two sets are equal if they have the same elements.*

Axiom 13.2 *(Pairing) If x and y are sets, then $\{x, y\}$ is a set.*

Axiom 13.3 *(Union) If x is a set, then so is $\cup x$.*

Axiom 13.4 *(Separation) For any formula φ and any set x, $\{y \in x : \varphi(y)\}$ is a set.*

Axiom 13.5 *(Infinity) There is a set N with $\varnothing \in N$, and if $x \in N$, then $x \cup \{x\} \in N$, i.e. the natural numbers form a set.*

Axiom 13.6 *(Power Set) If x is a set, then the collection of all subsets of x is also a set.*

Axiom 13.7 *(Replacement) If the domain of a function is a set, then the range of the function is also a set.*

Axiom 13.8 *(Regularity) Every nonempty set has a minimal element with respect to $\in$, i.e. if $x \neq \varnothing$, then there exists $y \in x$ such that $y \cap x = \varnothing$.*

An axiom that is also generally accepted, but does not enjoy the same consensus that it is obviously true, is the axiom of choice. The axiom of choice says that if we have any collection of nonempty sets, then it is possible to choose one point from each set. If we have five sets, then we can do this, but if the collection is infinite, then the specification of a *method* to choose the points becomes more problematic. Nevertheless, when we think of the usual axioms for set theory, we will include the axiom of choice. We consequently refer to the usual axioms as ZFC.

Axiom 13.9 *(Axiom of choice, AC) If $\mathcal{A}$ is a nonempty family of nonempty sets, then there is a function f whose domain is $\mathcal{A}$ such that $f(A) \in A$ for each $A \in \mathcal{A}$.*

We usually call the function whose existence is given by AC a *choice function* on $\mathcal{A}$. This is because it chooses a point from each member of $\mathcal{A}$.

There are many equivalent forms of AC. Many of them are used so commonly that we often don't even think about the fact that they are forms of AC. One such statement which we did use in part one is Zorn's lemma.

Theorem 13.10 *(Zorn's lemma) If $\mathbb{P}$ is a nonempty partially ordered set, and $\mathbb{P}$ has the property that every linearly ordered subset of $\mathbb{P}$ has an upper bound in $\mathbb{P}$, then $\mathbb{P}$ has a maximal element.*

Zorn's lemma is equivalent to the axiom of choice. Many people think of Zorn's lemma as a theorem, and it *is* if you believe in AC. It is perhaps instructive to look at the proof that Zorn's lemma implies the axiom of choice.

Proof. Suppose that Zorn's lemma is true, and suppose that $\mathcal{A}$ is a nonempty collection of nonempty sets. Let

$$\mathbb{P} = \{f : f \text{ is a function, } dom f \subset \mathcal{A}, \text{ and } f(x) \in x \text{ for each } x \in dom f\}$$

and order $\mathbb{P}$ by function extension, which is actually subset inclusion if we think of a function as a set of ordered pairs. Since $\mathcal{A}$ is nonempty, we choose $A_0 \in \mathcal{A}$. Now since $A_0 \neq \varnothing$, we choose $x \in A_0$. Since $\{(A_0, x)\} \in \mathbb{P}$, we see that $\mathbb{P}$ is nonempty. Now suppose that $\mathcal{K}$ is a linearly ordered subset of $\mathbb{P}$. We want to show that $\cup \mathcal{K} \in \mathbb{P}$.

Since each element of $\mathcal{K}$ is a set of ordered pairs, so is $\cup \mathcal{K}$. Suppose (x_1, y_1) and (x_2, y_2) are distinct elements of $\cup \mathcal{K}$. We choose f_1 and f_2 elements of $\mathcal{K}$ such that $(x_1, y_1) \in f_1$ and $(x_2, y_2) \in f_2$. Since $\mathcal{K}$ is linearly ordered, either $f_1 \subset f_2$ or $f_2 \subset f_1$, and in either case, (x_1, y_1) and (x_2, y_2) are elements of a function, so $x_1 \neq x_2$. Hence $\cup \mathcal{K}$ is a function. Now, for a set to be an element of $dom \cup \mathcal{K}$, it must be an element of some member of $\mathcal{K}$. Hence $dom \cup \mathcal{K} \subset \mathcal{A}$. Finally, if $x \in dom \cup \mathcal{K}$, then there is some $f \in \mathcal{K}$ such that $x \in domf$, and thus $(\cup \mathcal{K})(x) = f(x) \in x$. So we see that $\cup \mathcal{K} \in \mathbb{P}$, and thus the hypotheses of Zorn's lemma are satisfied

This means that $\mathbb{P}$ must have a maximal element, which we will call f. If there is a set $A \in \mathcal{A} \backslash domf$, then we can choose a point $x \in A$, and $f \cup \{(A, x)\}$ would contradict the maximality of f. Hence f must be the choice function that we desired. ∎

Another statement which is equivalent to AC is the well-ordering principle.

Theorem 13.11 *(Well-ordering principle) Every set can be well ordered.*

A common misconception by students upon first hearing of this result is that the well ordering should have something to do with some existing order. *This is not the case.* We know that the real numbers can be well ordered, but the well ordering will not be the usual order on the reals.

The ordinal numbers are the sets that represent the various order types of well orderings. For example, if you introduce a new order on the natural numbers by declaring that each even number precedes all the odd numbers, then the set of natural numbers with this order is still well ordered. However the order type is now different. In fact, it is now $\omega + \omega$, rather than ω. We will not say too much more about the ordinals here, but simply refer the reader to the student project in chapter 7, and recall that each ordinal is equal to the set of all smaller ordinals.

The concept of two sets having the same cardinality is defined in definition 1.17. From the axiom of choice, we know that for any set we can find a well ordering, and consequently for any set there is a bijection from that set onto some ordinal. This gives us the opportunity to assign an actual number to the cardinality of a set.

Definition 13.12 *If X is a set, then*

$$|X| = first \{\alpha : \alpha \text{ is an ordinal and } |X| = |\alpha|\} .$$

The ordinals that serve as cardinalities of sets are called the *cardinal numbers*, or just the cardinals. On a moment's reflection, it is clear that a cardinal is an ordinal which is not equipotent with any smaller ordinal. These are sometimes referred to as *initial ordinals*. The class of all cardinals is denoted by *Card*.

We have used the term *class* a couple of times. A few words of explanation are in order. The axiom of regularity prohibits a set from being an element of itself. This leads to the paradox of the set of all sets. Would not this set be an element of itself? To avoid this paradox, what is done in Zermelo-Fraenkel set theory is to take the undefined object to be a *class*, and *sets* are classes which are "small enough" that they are elements of other classes. Classes that are not sets are called *proper classes*.

We need to have some small introduction to cardinal arithmetic. This will be very brief; again we refer the reader to any standard introductory set theory book for the details. (Roitman, Monk, and Kunen all have nice presentations.)

Definition 13.13 *Suppose κ and λ are cardinals.*

1. $\kappa + \lambda = |(\kappa \times \{0\}) \cup (\lambda \times \{1\})|$, *that is, $\kappa + \lambda$ is the cardinality of the disjoint union of two sets, one of cardinality κ and one of cardinality λ.*

2. $\kappa \cdot \lambda = |\kappa \times \lambda|$.

3. $\kappa^\lambda = |\{f : f$ *is a function with domain λ and range contained in κ}$|$.

Since $2 = \{0, 1\}$, we can identify any subset of a given set X with the function that takes value 1 on the subset and the value 0 on the complement. So, we see that the cardinality of the power set is 2 raised to the cardinality of the set. It is customary to denote the power set of X by $\mathcal{P}(X)$.

Lemma 13.14 *For any set X, $2^{|X|} = |\mathcal{P}(X)|$.*

The relationship between κ and 2^κ is one of the most interesting things in set theory. Cantor showed by a very clever argument, which we will repeat below, that κ is strictly smaller than 2^κ.

Theorem 13.15 *For any set X, we have $|X| < 2^{|X|}$.*

Proof. Clearly, $|X| \leq 2^{|X|}$, since the mapping $x \mapsto \{x\}$ is a one-to-one function of X into the power set of X. Now suppose that there exists a function $f : X \to \mathcal{P}(X)$ which is onto. Consider the set $A = \{x : x \notin f(x)\}$. Now $A \in \mathcal{P}(X)$, so there must be a point $z \in X$ with $f(z) = A$. From the definition of A, we see that $z \in A$ if and only if $z \notin A$, a contradiction. So no such function can exist, and the result is proved. ∎

The amount of the gap between the two cardinal numbers κ and 2^κ is the topic of the continuum hypothesis (when $\kappa = \omega$) and the generalized continuum hypothesis.

Continuum hypothesis. $\omega_1 = 2^\omega$.

From the student project in chapter 7, we recall that ω_1 is the first uncountable ordinal, and thus the first cardinal larger than ω. We usually denote the continuum hypothesis by CH.

Generalized continuum hypothesis. For any infinite cardinal κ, $\kappa^+ = 2^\kappa$, where κ^+ is the first cardinal strictly larger than κ.

Volumes have been written on the continuum hypothesis and related matters. We will be satisfied with just a few additional remarks here. Gödel showed in 1943 that in the "constructible universe," which is in some sense the smallest model of set theory, both the continuum hypothesis and the generalized continuum hypothesis are theorems. Hence one need not fear causing a contradiction by using these results. On the other hand, Cohen showed in 1963 that, in certain large models of set theory, both of these statements can be false. Hence one need not fear causing a contradiction by assuming that these statements are false. These are examples of statements which can be neither proved nor disproved by the axioms of ZFC. Many such statements have been discovered in the years since Cohen's result, and the occurrence of such things has been frequent and tantalizing in the field of set-theoretic topology for the last forty years.

Since any number in $[0, 1]$ can be written as $\sum_{n=1}^{\infty} \frac{a_n}{2^n}$ for an appropriately chosen sequence $(a_n : n \in \mathbb{N})$ consisting of 0's and 1's, it is clear that $|[0, 1]| = 2^\omega$. From this it follows that $|\mathbb{R}| = 2^\omega$. Hence we can reformulate the continuum hypothesis as a statement about the real numbers.

Theorem 13.16 *CH is true if and only if whenever A is an infinite subset of $\mathbb{R}$, either $|A| = |\mathbb{N}|$ or $|A| = |\mathbb{R}|$.*

An interesting result, which has been known for almost 100 years, but seems to have escaped the notice of many in the mathematics community, is that when the set A in the above is required to be a Borel set, this statement is a theorem of ZFC. The class of Borel sets is the smallest subset of $\mathcal{P}(\mathbb{R})$ which contains the open sets and is closed under both complements and countable unions. This is the smallest class needed to talk about measure on the real line. In particular, for most of the sets which arise in engineering, analysis, and probability, CH is "true."

There is a proper class function that maps the class of ordinals one-to-one and onto the class of all infinite cardinals. This function is called $\aleph$, which is the first letter of the Hebrew alphabet, *aleph*. It is defined as follows.

Definition 13.17 *We define $\aleph$ by recursion as follows:*

1. $\aleph_0 = \omega$.

2. $\aleph_{\beta+1} = (\aleph_\beta)^+$.

3. If α is a limit ordinal, then $\aleph_\alpha = \bigcup_{\beta<\alpha} \aleph_\beta$.

Using this notation, we could formulate the continuum hypothesis as "$\aleph_1 = 2^{\aleph_0}$," and the generalized continuum hypothesis as "$\aleph_{\beta+1} = 2^{\aleph_\beta}$ for all ordinals β."

We include, without proof, the following theorem, which gives us what we need about cardinal arithmetic. The reader is referred again to any standard introductory text on set theory. Kunen's *Set Theory* has a nice concise treatment. We also remind the reader that we are assuming in this theorem that we are dealing with *infinite* cardinals. These results are not all true for the finite cardinals, i.e. the natural numbers.

Theorem 13.18 *Suppose κ and λ are infinite cardinals.*

1. $\kappa + \lambda = \max\{\kappa, \lambda\}$.

2. $\kappa \cdot \lambda = \max\{\kappa, \lambda\}$.

3. if $\kappa \leq \lambda$, then $\kappa^\lambda = 2^\lambda$.

4. $(2^\kappa)^\lambda = 2^{\kappa \cdot \lambda}$.

Our last result of this chapter is another for which we will refer the reader to a set theory book, such as Kunen. This result is very useful in studying spaces of ordinals. While the result is true more generally, we will state it only for ω_1 since that is the ordinal space where we will use it. Before stating this result, we need one more definition.

Definition 13.19 *A subset $S \subset \omega_1$ is called* **stationary** *provided that $S \cap A \neq \varnothing$ whenever A is a closed unbounded subset of ω_1.*

Theorem 13.20 *(Pressing down lemma) If S is a stationary subset of ω_1, and $f : S \to \omega_1$ has the property that $f(\alpha) < \alpha$ for all $\alpha \in S$, then there is a point $\beta \in \omega_1$ such that $f^{-1}(\beta)$ is stationary.*

Exercises

1. (Every vector space has a basis.) By a basis for a vector space V, we mean a subset $B \subset V$ that is *linearly independent* in the sense that there is no nontrivial finite linear combination of elements of B that produces the zero vector and that *spans* V in the sense that every vector in V can be written (uniquely) as a finite linear combination of elements of B. Use Zorn's lemma to show that every vector space contains a maximal linearly independent subset, and use the maximality to show that such a set must be a basis.

2. (Perfect spaces.) Sets that can be written as the intersection of some countable family of open sets are called G_δ-sets. A space X is called a *perfect space* if every closed subset of X is a G_δ-set.

   (a) Show that the set L of limit ordinals in ω_1 is a stationary set, and use the pressing down lemma to show that if U is an open subset of ω_1 which contains L, then $\omega_1 \backslash U$ must be countable. Use this result to show that L cannot be written as the intersection of any countable family of open sets in ω_1. Thus ω_1 is not perfect.

   (b) Prove that every metric space is perfect.

   (c) Prove that if X is a regular hereditarily Lindelöf space, then X is perfect.

   (d) Prove that if X has the property that for every closed subset $A \subset X$, there is a continuous $f : X \to [0,1]$ with $A = f^{-1}(0)$, then X is perfect.

   (e) Use the Baire category theorem to show that $\mathbb{Q}$ is not a G_δ-set in $\mathbb{R}$. Use this result to show that the Michael line $\mathbb{M}$ is not perfect.

A word of caution: an appealing word like *perfect* is bound to have other usages. One of the most common is that a space is perfect if every point of the space is a cluster point, i.e. there are no singleton open sets. In that usage, sometimes perfect sets are also required to be compact.

The origin of our usage of the word perfect is from the study of normality. If a space is normal and perfect, then it is *perfectly normal* in the sense that in the Urysohn's lemma (theorem 6.26) characterization of normality $A = f^{-1}(0)$ and $B = f^{-1}(1)$.

Chapter 14

Topological Spaces II

If part two of this book is to be used as a free standing one-semester course in general topology for graduate students, then some of the material from part one would need to be mentioned again. In this chapter, we remind the reader of many of the basic properties of topological spaces, and we reexamine several of the most interesting examples and some new examples. The treatment is brief, containing few proofs, so the students may need to review relevant portions of part one.

Definition 14.1 *A **topology** on a set X is a subcollection $\tau \subset \mathcal{P}(X)$ such that*

  1. *$\varnothing \in \tau$,*

  2. *$X \in \tau$,*

  3. *if $U \in \tau$ and $V \in \tau$, then $U \cap V \in \tau$, and*

  4. *if $\mathcal{U} \subset \tau$, then $\cup \mathcal{U} \in \tau$.*

We refer to the pair (X, τ) as a topological space. Often the topology is clear from the context, and then we usually refer to the topological space X, or just the space X. The subsets of X which are elements of the topology are called the *open sets* of the topological space X.

Example 14.2 *Here we remind the reader of some familiar examples from part one. See example 4.4.*

  1. *For any set X, $\mathcal{P}(X)$ is a topology on X. We call this the discrete topology.*

2. If (X,d) is a metric space, then the topology generated by the metric consists of all those subsets $U \subset X$ such that for each $x \in U$ there exists $\varepsilon_x > 0$ such that $B(x, \varepsilon_x) \subset U$.

3. In particular, $\mathbb{R}$ is the space of real numbers with the topology generated by the Euclidean metric, i.e. $d(x,y) = |x - y|$.

4. For any natural number n, we denote by $\mathbb{R}^n$ the n-dimensional Euclidean space.

5. The Michael line, denoted by $\mathbb{M}$, is the space obtained by changing the usual topology on $\mathbb{R}$ by isolating all the irrationals. That is $\mathbb{M} = (\mathbb{R}, \tau_M)$ where

$$\tau_M = \{A \subset \mathbb{R} : A = U \cup F \text{ where } U \text{ is open in } \mathbb{R} \text{ and } F \subset \mathbb{R}\backslash\mathbb{Q}\}.$$

The reader should verify that these topologies really do satisfy the definition.

We regard two topological spaces to be the same if there is a one-to-one correspondence between the points of the two underlying sets and the same correspondence induces a one-to-one correspondence between the open sets. That is, we will regard two spaces as equal if they are homeomorphic (see chapter 5).

Definition 14.3 *A continuous, open, one-to-one, and onto mapping is called a* **homeomorphism**. *If there is a homeomorphism f mapping X onto Y, then we say X and Y are homeomorphic.*

Definition 14.4 *If τ_1 and τ_2 are two topologies on the same set X, then we say τ_1 is* **weaker** *than τ_2 provided $\tau_1 \subset \tau_2$, i.e. the weaker topology has "fewer" open sets. We also sometimes say that τ_2 is* **stronger** *than τ_1. Other terms are also used; "**coarser**" for weaker, and "**finer**" for stronger.*

It is often more convenient to work with a base for the topology rather than the entire topology. We remind the reader of the definitions of base and subbase.

Definition 14.5 *A* **base** *for a topology τ on X is a subcollection $\mathcal{B} \subset \tau$ such that if $U \in \tau$, then there is a subcollection $\mathcal{B}_U \subset \mathcal{B}$ such that $U = \cup\mathcal{B}_U$.*

The elements of a particular base are called *basic open sets*.

Definition 14.6 *A* **subbase** *for a topology τ is a subcollection $\mathcal{S} \subset \tau$ such that $\{(\cap\mathcal{F}) \cap X : \mathcal{F}$ is a finite subset of $\mathcal{S}\}$ is a base for τ.*

The elements of a particular subbase are called *subbasic open sets*.

We often describe topological spaces by saying what a base for the topology is. In the examples above, we said that the topology generated by a metric has the collection of open balls as a base. On the other hand, also in the example above, we gave a full description of the topology for the Michael line. We could have described a base for this space as well by saying that the base consists of the all the usual open sets in the reals together with the singletons of all the irrationals.

In chapter 4, we proved theorems that characterize which collections of sets can serve as a base or a subbase for a topology. We repeat those theorems here.

Theorem 14.7 *(see theorem 4.19) If X is a set, and S is any collection of subsets of X, then S is a subbase for some topology on X.*

Theorem 14.8 *(see theorem 4.20) A family B of subsets of a set X is a base for a topology on X if and only if B has the following two properties:*

B1 $\cup B = X$

B2 *If $U \in B$, $V \in B$, and $x \in U \cap V$, then $\exists W \in B$ such that $x \in W \subset U \cap V$.*

Example 14.9 *Here we list some examples that are easily described in terms of bases.*

1. *If X is a set which has a linear order $<$, and we follow the custom that for any points $a, b \in X$, we have $(a, b) = \{x \in X : a < x < b\}$, $(\leftarrow, b) = \{x \in X : x < b\}$, and $(a, \rightarrow) = \{x \in X : x > a\}$, then the collection $S = \{(\leftarrow, b) : b \in X\} \cup \{(a, \rightarrow) : a \in X\}$ is a subbase for a topology on X. With this topology, X is called a **Linearly Ordered Topological Space**, or a LOTS for short. The study of spaces of this type has been very fruitful and has generated many familiar spaces.*

2. *The usual topology on $\mathbb{R}$ is the LOTS topology generated by the usual order on the reals.*

3. *The ordinal space ω_1 is a LOTS. It is interesting to note that for any $\alpha \in \omega_1$, we have that $(\beta, \alpha + 1) = (\beta, \alpha]$, and if α is a limit ordinal, then $(\beta, \alpha) = \bigcup_{\gamma < \alpha} (\beta, \gamma]$. Thus we can also use $\{(\beta, \alpha] : \beta < \alpha\} \cup \{\{0\}\}$ as a base for ω_1.*

4. *The lexicographic square is the LOTS defined on $[0, 1] \times [0, 1]$ with the lexicographic, or dictionary order. That is $(a, b) < (c, d)$ if and only if either $(a < c)$ or $(a = c$ and $b < d)$. See the project in chapter 11.*

5. *The Sorgenfrey line is denoted by* $\mathbb{S}$, *and it is the space which has* $\mathbb{R}$ *as the underlying set and the base consisting of* $\{[a, b) : a < b\}$.

The reader should verify that the examples above satisfy the criterion to be a base so that the spaces mentioned are indeed well defined.

Definition 14.10 *If* X *is a topological space and* $F \subset X$, *then we say* F *is a* **closed** *set if and only if* $X \backslash F$ *is an open set.*

We see from DeMorgan's laws and the definition of topology that $\varnothing$ and X are closed, and that *finite* unions of closed sets are closed, while *arbitrary* intersections of closed sets are closed.

Definition 14.11 *Suppose* A *is a subset of a space* X. *We define the* **closure** *of* A *to be the set* $\overline{A}$ *where*

$$\overline{A} = \cap\{F \subset X : F \text{ is closed in } X \text{ and } A \subset F\}.$$

Theorem 14.12 *If* X *is a space and* $A \subset X$, *then the following are true:*

1. $\overline{A}$ *is a closed set.*

2. $A \subset \overline{A}$.

3. *If* $A \subset B \subset X$, *then* $\overline{A} \subset \overline{B}$.

4. *If* F *is a closed set with* $A \subset F$, *then* $\overline{A} \subset F$, *that is* $\overline{A}$ *is the smallest closed set containing* A.

Proof. Left to the reader. ■

We also note the Kuratowski closure axioms in the theorem below. These can be used as an approach to the notion of a topological space. By that we mean that if an operation on the subsets of a set X satisfies conditions $1, 2, 3, and 4$ of the following theorem, and we define the word closed by 5, then the complements of the closed sets form a topology.

Theorem 14.13 *(Kuratowski closure axioms) Suppose* X *is a topological space.*

1. *For every* $A \subset X$, $A \subset \overline{A}$.

2. *For every* $A \subset X$, $\overline{(\overline{A})} = \overline{A}$.

3. *For every $A \subset X$ and every $B \subset X$, $\overline{A \cup B} = \overline{A} \cup \overline{B}$.*

4. $\overline{\varnothing} = \varnothing$.

5. *For every $A \subset X$, A is closed if and only if $A = \overline{A}$.*

Proof. Left to the reader. ■

It is also perfectly reasonable, and some books do so, to denote the closure of a set $cl(A)$. This is especially useful if there is more than one space X in the discussion. In that case, we will often use $cl_X(A)$ to denote the closure of A in the space X. Dual to the idea of the closure of a set is the idea of interior of a set. Rather than intersecting from the outside as we did with closure, to obtain the interior we union from the inside.

Definition 14.14 *Suppose A is a subset of a space X. We define the interior of A to be the set $Int(A)$ where*

$$Int(A) = \cup \{U \subset X : U \text{ is open in } X \text{ and } U \subset A\}.$$

We sometimes write $IntA$ for $Int(A)$.

Theorem 14.15 *If X is a space and $A \subset X$, then the following are true:*

1. *$Int(A)$ is an open set.*

2. *$Int(A) \subset A$.*

3. *If $A \subset B \subset X$, then $Int(A) \subset Int(B)$.*

4. *If U is an open set with $U \subset A$, then $U \subset Int(A)$, that is $Int(A)$ is the largest open set contained in A.*

5. *$Int(A) = X \backslash \overline{(X \backslash A)}$.*

Proof. Left to the reader. ■

We also obtain easily the interior analogue of the Kuratowski closure axioms.

Theorem 14.16 *Suppose X is a topological space.*

1. *For every $A \subset X$, $Int(A) \subset A$.*

2. *For every $A \subset X$, $Int(Int(A)) = Int(A)$.*

3. *For every $A \subset X$ and every $B \subset X$, $Int(A \cap B) = Int(A) \cap Int(B)$.*

4. *$Int(X) = X$.*

5. *For every $A \subset X$, A is open if and only if $A = Int(A)$.*

We note that other popular notations exist for interior, for instance $\overset{\circ}{A}$ or A°. We will continue with $Int(A)$ or $IntA$. As with the closure operator, when there is more than one space in the discussion, we will use $Int_X(A)$ to denote the interior of A in the space X.

We have that the closure is the smallest closed set that contains A and the interior is the largest open set inside A. What lies between is called the *boundary* of A.

Definition 14.17 *Suppose A is a subset of a space X. We define the **boundary** of A to be the set $Bdry(A)$ where $Bdry(A) = \overline{A} \cap \overline{X \backslash A}$. We sometimes write $BdryA$ for $Bdry(A)$.*

Notice that $Bdry(A) = \overline{A} \backslash Int(A)$.

The boundary of a set is also called the *frontier* in some books. Other notations which appear in the literature for this notion are [in addition to $Bdry(A)$] $Fr(A)$ and $\partial(A)$. The boundary is an appealing term to use for this idea since these points are on the "edge" of A because they are in the closures of both A and $X \backslash A$.

Definition 14.18 *If X is a space and $A \subset X$, then we say A is a **neighborhood** of a point x provided $x \in Int(A)$. We often abbreviate the word neighborhood by "nbd."*

Lemma 14.19 *Suppose X is a space and $A \subset X$. The set A is a neighborhood of a point x if and only if there exists an open set U with $x \in U \subset A$.*

Proof. Left to the reader. ■

Lemma 14.20 *Suppose X is a space and $A \subset X$. The point $x \in \overline{A}$ if and only if for every neighborhood U of x we have $U \cap A \neq \varnothing$.*

Proof. Suppose $x \in \overline{A}$ and U is a neighborhood of x. Since $X \backslash Int(U)$ is a closed set, if it were the case that $U \cap A = \varnothing$, then we would have $\overline{A} \subset X \backslash Int(U)$, which would make it impossible to have $x \in \overline{A}$. Hence $U \cap A \neq \varnothing$. Now suppose every neighborhood of x intersects A. If $x \notin \overline{A}$, then $X \backslash \overline{A}$ would be a neighborhood of x which does not intersect A, a contradiction. Hence $x \in \overline{A}$. ∎

This lemma can be restated for open sets, and we do this below.

Lemma 14.21 *Suppose X is a space and $A \subset X$. A point $x \in \overline{A}$ if and only if whenever U is an open set in X with $x \in U$, we have $U \cap A \neq \varnothing$.*

Proof. Left to the reader. ∎

These lemmas provide a very handy way to determine when points are in the closure of a set, probably more commonly used than the definition itself. They also provide some motivation for the language. A point is in the closure of a set iff each neighborhood of the point intersects the set. So the points of the closure are points that are really close to the set even though they may not be elements of the set.

We also get a nice characterization of the boundary from these lemmas. That is, a point is in the boundary of a set provided each of its neighborhoods intersects both the set and its complement.

Lemma 14.22 *Suppose X is a space and $A \subset X$. The point $x \in Bdry(A)$ if and only if for every neighborhood U of x we have $U \cap A \neq \varnothing$ and $U \cap (X \backslash A) \neq \varnothing$.*

Example 14.23 *We refer the reader to chapters 2 and 4 for additional examples. Here we list a few.*

1. *In $\mathbb{R}$, $Int[0,1] = (0,1)$, $Bdry[0,1] = \{0,1\}$, and $\overline{[0,1]} = [0,1]$.*

2. *In $\mathbb{R}$, $Int\mathbb{Q} = \varnothing$, $Bdry\mathbb{Q} = \mathbb{R}$, and $\overline{\mathbb{Q}} = \mathbb{R}$.*

3. *In $\mathbb{M}$, the Michael line, $Int\mathbb{Q} = \varnothing$, $Bdry\mathbb{Q} = \mathbb{Q}$, and $\overline{\mathbb{Q}} = \mathbb{Q}$.*

Proof. The reader should prove these results. ∎

Notation 14.24 *For a space X and a point $x \in X$, we denote*

$$\mathcal{U}_x = \{U \subset X : U \text{ is a nbd of } x\}.$$

We call $\mathcal{U}_x$ the *complete neighborhood system* at x, sometimes just the *neighborhood system* at x, or sometimes the *neighborhood filter* at x.

Definition 14.25 *A **neighborhood base** at a point $x \in X$ is a subset $\mathcal{B}_x \subset \mathcal{U}_x$ such that if $U \in \mathcal{U}_x$, then there is an element $B \in \mathcal{B}_x$ such that $x \in B \subset U$.*

Definition 14.26 *A **local base** at x is a neighborhood base at x whose members are open sets.*

A word of caution: some authors require that neighborhoods be open sets. We have elected to not do that; in some cases it is convenient to be able to talk about closed neighborhoods or compact neighborhoods. Some authors also use the phrases "neighborhood base" and "local base" interchangeably.

Theorem 14.27 *Suppose that X is a set, and for each $x \in X$, $\mathcal{B}_x$ is a nonempty collection of subsets of X. Let $\tau = \{U \subset X : \forall x \in U \exists B \in \mathcal{B}_x \text{ such that } B \subset U\}$. The collection τ is a topology on X and for each x, $\mathcal{B}_x$ is a neighborhood base at x in the topology τ if and only if the following conditions are satisfied:*

1. *for each $x \in X$ and each $V \in \mathcal{B}_x$, $x \in V$*

2. *if $V_1, V_2 \in \mathcal{B}_x$, then there exists $V_3 \in \mathcal{B}_x$ such that $V_3 \subset V_1 \cap V_2$*

3. *if $V \in \mathcal{B}_x$, then there exists $V_0 \in \mathcal{B}_x$ such that for each $y \in V_0$ there is $W \in \mathcal{B}_y$ with $W \subset V$.*

Proof. First suppose that τ is a topology, and $\mathcal{B}_x$ is a neighborhood base at x for each $x \in X$. Since each element of $\mathcal{B}_x$ is a neighborhood of x, each must contain the point x. Hence 1 is true. To see 2, suppose $V_1, V_2 \in \mathcal{B}_x$, and note that we then have $x \in IntV_1 \cap IntV_2 = Int(V_1 \cap V_2)$. Hence $V_1 \cap V_2$ is a neighborhood of x, and thus there is $V_3 \in \mathcal{B}_x$ with $V_3 \subset V_1 \cap V_2$. Finally, let $V \in \mathcal{B}_x$, then $x \in IntV$. Hence we may choose $V_0 \in \mathcal{B}_x$ with $V_0 \subset IntV$. Now if $y \in V_0$, then $IntV$ is a neighborhood of y, so there exists $W \in \mathcal{B}_y$ with $y \in W \subset IntV \subset V$.

Now for the converse, suppose that we have $\mathcal{B}_x$ for each $x \in X$ such that $1, 2$, and 3 are satisfied, and τ is defined as indicated. We need to show that τ is a topology on X, and that each $B \in \mathcal{B}_x$ is a neighborhood of x. This will complete the proof since the definition of τ would then make $\mathcal{B}_x$ a neighborhood base at x.

It is clear from the definition that $\varnothing \in \tau$ and $X \in \tau$, since each $\mathcal{B}_x \neq \varnothing$. Suppose $\mathcal{G} \subset \tau$, and $x \in \cup\mathcal{G}$. Choose $G_0 \in \mathcal{G}$ with $x \in G_0$. Since $G_0 \in \tau$, we can choose $B \in \mathcal{B}_x$ with $B \subset G_0$, and thus $B \subset \cup\mathcal{G}$. Hence $\cup\mathcal{G} \in \tau$. Now suppose that $G_1, G_2 \in \tau$, and $x \in G_1 \cap G_2$. By the definition of τ, we choose $V_1, V_2 \in \mathcal{B}_x$ such that $V_1 \subset G_1$ and $V_2 \subset G_2$. Now by 3, choose $V_3 \in \mathcal{B}_x$ with $V_3 \subset V_1 \cap V_2$, and we have $V_3 \subset G_1 \cap G_2$. Hence $G_1 \cap G_2 \in \tau$, and we have shown that τ is a topology.

Now we need to show that with respect to the topology τ, each $B \in \mathcal{B}_x$ is a neighborhood of x. Suppose $x \in X$ and $B \in \mathcal{B}_x$. Define G as

$$G = \{y \in B : \exists V_y \in \mathcal{B}_y \text{ with } V_y \subset B\}.$$

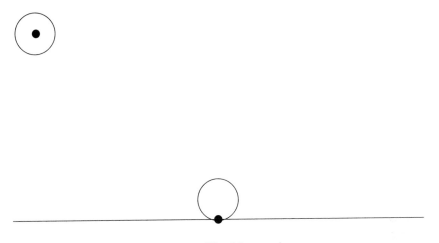

Figure 14.1: The Moore plane.

Since $B \subset B$, we see that $x \in G$. We claim that $G \in \tau$. Suppose $y \in G$, and choose $V \in \mathcal{B}_y$ with $V \subset B$. Now let $V_0 \in \mathcal{B}_y$ be the set whose existence is given by 3. Suppose $z \in V_0$, and by 3, there is $W \in \mathcal{B}_z$ with $W \subset V$. Since $V \subset B$, we have that $W \subset B$, and thus $z \in G$. Hence we have shown that $V_0 \subset G$, and since this is true for each $y \in G$, $G \in \tau$. Hence $x \in IntB$, and the result is proved. ∎

This result is important because it gives us easy criteria to determine when we have made a valid assignment of neighborhoods as we are trying to construct examples of topological spaces.

Example 14.28 *(Moore plane) We call this example Γ. The underlying set for Γ is the set $\mathbb{R} \times [0, \infty)$, i.e. the upper half plane. As usual, for (a, b) with $b > 0$ we denote the ball centered at (a, b) of radius ε by $B((a, b), \varepsilon) = \{(x, y) : (x - a)^2 + (y - b)^2 < \varepsilon^2\}$. In the case that $b = 0$, we introduce a new notation for the ball tangent to the x-axis at the point $(a, 0)$ of radius ε, namely $B^T((a, 0), \varepsilon) = \{(a, 0)\} \cup B((a, \varepsilon), \varepsilon)$. We now assign neighborhood bases at each of the points of Γ. If $b > 0$, then $\mathcal{B}_{(a,b)} = \{B((a, b), \varepsilon) : 0 < \varepsilon < b\}$, and $\mathcal{B}_{(a,0)} = \{B^T((a, 0), \varepsilon) : \varepsilon > 0\}$. We check that this is a valid assignment of neighborhood bases by verifying theorem 14.27.*

Proof. Suppose $x \in \Gamma$. Clearly $x \in B$ for each $B \in \mathcal{B}_x$. If $V_1, V_2 \in \mathcal{B}_x$, then $V_1 \cap V_2$ is the one of V_1 and V_2 with the smaller radius. So we can choose $V_3 = V_1 \cap V_2$. Finally, for each $V \in \mathcal{B}_x$, it is the case that for each point $y \in V$ there is $W \in \mathcal{B}_y$ with $W \subset V$. So we can take $V_0 = V$ for part 3. Hence parts $1, 2$, and 3 of theorem 14.27 are all satisfied and we have a topology. ∎

This example is very important in the study of some of the most interesting questions in topology. As a result, it has been discovered many times. Γ is also known as the Moore half plane, the Niemytski plane, the tangent bubble space, the tangent disk space, and the tangent ball space.

Now that we have the ideas of base and neighborhood base, we can introduce the first two countability conditions on a space.

Definition 14.29 *A space X is called **first countable**, or is said to satisfy the first axiom of countability, if and only if at each point of X there is a countable neighborhood base.*

Definition 14.30 *A space X is called **second countable**, or is said to satisfy the second axiom of countability, if and only if there is a countable base for the topology on X.*

Both of these properties were introduced in part one. Readers who have not already done so should read that material.

Example 14.31 *The following are examples regarding first and second countability.*

1. *All second-countable spaces are first countable.*

2. *All countable first-countable spaces are second countable.*

3. $\mathbb{R}$ *is second countable.*

4. $\mathbb{S}$ *is first countable and not second countable.*

5. $\mathbb{M}$ *is first countable and not second countable.*

6. Γ *is first countable and not second countable.*

7. ω_1 *is first countable and not second countable.*

8. $\omega_1 + 1$ *is not first countable.*

Proof. The proofs that these examples have the properties claimed are left to the reader. ∎

We now turn to the basic constructions. By that we mean the construction of subspaces, quotients, and products. We covered subspaces in chapter 5, but we repeat the main points here. We refer the reader to chapter 5 for proofs.

Definition 14.32 *Suppose X is a topological space and $Y \subset X$. The **subspace topology** on Y is the collection $\{U \cap Y : U \text{ is open in } X\}$.*

This is a topology on Y. The subspace topology on Y is also often called the *relative topology* on Y. When Y has this topology, we say Y is a *subspace* of X.

Theorem 14.33 *Suppose X is a topological space, and $A \subset X$ is a subspace of X. The following are true:*

1. *$H \subset A$ is open in A iff $\exists G$ open in X with $H = A \cap G$.*

2. *$F \subset A$ is closed in A iff $\exists K$ closed in X with $F = A \cap K$.*

3. *If $E \subset A$, then $cl_A E = A \cap cl_X E$.*

4. *If $x \in A$, then V is a neighborhood of x in A iff $V = U \cap A$ where U is a neighborhood of x in X.*

5. *If $x \in A$ and $\mathcal{B}_x$ is a neighborhood base at x in X, then $\{B \cap A : B \in \mathcal{B}_x\}$ is a neighborhood base at x in A.*

6. *If $\mathcal{B}$ is a base for X, then $\{B \cap A : B \in \mathcal{B}\}$ is a base for A.*

When a property is such that whenever X has the property, then so does every subspace of X, we say that the property is hereditary. From part 6 of the theorem above, it is clear that the property of being second countable is hereditary. It is also clear from 5 that being first countable is hereditary.

An odd thing about subspace topologies is how they interact with subspace orders. If X is a *LOTS*, and $A \subset X$, then A inherits a linear order from X. This order generates a *LOTS* topology on A, but that topology may not agree with the subspace topology on A.

Example 14.34 *The top line (minus the endpoints) of the lexicographic square has the usual order on $(0,1)$, and thus the LOTS topology on the subspace is homeomorphic to the real line $\mathbb{R}$. On the other hand, the subspace topology on this subspace is homeomorphic to $\mathbb{S}$, the Sorgenfrey line.*

Definition 14.35 *We say a space X is a **generalized ordered space**, or GO-space, provided that X is homeomorphic to a subspace of a LOTS.*

Such spaces are also called *suborderable*.

Example 14.36 *Both $\mathbb{M}$ and $\mathbb{S}$ are GO-spaces.*

Proof. For the Michael line, consider the space $X = (\mathbb{Q} \times \{0\}) \cup ((\mathbb{R} \backslash \mathbb{Q}) \times (-1, 1))$ and order lexicographically. We leave it as an exercise to show that $\mathbb{M} = \mathbb{R} \times \{0\}$ with the subspace topology from X. ∎

Exercises

1. Prove that the examples of topologies listed in example 14.2 are actually topologies.

2. Prove that the examples of bases listed in example 14.9 do satisfy the criteria to be a base.

3. Prove theorem 14.12 and theorem 14.13.

4. Prove theorem 14.15 and theorem 14.16.

5. Prove lemma 14.19.

6. Prove lemma 14.21 and lemma 14.22.

7. Verify the results in example 14.23.

8. (Second-countable spaces.) Prove that if X is a second-countable space, then for any base $\mathcal{B}$ for X there is a countable $\mathcal{B}_0 \subset \mathcal{B}$ such that $\mathcal{B}_0$ is a base for X. In other words, if there is a countable base for a space, then you can find one within any given base. So you can specify the form of the countable base elements. Prove that if X is a second-countable space, then X can have at most countably many isolated points. Prove that if X is a second-countable space, then X cannot have an uncountable closed discrete subspace. Use the results you have just established to prove the results in example 14.31.

9. Complete the proof that $\mathbb{M}$ is a GO-space.

Chapter 15

Quotients and Products

A quotient in topology is generated by a certain kind of function, just as it is in algebra. Readers who have had an algebra course will recall from the first isomorphism theorem that the quotient of a group is simply the image of the group under a homomorphism, or a "group-structure-preserving" function. Here we define a quotient mapping as a kind of "topology-preserving" function.

Definition 15.1 *Suppose X is a topological space where $f : X \to Y$ and f is onto Y. The **quotient topology** on Y is given by $\tau_f = \{U \subset Y : f^{-1}(U) \text{ is open in } X\}$.*

Definition 15.2 *When τ_f is the topology on Y, we say f is a **quotient mapping**, and we say Y is a **quotient space** of X. We sometimes write $Y = X/f$, which is read "X mod f."*

We recall from part one that a function $f : X \to Y$ is continuous if and only if $f^{-1}(U)$ is open in X for each open $U \subset Y$. So quotient mappings are always continuous, but more is true. The quotient topology on Y is the strongest topology on Y which will make f be a continuous map. There are interesting questions about what additional conditions will make a continuous function be a quotient mapping. First let us recall the following definition from part one.

Definition 15.3 *Suppose $f : X \to Y$. We say f is **closed** if and only if $f(H)$ is a closed subset of Y whenever H is a closed subset of X. We say f is **open** if and only if $f(G)$ is an open subset of Y whenever G is an open subset of X.*

Theorem 15.4 *If $f : X \to Y$ is continuous and onto Y, then f is a quotient mapping if f is either open or closed.*

Proof. Let τ denote the topology on Y. We want to show that $\tau = \tau_f$. Since f is continuous, we know that $\tau \subset \tau_f$. Suppose that f is an open mapping, and let $U \in \tau_f$. By the definition of τ_f, we know that $f^{-1}(U)$ is open in X, and since f is onto, $f(f^{-1}(U)) = U$. Now since f is an open mapping, $f(f^{-1}(U)) \in \tau$. Thus we have that $\tau = \tau_f$, and f is a quotient mapping.

Now suppose that f is a closed mapping, and let $U \in \tau_f$. We know $X \backslash f^{-1}(U)$ is closed in X, so $f(X \backslash f^{-1}(U))$ is closed in (Y, τ). We will show that $U = Y \backslash f(X \backslash f^{-1}(U))$, and thus $U \in \tau$. Suppose that $y \in U$, and $x \in X$ with $f(x) = y$. We see that $x \in f^{-1}(U)$, and in particular, $x \notin X \backslash f^{-1}(U)$. Since this is true for every such x, $y \notin f(X \backslash f^{-1}(U))$. Hence $U \subset Y \backslash f(X \backslash f^{-1}(U))$. Now suppose that $y \in Y \backslash f(X \backslash f^{-1}(U))$. Since f is onto Y, there does exist $x \in X$ with $f(x) = y$, and this $x \notin X \backslash f^{-1}(U)$. Hence $x \in f^{-1}(U)$, and thus $y = f(x) \in U$. Thus $Y \backslash f(X \backslash f^{-1}(U)) \subset U$. This gives us that $Y \backslash f(X \backslash f^{-1}(U)) = U$, and the proof is complete. ∎

Example 15.5 *Consider $X = [0, 2\pi]$ and $Y = \{(x, y) : x^2 + y^2 = 1\}$. Define $f : X \to Y$ by $f(\theta) = (\cos \theta, \sin \theta)$. Note that f is a quotient mapping, in fact, a closed, continuous mapping. Hence the circle, with its usual topology, is a quotient of the interval.*

Just as a quotient group can be formed by decomposing a given group into cosets, a quotient space can be formed by decomposing a space into equivalence classes under some equivalence relation.

Definition 15.6 *Suppose X is a topological space and $\sim$ is an equivalence relation on X. The set Y of equivalence classes determined by $\sim$ is a partition of X. We define a topology on Y by declaring a subset $V \subset Y$ to be open in Y if and only if the subset $\cup V \subset X$ is open in X. With this topology, Y is called a **decomposition space** of X. We usually denote the space $Y = X / \sim$.*

The fact that decomposition spaces and quotients are really the same is given by the following theorem.

Theorem 15.7 *Suppose X is a topological space.*

1. *If $\sim$ is an equivalence relation on X, then the mapping $\pi : X \to X / \sim$, where $\pi(x)$ is the equivalence class containing x, is a quotient mapping.*

2. *If $f : X \to Y$ is a quotient mapping, then the relation $x \sim y$ if and only if $f(x) = f(y)$ is an equivalence relation on X, and Y is homeomorphic to $X / \sim$.*

Proof. For 1, note that $\pi^{-1}(V) = \cup V$ for any $V \subset X/\sim$.

Let us prove 2. To see that $\sim$ is an equivalence relation, we need to check that $\sim$ is reflexive, symmetric, and transitive. Since $f(x) = f(x)$, $\sim$ is reflexive. Similarly, since the relation $=$ on X is symmetric and transitive, so is $\sim$. Thus we have an equivalence relation, and the equivalence classes are the fibers of the mapping, that is, $[x] = f^{-1}(f(x))$. The mapping $\varphi : Y \to X/\sim$ given by $\varphi(y) = f^{-1}(y)$ is clearly one-to-one and onto since f is a function and f is onto.

Suppose $V \subset X/\sim$, and V is open. Let $A = \varphi^{-1}(V)$, and note that $f^{-1}(A) = \{x \in X : f(x) \in A\} = \{x \in X : \varphi(f(x)) \in V\} = \cup V$. Thus $f^{-1}(A)$ is open, and since f is a quotient map, A is open. Hence φ is continuous.

Now suppose that $U \subset Y$, and U is open. Now $\varphi(U) = \{f^{-1}(y) : y \in U\}$, and thus $\cup\varphi(U) = f^{-1}(U)$ which is open in X. Hence $\varphi(U)$ is an open set in $X/\sim$. Hence φ is the homeomorphism we seek. ∎

Notice that what we have shown above is that the range of a quotient mapping is homeomorphic to the decomposition space of fibers of the mapping, just as the range of a group homomorphism is isomorphic to the quotient group of cosets of its kernel. Also note that the cosets of the kernel of a group homomorphism are the fibers of the homomorphism.

Example 15.8 *Look again at example 15.5. Notice that in the equivalence relation defined above, the equivalence classes are all singleton sets except for $f^{-1}((1,0))$ which is the two point set $\{0, 2\pi\}$.*

The space $X/\sim$ is often called an *identification space*. This is because it *identifies* the points of each equivalence class together so that they are now one point. Usually, we only talk about identifying the equivalence classes with more than one element. So what we would often say in example 15.8 is that the circle is obtained from the interval by *identifying the endpoints*.

The idea of identifying points is used to create many interesting and important examples. Suppose we look at a two-dimensional version of example 15.5.

Example 15.9 *Start with the box $[0, 2\pi] \times [0, 2\pi]$ viewed as a subspace of $\mathbb{R}^2$. We identify the right and left edges, i.e., $(0, x) \sim (2\pi, x)$ for each x. We obtain the cylinder $S^1 \times [0, 2\pi]$. If we now identify the top and bottom circles, we obtain the torus, or "doughnut."*

If we do something similar to the example above, but change it slightly, we obtain another useful and interesting example.

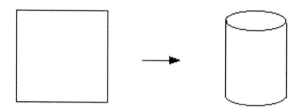

Example 15.10 *Start with the box* $[0, 2\pi] \times [0, 2\pi]$ *viewed as a subspace of* $\mathbb{R}^2$. *We again identify the right and left edges only this time we identify* $(0, x) \sim (2\pi, 2\pi - x)$. *We have reversed the orientation of the edges; that is, we put a half twist in the interval before attaching the edges. This is called a Möbius strip.*

The easiest way to see this is to make one from a strip of paper. Start with a strip of paper; about 1 inch by 8 inches will work better than a square. Twist one-half turn, and attach the ends. This surface in $\mathbb{R}^3$ has only one side. To see this, start at some point on the surface and trace around the surface with your finger until you are directly opposite the point where you started.

This shape is used in industrial applications. If you walk into any factory which uses conveyor belts and look below the roller bed which supports the belt, you will see that there is a half twist in the belt. This creates a Möbius strip and causes the belt to wear evenly on both sides, which both extends the life of the belt and decreases the probability that the belt will crack.

Theorem 15.11 *Suppose* $f : X \to Y$ *is a continuous mapping onto* Y. *The mapping* f *is closed if and only if whenever* $y \in Y$ *and* U *is an open subset of* X *with* $f^{-1}(y) \subset U$, *there is an open set* $V \subset Y$ *with* $f^{-1}(y) \subset f^{-1}(V) \subset U$.

Proof. First suppose f is a closed mapping. Let $y \in Y$ and let U be an open subset of X with $f^{-1}(y) \subset U$. Let $V = Y \backslash f(X \backslash U)$. Clearly, V is open, and $y \in V$. If $x \in f^{-1}(V)$, then $f(x) \notin f(X \backslash U)$, and thus $x \notin X \backslash U$, i.e. $x \in U$. Hence $f^{-1}(V) \subset U$.

Now suppose that f has the property of the theorem, and let H be a closed subset of X. Suppose that $y \in Y \backslash f(H)$, and note that we then have $f^{-1}(y) \subset X \backslash H$. By the property, choose an open set $V \subset Y$ such that $f^{-1}(y) \subset f^{-1}(V) \subset X \backslash H$. From this we have that $V \cap f(H) = \varnothing$, and thus there is a neighborhood of y that is disjoint from $f(H)$. Hence $Y \backslash f(H)$ is open and $f(H)$ is closed, so we have shown that f is a closed mapping. ∎

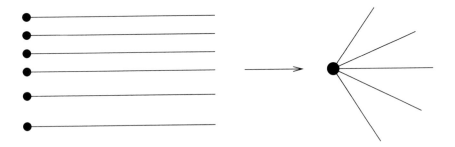

Figure 15.1: Identifying a closed set to a point.

When we translate the last result into the language of decompositions, we obtain a result that is of importance in many classical results in general topology, geometric topology, and continuum theory. A *saturated set* in the underlying space of a decomposition is a set that equals the union of fibers of the decomposition. Suppose a decomposition space has the property that if a fiber of the decomposition is contained in an open set of the original space, then there is an open set between them which is saturated. Such a decomposition is called an *upper semi-continuous decomposition*. What we showed is that the upper semi-continuous decompositions correspond exactly to the quotient mappings that are closed and continuous.

Example 15.12 *One situation that gives rise to an upper semi-continuous decomposition is when we identify a single closed set to a point. To be more specific, suppose X is a space and A is a closed subset of X. We define $\sim_A$ on X by $x \sim_A y$ if $\{x, y\} \subset A$. For all other points $x \sim_A y$ if and only if $x = y$. The equivalence classes under this equivalence relation are the set A and the singletons of all points of $X \backslash A$. The space $X / \sim_A$ is an upper semi-continuous decomposition of X. We usually denote this space by X/A.*

Proof. We leave the proof that this decomposition is upper semi-continuous as an exercise. ∎

It is an easy exercise to show that the composition of two quotient maps is also a quotient mapping, and that the composition of two closed, continuous mappings is again closed and continuous. Consequently, by induction, the example above can be extended to any collapsing of finitely many pairwise disjoint closed sets to points.

We close out this discussion of quotients with one more theorem about compositions.

Theorem 15.13 *Suppose $f : X \to Y$ is a quotient mapping onto Y. A mapping $g : Y \to Z$ is continuous if and only if $g \circ f : X \to Z$ is continuous.*

Proof. The "only if" portion is trivial, since any composition of two continuous maps is continuous. Suppose that $g \circ f$ is continuous. We show that g is continuous. Let U be an open subset of Z. By continuity of $g \circ f$, we know that $(g \circ f)^{-1}(U)$ is open in X, but $(g \circ f)^{-1}(U) = f^{-1}(g^{-1}(U))$. Since Y has the quotient topology, we then know that $g^{-1}(U)$ is open in Y. Hence g is continuous. ■

$$\begin{array}{ccccc} X & & \xrightarrow{f} & & Y \\ & \overset{g\circ f}{\searrow} & & \overset{g}{\swarrow} & \\ & & Z & & \end{array}$$

We turn now to a discussion of products. In Part One we discussed the topology on finite and countable Cartesian products. Now we will extend those constructions to arbitrary products.

Definition 15.14 *Suppose $\{X_\alpha : \alpha \in A\}$ is a collection of sets. The* ***Cartesian product*** *of these sets is denoted by $\prod_{\alpha\in A} X_\alpha$ and is defined by*

$$\prod_{\alpha\in A} X_\alpha = \left\{x : x \text{ is a function from } A \text{ into } \bigcup_{\alpha\in A} X_\alpha \text{ and } x(\alpha) \in X_\alpha \forall\alpha\right\}.$$

Notation 15.15 *There are several easy observations and notational conventions associated with products that need to be mentioned.*

1. *If $x \in \prod_{\alpha\in A} X_\alpha$, we denote $x(\alpha) = x_\alpha$ and call this the αth* ***coordinate*** *of x.*
2. *The set X_β is the called the βth* ***factor*** *of the product $\prod_{\alpha\in A} X_\alpha$.*
3. *If $X_n = X$ for all $n \in \omega$, then $\prod_{n\in\omega} X_n$ is the set of all sequences in X. We denote this X^ω.*
4. *Following* 1 *and our conventions for sequences, we often write $x = (x_\alpha : \alpha \in A)$.*
5. *If $A = \{1,2\}$, we usually write $X_1 \times X_2$ rather than $\prod_{\alpha\in\{1,2\}} X_\alpha$. We also write $x = (x_1, x_2)$ rather than $(x_i : i \in \{1,2\})$. Note that $(x_i : i \in \{1,2\}) = \{(1,x_1),(2,x_2)\}$ so this is not* really *equal to (x_1,x_2). However, no confusion usually arises.*

6. $\mathbb{R} \times \mathbb{R}$ *is* $\mathbb{R}^2$.

7. *When each* X_α *is the same set* X, *we denote the product* $\prod_{\alpha \in A} X_\alpha$ *by* X^A.

8. $X \times Y$ *is the same as* $X_1 \times X_2$ *where* $X = X_1$ *and* $Y = X_2$.

9. *We often write* $\prod_{k=1}^n X_k$ *for* $\prod_{k \in \{1,2,\ldots,n\}} X_k$, *and* $\prod_{n=0}^\infty X_n = \prod_{n \in \omega} X_n$.

10. *The mapping* $\pi_\beta : \prod_{\alpha \in A} X_\alpha \to X_\beta$ *defined by* $\pi_\beta(x) = x_\beta$ *is called the* **projection mapping** *onto the* β^{th} *factor, or sometimes just the* β^{th} *projection.*

11. *If the product space is nonempty, then the projection mappings are always onto mappings.*

12. *Each point* $x \in \prod_{\alpha \in A} X_\alpha$ *is a choice function on the collection* $\{X_\alpha : \alpha \in A\}$. *By assuming that the product* $\prod_{\alpha \in A} X_\alpha$ *is nonempty whenever each factor is, we are assuming the axiom of choice to be true.*

We leave it as an exercise to prove part 11.

We would like to define a topology on the product space $\prod_{\alpha \in A} X_\alpha$. Before defining the product topology, let us first look at the finite and countable products that we talked about in chapter 5. The topology on the product of two spaces $X \times Y$ is very natural. We take as a base the collection of boxes $\{U \times V : U \text{ is open in } X \text{ and } V \text{ is open in } Y\}$. With this topology, the projection mappings are open continuous mappings. Thus each factor is a quotient of the product space. This is certainly a satisfying result.

The topology for the countably infinite product, which we defined in chapter 5, seemed a bit more complicated. We claim that it is still natural. Following the familiar idea of decimal representations, we say that two sequences are close together if the first finitely many terms are close together, and after that we don't care. Using this idea, we defined a topology on countable products which extended the idea of the topology on finite products and had many nice characteristics. For example, the projections were still open and continuous, so that each factor space was a quotient of the product. Also, the product of countably many metric spaces was again a metric space, and the metric was given by a nice geometric series combination of the metrics on the factors. This result was crucial to the student project at the end of chapter 6.

We will define the topology on arbitrary products in a similar way; we will make sure that for points to be close, they must be close on a finite number of factors.

Definition 15.16 *The **product topology**, sometimes called the **Tychonoff product topology**, on the space $\prod_{\alpha \in A} X_\alpha$ is the topology having as a base all sets of
form $\prod_{\alpha \in A} U_\alpha$ where for each $\alpha \in A$, U_α is open in X_α, and $U_\alpha = X_\alpha$ for all but
finitely coordinates.*

We should check that this really is a topology. The easiest way to see this is
to note that if the set of coordinates for which $U_\alpha \neq X_\alpha$ is the set $\{\alpha_1, \alpha_2, ..., \alpha_n\}$,
then

$$\prod_{\alpha \in A} U_\alpha = \bigcap_{k=1}^{n} \pi_{\alpha_k}^{-1}(U_{\alpha_k}).$$

Hence the topology we have defined is exactly the topology with the following set
as a subbase,

$$\left\{\pi_\alpha^{-1}(U_\alpha) : \alpha \in A \text{ and } U_\alpha \text{ is open in } X_\alpha\right\}.$$

Thus we do indeed have a topology.

Many people prefer to write the basic open sets in the product as this finite
intersection of inverse projections, and there are times when this is very convenient.
There are also times when it is very convenient to write basic open sets as "boxes,"
where all by finitely many factors are the entire factor space. This is largely a
matter of taste, and by making a big fuss about these two notations being the
same object, perhaps we can remove some of the mystique of the messy notation.

Theorem 15.17 *If the product space $\prod_{\alpha \in A} X_\alpha$ is nonempty, then each projection
mapping $\pi_\beta : \prod_{\alpha \in A} X_\alpha \to X_\beta$ is an open continuous mapping onto X_β.*

Proof.   That the projections are open follows since $\pi_\beta(\prod_{\alpha \in A} U_\alpha) = U_\beta$ and
any open set is the union of basic open sets. The continuity of each π_β follows
immediately from the definition of the topology. ∎

Theorem 15.18 *The product topology on $\prod_{\alpha \in A} X_\alpha$ is the weakest topology that
makes all projections continuous.*

Proof.   If τ is a topology on $\prod_{\alpha \in A} X_\alpha$ that makes all projections continuous,
then each of the subbasic open sets for the product topology would have to be an
element of τ. Hence the product topology would have to be a subset of τ, since
any topology is closed under finite intersections and arbitrary unions. ∎

From this point forward, whenever we write $\prod_{\alpha \in A} X_\alpha$, we will assume that the
topology on this set is the one we have just defined, the Tychonoff product topology.

There is another topology on the product $\prod_{\alpha \in A} X_\alpha$ which seems more natural a first glance. This is the topology generated by taking as a base the set of all "boxes," that is, the topology having the set

$$\left\{ \prod_{\alpha \in A} U_\alpha : U_\alpha \text{ is open in } X_\alpha \right\}$$

as a base. This topology is called the *box topology*. There are still interesting and basic questions about box topologies that are unsolved, and box topology is an object of study for quite a number of people. The characteristics of the box topology are generally bad. For instance, nontrivial infinite box products are never compact, and there is a countable family of complete separable metric spaces whose box product is not normal. For the interested reader, we mention that Scott Williams has written a very nice survey on box products in the *Handbook of Set-Theoretic Topology*.

Theorem 15.19 *A function $f : X \to \prod_{\alpha \in A} X_\alpha$ is continuous if and only if for each $\beta \in A$ the function $\pi_\beta \circ f : X \to X_\beta$ is continuous.*

Proof. The "only if" is clear since the composition of two continuous functions is continuous. Now suppose that $\pi_\beta \circ f$ is continuous for each $\beta \in A$. Let B be a basic open set in $\prod_{\alpha \in A} X_\alpha$. Choose a finite set $\{\alpha_1, \alpha_2, ..., \alpha_n\} \subset A$ such that $B = \bigcap_{k=1}^n \pi_{\alpha_k}^{-1}(U_{\alpha_k})$ where U_{α_k} is open in X_{α_k} for each k. Now $f^{-1}(B) = f^{-1}(\bigcap_{k=1}^n \pi_{\alpha_k}^{-1}(U_{\alpha_k})) = \bigcap_{k=1}^n f^{-1}(\pi_{\alpha_k}^{-1}(U_{\alpha_k})) = \bigcap_{k=1}^n (\pi_{\alpha_k} \circ f)^{-1}(U_{\alpha_k})$, and this set is open by the continuity of each of the functions $\pi_{\alpha_k} \circ f$, $k = 1, 2, ..., n$. Hence f is continuous. ∎

$$
\begin{array}{ccc}
X & \xrightarrow{\ f\ } & \prod_{\alpha \in A} X_\alpha \\
& \searrow^{\pi_\beta \circ f} \quad \swarrow_{\pi_\beta} & \\
& X_\beta &
\end{array}
$$

This result is characteristic of weak topologies. Let us make this a bit more formal.

Definition 15.20 *Let X be a set and for each $\alpha \in A$, $f_\alpha : X \to X_\alpha$ is a function into a topological space X_α. The **weak topology induced by the family of mappings** $\{f_\alpha : \alpha \in A\}$ is the weakest topology on X that makes each of the functions f_α continuous.*

It is clear, after a moment's thought, that this is the topology having

$$\left\{ f_\alpha^{-1}(U_\alpha) : \alpha \in A \text{ and } U_\alpha \text{ is open in } X_\alpha \right\}$$

as a subbase.

Example 15.21 *The product topology is the weak topology induced by the family of projections.*

We also get the analogue of theorem 15.19 with virtually no change in the proof.

Theorem 15.22 *If X has the weak topology induced by a collection $f_\alpha : X \to X_\alpha$, for $\alpha \in A$, then a mapping $f : Y \to X$ is continuous if and only if $f_\alpha \circ f$ is continuous for all $\alpha \in A$.*

We see in our next group of results that weak topologies induced by collections of functions are often products. Indeed, if we have enough functions to distinguish among the points of X, then the space with the weak topology will be embedded into a product. First, we need a couple of definitions.

Definition 15.23 *Let X be a set and for each $\alpha \in A$, $f_\alpha : X \to X_\alpha$ is a function into a space X_α. We say the family $\{f_\alpha : \alpha \in A\}$ **separates points** in X if and only if whenever $x, y \in X$ and $x \neq y$, there is $\beta \in A$ with $f_\beta(x) \neq f_\beta(y)$. We say the family $\{f_\alpha : \alpha \in A\}$ **separates points from closed sets** in X if and only if whenever H is a closed subset of X and $x \in X \backslash H$, there is $\beta \in A$ such that $f_\beta(x) \notin \overline{f_\beta(H)}$.*

Definition 15.24 *Let X be a set and for each $\alpha \in A$, $f_\alpha : X \to X_\alpha$ be a function into a set X_α. The **evaluation map** is the function $e : X \to \prod_{\alpha \in A} X_\alpha$ defined by $[e(x)]_\alpha = f_\alpha(x)$, that is $e(x) = (f_\alpha(x) : \alpha \in A) \in \prod_{\alpha \in A} X_\alpha$.*

$$
\begin{array}{ccc}
X & \xrightarrow{\ e\ } & \prod_{\alpha \in A} X_\alpha \\
\ _{f_\alpha}\searrow & & \swarrow_{\ \pi_\alpha \circ e} \\
& X_\alpha &
\end{array}
$$

Definition 15.25 *An **embedding** is a function $f : X \to Y$ that is a homeomorphism onto its range. That is, f is a one-to-one, continuous, open (onto the subspace $f(X)$) mapping.*

If an embedding, as above, exists, then X is homeomorphic to a subspace of Y, namely $f(X)$.

Lemma 15.26 *For each $\alpha \in A$, let $f_\alpha : X \to X_\alpha$. The evaluation map $e : X \to \prod_{\alpha \in A} X_\alpha$ is an embedding if and only if X has the weak topology induced by the family $\{f_\alpha : \alpha \in A\}$, and the family $\{f_\alpha : \alpha \in A\}$ separates points in X.*

Proof. Notice that from the definition of the evaluation mapping we know $\pi_\alpha \circ e = f_\alpha$ for each $\alpha \in A$.

First we suppose that e is an embedding, that is, $e : X \to e(X)$ is a homeomorphism. Suppose $x \neq y$ for $x, y \in X$. Since $e(x) \neq e(y)$, there is some coordinate β where $[e(x)]_\beta \neq [e(y)]_\beta$, but this means that $f_\beta(x) \neq f_\beta(y)$. Thus the family $\{f_\alpha : \alpha \in A\}$ separates points in X.

Since e is an embedding, e is continuous, and thus each $\pi_\alpha \circ e$ is continuous. Hence we have that each f_α is continuous. Thus the topology on X includes the weak topology induced by $\{f_\alpha : \alpha \in A\}$.

Suppose that U is open in X, and $x \in U$. Since e is an embedding, $e(U)$ is open in $e(X)$, and $e(x) \in e(U)$. Choose an open set $V \subset \prod_{\alpha \in A} X_\alpha$ such that $e(U) = V \cap e(X)$. Now choose a basic open set B in the product space with $e(x) \in B \subset V$. So we can choose a finite number of coordinates $\alpha_1, \alpha_2, ..., \alpha_n$ and open sets U_{α_k} in X_{α_k} with $B = \bigcap_{k=1}^n \pi_{\alpha_k}^{-1}(U_{\alpha_k})$.

We claim that $x \in \bigcap_{k=1}^n f_{\alpha_k}^{-1}(U_{\alpha_k}) \subset U$. Since $e(x) \in B$ for each k, $f_\alpha(x) = \prod_{\alpha_k}(e(x)) \in U_{\alpha_k}$, and thus $x \in f_{\alpha_k}^{-1}(U_{\alpha_k})$. Hence $x \in \bigcap_{k=1}^n f_{\alpha_k}^{-1}(U_{\alpha_k})$. On the other hand, suppose that $y \in \bigcap_{k=1}^n f_{\alpha_k}^{-1}(U_{\alpha_k})$. For each k, $f_{\alpha_k}(y) \in U_{\alpha_k}$, and thus $[e(y)]_{\alpha_k} \in U_{\alpha_k}$, but this means that $e(y) \in B \cap e(X) \subset e(U)$. Hence $y \in U$, and we have our claim. This shows that every open set in X is in the topology having $\{f_\alpha^{-1}(U_\alpha) : \alpha \in A \text{ and } U_\alpha \text{ is open in } X_\alpha\}$ as a subbase, i.e. X has the weak topology induced by $\{f_\alpha : \alpha \in A\}$.

Now for the converse, suppose that X has the weak topology induced by the family $\{f_\alpha : \alpha \in A\}$, and the family $\{f_\alpha : \alpha \in A\}$ separates points in X. We need to show that e is one-to-one, that e is continuous, and that e^{-1} is continuous. Suppose $x, y \in X$ and $x \neq y$. Since the family $\{f_\alpha : \alpha \in A\}$ separates points in X, choose $\beta \in A$ with $f_\beta(x) \neq f_\beta(y)$. We have then that $[e(x)]_\beta \neq [e(y)]_\beta$ and thus $e(x) \neq e(y)$. So e is one-to-one.

Since each f_α is continuous, then for each $\alpha \in A$, $\pi_\alpha \circ e$ is continuous. Hence, by theorem 15.19, e is continuous. Now $e^{-1} : e(X) \to X$ and if we suppose $y \in e(X)$, then for $x = e^{-1}(y)$, we have for each α, $f_\alpha(e^{-1}(y)) = f_\alpha(x) = \pi_\alpha(e(x)) = \pi_\alpha(y)$. Hence, for each α, $f_\alpha \circ e^{-1} = \pi_\alpha$ (restricted to the subspace $e(X)$), which is a continuous function. Hence, by theorem 15.22, e^{-1} is continuous, and the proof is complete. ∎

We can phrase this last result in a slightly nicer way by using the idea of separating points from closed sets. The first result toward that end is that the existence of such families gives us a weak topology.

Lemma 15.27 *A family* $\{f_\alpha : \alpha \in A\}$ *of continuous functions on a space* X *separates points from closed sets in* X *if and only if*

$$\{f_\alpha^{-1}(U_\alpha) : \alpha \in A \text{ and } U_\alpha \text{ is open in } X_\alpha\}$$

forms a base for the topology on X.

Proof. Suppose the family separates points from closed sets and $U \subset X$ is open with $x \in U$. Choose $\alpha \in A$ with $f_\alpha(x) \in X_\alpha \backslash \overline{f_\alpha(X \backslash U)}$. Since this set is open in X_α, we let $U_\alpha = X_\alpha \backslash \overline{f_\alpha(X \backslash U)}$. Clearly, $x \in f_\alpha^{-1}(U_\alpha)$, and if $y \in f_\alpha^{-1}(U_\alpha)$, then $f_\alpha(y) \notin \overline{f_\alpha(X \backslash U)}$, and thus $y \in U$. Hence, $x \in f_\alpha^{-1}(U_\alpha) \subset U$. Thus $\{f_\alpha^{-1}(U_\alpha) : \alpha \in A \text{ and } U_\alpha \text{ is open in } X_\alpha\}$ forms a base for the topology on X.

For the converse, suppose that this collection does form a base for X. Suppose that H is a closed subset of X, and $x \in X \backslash H$. Choose $\alpha \in A$ and U_α open in X_α with $x \in f_\alpha^{-1}(U_\alpha) \subset X \backslash H$. Now $f(x) \in U_\alpha$ and $U_\alpha \cap f_\alpha(H) = \varnothing$. Thus $f_\alpha(x) \notin \overline{f_\alpha(H)}$. Hence the family of functions separates points from closed sets. ∎

Corollary 15.28 *If a family* $\{f_\alpha : \alpha \in A\}$ *of continuous functions on a space* X *separates points from closed sets in* X, *then* X *has the weak topology induced by the family* $\{f_\alpha : \alpha \in A\}$.

Of course, if the singleton sets in X are themselves closed sets, then a family of functions that separate points from closed sets must also separate points. As we recall from part one, spaces that have the property that all singleton sets are closed are called T_1-spaces. Combining these last several results we have the following major theorem.

Theorem 15.29 *(Embedding lemma) If* X *is a* T_1-space *and* $\{f_\alpha : \alpha \in A\}$ *is a family of continuous functions, with* $f_\alpha : X \to X_\alpha$, *which separates points from closed sets, then the evaluation mapping* $e : X \to \prod_{\alpha \in A} X_\alpha$ *is an embedding.*

Proof. This is immediate from corollary 15.28 and lemma 15.26. ∎

Example 15.30 *In the project in chapter 6, you used the idea of the embedding lemma to show that every regular second-countable space can be embedded in a product of countably many closed intervals.*

We close out this section with one more construction. This is called the disjoint union or sometimes the free union of spaces.

Definition 15.31 *Suppose that X_α is a topological space for each $\alpha \in A$. We denote by $X_\alpha^* = X_\alpha \times \{\alpha\} = \{(x, \alpha) : x \in X_\alpha\}$, with the product topology. The **disjoint union** of these spaces is denoted $\sum_{\alpha \in A} X_\alpha$ and is defined by $\sum_{\alpha \in A} X_\alpha = \bigcup_{\alpha \in A} X_\alpha^*$, where a subset U of $\sum_{\alpha \in A} X_\alpha$ is open if and only if $U \cap X_\alpha^*$ is open in X_α^* for every $\alpha \in A$.*

Notice what this construction does. Since there is only one topology on $\{\alpha\}$, each X_α^* is a well-defined topological space homeomorphic to the corresponding X_α. Passing from X_α to X_α^* simply ensures that the spaces are pairwise disjoint. We have defined the topology so that each X_α^* is an open subspace of the union, and each is homeomorphic in the subspace topology to the corresponding X_α.

Exercises

1. Prove that the result in example 15.12 is true. That is, prove that the decomposition space obtained by collapsing a single closed set to a point is upper semi-continuous.

2. (Open mappings and closed mappings.)

   (a) Show that the projection $\pi_1 : \mathbb{R}^2 \to \mathbb{R}$ is not a closed mapping. Hence there are open continuous mappings that are not closed.

   (b) Show that the natural quotient mapping of the decomposition space obtained by shrinking $[0, 1]$ to a point in $\mathbb{R}$ is a closed mapping that is not open. Hence there are closed continuous mappings that are not open.

   (c) Show that the projection $\pi_2 : [0, 1] \times \mathbb{R} \to \mathbb{R}$ is a closed mapping.

3. (Quotients and Subspaces.) Any partition of a space X induces a partition on a subset $Y \subset X$. If we let $\sim$ be the equivalence relation generated by the partition, then we have two natural objects which could be thought of as corresponding to Y in the decomposition space. When Y has the subspace topology, then the decomposition space $Y/\sim$ is well defined. Also if we look at the subspace $Y^* = \{A \in X/\sim : A \cap Y \neq \varnothing\}$ of the decomposition space $X/\sim$, then there is a one-to-correspondence between the points of Y^* and the points of $Y/\sim$. Are these two topological spaces homeomorphic? If not,

then can you find a nice condition on Y which will make sure that they are homeomorphic?

4. From the space $\mathbb{R}$ of real numbers, form the identification space $\mathbb{R}/\sim$ by identifying $\frac{1}{n} \sim n$ for every $n \in \mathbb{N}$. Describe a typical neighborhood of 0 in the space $\mathbb{R}/\sim$. Look at the subset $A = \bigcup_{n=1}^{\infty}(n, n+1)$. Show that $0 \in \overline{A}$, but no sequence in A converges to 0.

5. In the space $\mathbb{R}$ of real numbers, identify the integers to a point. This is the space $\mathbb{R}/\mathbb{Z}$. Describe a typical neighborhood of $\mathbb{Z}$ in the decomposition space.

6. If $X_\alpha = X$ for all $\alpha \in A$, then prove that $\sum_{\alpha \in A} X_\alpha$ is homeomorphic to the product space $X \times A$ where A is given the discrete topology.

7. Suppose that $\prod_{\alpha \in A} X_\alpha$ is a nonempty product space.

   (a) Prove that if V is a nonempty open subset of $\prod_{\alpha \in A} X_\alpha$, then $\pi_\alpha(V) = X_\alpha$ for all but finitely many α.

   (b) If $c_\alpha \in X_\alpha$ for each $\alpha \in A$, then for each β, X_β is homeomorphic to the subspace $X_\beta^* = \{x \in \prod_{\alpha \in A} X_\alpha : x_\alpha = c_\alpha \text{ for all } \alpha \neq \beta\}$. Hence each factor space is homeomorphic to subspace of the product.

8. (Closure and interior in finite products.) Suppose X and Y are spaces, and $A \subset X$ and $B \subset Y$.

   (a) Prove that $\overline{A \times B} = \overline{A} \times \overline{B}$.

   (b) Prove that $Int(A \times B) = (IntA) \times (IntB)$.

9. (Closure and interior in arbitrary products.)

   (a) Prove that if $H_\alpha \subset X_\alpha$ for all $\alpha \in A$, then $\overline{\prod_{\alpha \in A} H_\alpha} = \prod_{\alpha \in A} \overline{H_\alpha}$. In particular, any product of closed sets will again be a closed set.

   (b) Prove that in the product space $\mathbb{R}^\omega$, $Int((0, 1)^\omega) = \varnothing$. Hence the corresponding result for interiors fails.

Chapter 16

Convergence

In beginner-level analysis courses the closure of a set is often defined to be the set together with the limit points of all the convergent sequences drawn from the set. Moreover, we proved in chapter 2 that this definition is equivalent to the one we have given for metric spaces. It was an exercise in chapter 6 to show that if a point is the limit of a convergent sequence drawn from a subset, then that point is in the closure of the subset. The second part of that exercise was to find an example of a point that is in the closure of a subset, but no sequence from the subset converges to that point. This phenomenon is often referred to as "the inadequacy of sequences." We will describe two different, but equivalent, methods of building convergence structures that are adequate to describe closure.

First let us recall what we mean by a convergent sequence in a topological space.

Definition 16.1 *(Recall definition 6.10.) Suppose X is a topological space and $(x_n : n \in \omega)$ is a sequence in X. We say the sequence $(x_n : n \in \omega)$ **converges** to a point $x \in X$ if and only if for every neighborhood U of x there exists $N \in \omega$ such that if $n \geq N$, then $x_n \in U$. We denote this by $x_n \to x$.*

The first result is that for first-countable spaces, sequences are adequate to describe closure.

Theorem 16.2 *Suppose X is first countable and $A \subset X$. A point $x \in \overline{A}$ if and only if there is a sequence $(x_n : n \in \omega)$ consisting of points of A with $x_n \to x$.*

Proof. Suppose we have such a sequence, then every neighborhood of x will contain points, in fact an entire tail, of the sequence. Hence every neighborhood of x will intersect A, and we see that $x \in \overline{A}$. For the converse, suppose that $x \in \overline{A}$, and let $\{G_n : n \in \omega\}$ be a countable local base at x. For each $n \in \omega$, we let $B_n = \bigcap_{k=0}^{n} G_k$ and note that $\{B_n : n \in \omega\}$ is also a countable local base at x, and this one has the property that it is decreasing. Since $x \in \overline{A}$, for each $n \in \omega$, we can choose $x_n \in A \cap B_n$. Now, the sequence $(x_n : n \in \omega)$ consists entirely of points of A. Suppose U is a neighborhood of x. Choose $N \in \omega$ such that $x \in B_N \subset U$. Now if $n \geq N$, then $x_n \in B_n \subset B_N \subset U$. Hence we have that $x_n \to x$. ∎

In this proof, the reader will notice that we did not need to have a countable base at every point, just at the one where we were trying to build the convergent sequence. So the following, nominally stronger, result is actually what we proved.

Corollary 16.3 *Suppose X is a topological space, $x \in X$, and there is a countable local base at x. For any subset $A \subset X$, we have $x \in \overline{A}$ if and only if there is a sequence in A which converges to x.*

We turn now to spaces where the conclusion of theorem 16.2 fails. Obviously these are not first countable spaces.

Example 16.4 *We list two examples where sequences are inadequate. There are many such examples. I like these because they are in some sense opposite extremes. In the first there is only one point where the space fails to be first countable. In the second, no point has a countable neighborhood base.*

1. *In the space $\omega_1 + 1$ with the order topology, the point ω_1 is clearly in the closure of the set $[0, \omega_1)$, but if $(x_n : n \in \omega)$ is any sequence in $[0, \omega_1)$, then there is a point $\beta \in \omega_1$ with $x_n \leq \beta$ for all $n \in \omega$. Hence $(\beta, \omega_1] \cap \{x_n : n \in \omega\} = \varnothing$, and thus $x_n \not\to \omega_1$.*

2. *Let $\mathfrak{c}$ be the cardinality of $\mathbb{R}$. Consider the product space $\{0,1\}^\mathfrak{c}$. Let S be the functions with finite support, i.e.*

$$S = \{x \in \{0,1\}^\mathfrak{c} : |\{\alpha \in \mathfrak{c} : x_\alpha \neq 0\}| < \omega\}.$$

The point $\overline{1}$, i.e., the point of the product with all coordinates equal to 1, is in the closure of S, but no sequence in S converges to $\overline{1}$.

Proof. To see that 2 is true, suppose that B is a basic open set containing $\overline{1}$, and let F be the finite set of coordinates α such that $\pi_\alpha(B) \neq \{0,1\}$. Define $x = (x_\alpha : \alpha \in \mathfrak{c})$ by $x_\alpha = 1$ if $\alpha \in F$ and $x_\alpha = 0$ otherwise. It is clear that $x \in B \cap S$. Hence $\overline{1} \in \overline{S}$. On the other hand, if $(x_n : n \in \omega)$ is any sequence in

S, then if we let F_n be the set of coordinates on which x_n takes nonzero values, and we see that $\bigcup_{n \in \omega} F_n$ is a countable subset of $\mathfrak{c}$. Now choose $\alpha \in \mathfrak{c} \setminus \bigcup_{n \in \omega} F_n$, and $\pi_\alpha^{-1}(\{1\})$ is a basic open set about $\bar{1}$ which contains no terms of the sequence. Hence $x_n \nrightarrow \bar{1}$. $\blacksquare$

The problem of discovering convergence structures that *are* adequate to describe closure and still have the comfort of *feeling* like sequences has been solved in two different, but equivalent, ways. The first which we will discuss is the notion of a net. Nets look very much like sequences, and for this reason, students often like nets very much. Nets differ from sequences in that the index set is a more general set than the natural numbers.

Definition 16.5 *A **directed set** is a nonempty set Λ with a binary relation $\leq$ that satisfies the following conditions:*

1. *for each $\lambda \in \Lambda$, $\lambda \leq \lambda$,*

2. *if $\lambda_1 \leq \lambda_2$ and $\lambda_2 \leq \lambda_3$, then $\lambda_1 \leq \lambda_3$,*

3. *if $\lambda_1, \lambda_2 \in \Lambda$, then there is $\lambda_3 \in \Lambda$ with both $\lambda_1 \leq \lambda_3$ and $\lambda_2 \leq \lambda_3$.*

We often call the order relation on a directed set a *direction*, and we say that Λ is *directed* by $\leq$. The natural numbers with the usual order do form a directed set. A more typical example of a directed set is the next example.

Example 16.6 *Suppose that X is a topological space and $x \in X$. The complete neighborhood system $\mathcal{U}_x$ at the point x is a directed set when given the direction $\leq$ defined by $U_1 \leq U_2$ if and only if $U_1 \supset U_2$.*

In fact, any *neighborhood base* at a point is also a directed set when given this same direction.

Definition 16.7 *A **net** in a set X is a function $\nu : \Lambda \to X$, where Λ is a directed set. We usually denote $\nu(\lambda) = x_\lambda$, and we call x_λ the λth term of the net. We usually write $\nu = (x_\lambda : \lambda \in \Lambda)$. Indeed, the most common usage is to abandon the function notation altogether and simply refer to the net $(x_\lambda : \lambda \in \Lambda)$, just as we did with sequences.*

Since the natural numbers form a directed set, all sequences are nets. The converse is not true, as we leave the reader to discover. For this reason, nets are sometimes called *generalized sequences*.

Definition 16.8 *A **subnet** of a net $\nu = (x_\lambda : \lambda \in \Lambda)$ in a set X is a composition $\nu \circ \varphi$ where φ is an increasing and cofinal mapping from a directed set Σ into Λ. That is, $\varphi : \Sigma \to \Lambda$ satisfies the following:*

1. *if $\sigma_1 \leq \sigma_2$, then $\varphi(\sigma_1) \leq \varphi(\sigma_2)$, [$\varphi$ is increasing]*

2. *if $\lambda \in \Lambda$, then there exists $\sigma \in \Sigma$ with $\lambda \leq \varphi(\sigma)$. [$\varphi$ is cofinal]*

We easily see that a subsequence of a sequence is also a subnet of the sequence. Also following the subsequence customs, we often denote $\varphi(\sigma) = \lambda_\sigma$, and write $\nu \circ \varphi = (x_{\lambda_\sigma} : \sigma \in \Sigma)$ for the subnet.

Definition 16.9 *Suppose $(x_\lambda : \lambda \in \Lambda)$ is a net in a topological space X, and $x \in X$. We say the net $(x_\lambda : \lambda \in \Lambda)$ **converges** to the point x if and only if for every neighborhood U of x there exists $\lambda_0 \in \Lambda$ such that $\lambda \geq \lambda_0 \implies x_\lambda \in U$. We denote this by $x_\lambda \to x$, and we often say x is a **limit point** of the net $(x_\lambda : \lambda \in \Lambda)$.*

Again, compare this with what we know for sequences. If the net under consideration is actually a sequence, then this is exactly the definition of sequential convergence.

Other terminology which is popular is given by thinking of "tails." If $(x_\lambda : \lambda \in \Lambda)$ is a net, then the set $\{x_\lambda : \lambda \geq \lambda_0\}$ is called the *tail* of the net determined by λ_0. So we have $x_\lambda \to x$ if and only if every neighborhood of x contains a tail of the net. This is also sometimes described by saying that for every neighborhood U of x, the net is *eventually* in U.

A related idea is that of being frequently, or cofinally, in a neighborhood. We say a net is *frequently*, or *cofinally*, in a set U provided that for every $\lambda \in \Lambda$, there is a point $\mu \in \Lambda$ with $\mu \geq \lambda$ and $x_\mu \in U$. In the language of tails, this says that U intersects every tail of the net. As with sequences, this corresponds to the idea of cluster point.

Definition 16.10 *Suppose $(x_\lambda : \lambda \in \Lambda)$ is a net in a topological space X, and $x \in X$. We say x is a **cluster point** of the net $(x_\lambda : \lambda \in \Lambda)$ if and only if for every neighborhood U of x and every $\lambda \in \Lambda$, there exists $\mu \in \Lambda$ with $\mu \geq \lambda$ and $x_\mu \in U$.*

Hearkening back to the tail idea, this says that a point is a cluster point of a net if and only if every neighborhood of the point intersects every tail of the net, or equivalently every neighborhood contains *cofinally many* terms of the net. This also reminds us of sequences since a point is a cluster point of a sequence if and only if every neighborhood of the point contains infinitely many terms of the sequence.

We have analogues of the familiar theorems for sequences.

Theorem 16.11 *If* $(x_\lambda : \lambda \in \Lambda)$ *is a net in a space* X *and* $x_\lambda \to x \in X$, *then for any subnet* $(x_{\lambda_\sigma} : \sigma \in \Sigma)$ *we have* $x_{\lambda_\sigma} \to x$.

Proof. Suppose U is a neighborhood of x. Choose $\lambda_0 \in \Lambda$ such that $\lambda \geq \lambda_0 \implies x_\lambda \in U$. By cofinality, choose $\sigma_0 \in \Sigma$ such that $\lambda_{\sigma_0} \geq \lambda_0$. Now if $\sigma \geq \sigma_0$, then $\lambda_\sigma \geq \lambda_{\sigma_0}$, and thus $\lambda_\sigma \geq \lambda_0$. Hence $x_{\lambda_\sigma} \in U$. Thus we have $x_{\lambda_\sigma} \to x$. ∎

Theorem 16.12 *Suppose* $(x_\lambda : \lambda \in \Lambda)$ *is a net in a space* X, *and* $x \in X$. *The point* x *is a cluster point of* $(x_\lambda : \lambda \in \Lambda)$ *if and only if there is a subnet of* $(x_\lambda : \lambda \in \Lambda)$ *which converges to* x.

Proof. First suppose that x is a cluster point of the net $\nu = (x_\lambda : \lambda \in \Lambda)$. We define $\Sigma = \{(\lambda, U) : U \text{ is a neighborhood of } x \text{ and } x_\lambda \in U\}$. We define a direction on Σ by $(\lambda_1, U_1) \leq (\lambda_2, U_2)$ if and only if both $\lambda_1 \leq \lambda_2$ and $U_1 \supset U_2$.

We show that this is a direction. Clearly, $(\lambda, U) \leq (\lambda, U)$ for any $(\lambda, U) \in \Sigma$. Suppose $(\lambda_1, U_1) \leq (\lambda_2, U_2)$ and $(\lambda_2, U_2) \leq (\lambda_3, U_3)$. Since Λ is directed, $\lambda_1 \leq \lambda_3$, and by properties of subset, $U_1 \supset U_3$. Thus $(\lambda_1, U_1) \leq (\lambda_3, U_3)$. Now suppose $(\lambda_1, U_1), (\lambda_2, U_2) \in \Sigma$. Choose $\lambda_0 \in \Lambda$ with $\lambda_1 \leq \lambda_0$ and $\lambda_2 \leq \lambda_0$. Since U_1 and U_2 are neighborhoods of x, so is $U_1 \cap U_2$. Since x is a cluster point of $(x_\lambda : \lambda \in \Lambda)$, we can choose $\lambda_3 \geq \lambda_0$ with $x_{\lambda_3} \in U_1 \cap U_2$. Thus $(\lambda_3, U_1 \cap U_2) \in \Sigma$, and clearly $(\lambda_1, U_1) \leq (\lambda_3, U_1 \cap U_2)$ and $(\lambda_2, U_2) \leq (\lambda_3, U_1 \cap U_2)$.

To define the subnet, let $\varphi : \Sigma \to \Lambda$ be defined by $\varphi((\lambda, U)) = \lambda$. That φ is increasing is obvious, and φ is cofinal since $(\lambda, X) \in \Sigma$ for every $\lambda \in \Lambda$.

Suppose that U is a neighborhood of x. Choose $\lambda_0 \in \Lambda$ such that $x_{\lambda_0} \in U$. Now if $(\lambda, V) \in \Sigma$ and $(\lambda, V) \geq (\lambda_0, U)$, then $x_\lambda \in V \subset U$, and $x_\lambda = \nu \circ \varphi((\lambda, V))$. Hence the subnet converges to x.

Now for the converse, suppose that $(x_\lambda : \lambda \in \Lambda)$ has a subnet $(x_{\lambda_\sigma} : \sigma \in \Sigma)$ which converges to x, and let U be a neighborhood of x. Fix $\lambda_0 \in \Lambda$. Choose $\sigma_1 \in \Sigma$ such that $\lambda_{\sigma_1} \geq \lambda_0$. Now choose $\sigma_2 \in \Sigma$ such that $\sigma \geq \sigma_2 \implies x_{\lambda_\sigma} \in U$. Finally, we choose $\sigma_3 \in \Sigma$ such that $\sigma_1 \leq \sigma_3$ and $\sigma_2 \leq \sigma_3$. Now $x_{\lambda_{\sigma_3}} \in U$, and $\lambda_{\sigma_3} \geq \lambda_0$. Hence x is a cluster point of $(x_\lambda : \lambda \in \Lambda)$. ∎

Example 16.13 $(\cos \lambda : \lambda \in \mathbb{R})$ *is a net in* $\mathbb{R}$ *that clusters at every point in* $[-1, 1]$. *We leave it as an exercise to find subnets that converge to some of these points, say* 0 *and* 1.

Since a subnet of a subnet is a subnet of the original net, and every net is a subnet of itself, we have the following result.

Corollary 16.14 *A point x is a cluster point of a subnet of the net $(x_\lambda : \lambda \in \Lambda)$ in X if and only if x is a cluster point of $(x_\lambda : \lambda \in \Lambda)$.*

We now prove the result that we have been leading toward, namely that nets describe closure.

Theorem 16.15 *Suppose X is a space, $x \in X$, and $A \subset X$. The point $x \in \overline{A}$ if and only if there is a net $(x_\lambda : \lambda \in \Lambda)$ in A with $x_\lambda \to x$.*

Proof. First suppose there is a net $(x_\lambda : \lambda \in \Lambda)$ in A that converges to x. Let U be a neighborhood of x. Choose $\lambda_0 \in \Lambda$ such that $x_\lambda \in U$ for all $\lambda \geq \lambda_0$. Now $x_{\lambda_0} \in U \cap A$. Hence every neighborhood of x intersects A, and thus $x \in \overline{A}$.

Now suppose that $x \in \overline{A}$, and construct a net in A that converges to x as follows. We define $\Lambda = \{(y, U) : U \in \mathcal{U}_x \text{ and } y \in U \cap A\}$, and we direct Λ by $(y, U) \leq (z, V)$ if and only if $U \supset V$, i.e. reverse inclusion on the second term.

We show that this is a direction on Λ. First note that $(y, U) \leq (y, U)$ for all $(y, U) \in \Lambda$, and transitivity is clear. Suppose that (y, U) and (z, V) are elements of Λ. Since U and V are neighborhoods of x, so is $U \cap V$. Hence there is a point $t \in A \cap (U \cap V)$. Thus $(t, U \cap V) \in \Lambda$, and we clearly have $(y, U) \leq (t, U \cap V)$ and $(z, V) \leq (t, U \cap V)$. Hence we have a directed set. We claim that the net defined by $\nu((y, U)) = y$ converges to x. Suppose G is a neighborhood of x. Choose $y_0 \in G \cap A$, and note that $(y_0, G) \in \Lambda$. If $(y, U) \geq (y_0, G)$, then $x_{(y,U)} = y \in U \subset G$. Hence the net ν converges to x, and it is clear from the definition of Λ that the terms of this net are elements of A. ∎

We can also give a characterization of continuity similar to the one which we gave for metric spaces using sequences.

Theorem 16.16 *Suppose X and Y are spaces, $f : X \to Y$, and $x \in X$. The function f is continuous at x if and only if whenever a net $(x_\lambda : \lambda \in \Lambda)$ converges to x in X, we have that the net $(f(x_\lambda) : \lambda \in \Lambda)$ converges to $f(x)$ in Y.*

Proof. Suppose that f is continuous at x, i.e. for every neighborhood V of $f(x)$ there is a neighborhood U of x such that $f(U) \subset V$ (recall definition 5.5). Suppose $(x_\lambda : \lambda \in \Lambda)$ is a net in X that converges to x and V is a neighborhood of $f(x)$. Choose a neighborhood U of x with $f(U) \subset V$. Since $x_\lambda \to x$, we can choose $\lambda_0 \in \Lambda$ such that $\lambda \geq \lambda_0 \implies x_\lambda \in U$. Now if $\lambda \geq \lambda_0$, then $f(x_\lambda) \in f(U) \subset V$. Hence $f(x_\lambda) \to f(x)$.

For the converse, suppose f is not continuous at x. Choose a neighborhood V_0 of $f(x)$ such that for every neighborhood U of x, $f(U)\backslash V_0 \neq \varnothing$. Let $\Lambda = \{(z, U) : U \in \mathcal{U}_x, z \in U, \text{ and } f(z) \notin V_0\}$. Direct Λ by reverse inclusion on the second term. The verification that this is a direction is similar to some we have done before, so we leave it as an exercise. We define a net on Λ by $x_{(z,V)} = z$. It is clear that $x_{(z,V)} \to x$, but $f(x_{(z,V)}) \nrightarrow f(x)$, so the proof is complete. ∎

We now characterize convergence of nets in product spaces.

Theorem 16.17 *Suppose $(x_\lambda : \lambda \in \Lambda)$ is a net in the product space $\prod_{\alpha \in A} X_\alpha$. The net $(x_\lambda : \lambda \in \Lambda)$ converges in $\prod_{\alpha \in A} X_\alpha$ if and only if the net $(\pi_\alpha(x_\lambda) : \lambda \in \Lambda)$ converges in the space X_α for every $\alpha \in A$.*

Proof. The "only if" direction follows from the continuity of the projections. Suppose we have a net $(x_\lambda : \lambda \in \Lambda)$ in the product space so that for each $\alpha \in A$, $\pi_\alpha(x_\lambda) \to y_\alpha \in X_\alpha$. Consider the point $y = (y_\alpha : \alpha \in A)$ in the product space $\prod_{\alpha \in A} X_\alpha$. Suppose that B is a basic open neighborhood of y, and choose $\alpha_1, \alpha_2, \ldots, \alpha_n$ in A such that $B = \bigcap_{k=1}^n \pi_{\alpha_k}^{-1}(U_{\alpha_k})$ where each U_{α_k} is an open set in X_{α_k}. For each k, $y_{\alpha_k} \in U_{\alpha_k}$, so we choose $\lambda_k \in \Lambda$ such that $\lambda \geq \lambda_k \implies \pi_{\alpha_k}(x_\lambda) \in U_{\alpha_k}$. Since Λ is directed, it follows by induction that there exists $\lambda_0 \in \Lambda$ such that $\lambda_k \leq \lambda_0$ for $k = 1, 2, \ldots, n$. Now, if $\lambda \geq \lambda_0$, then $\lambda \geq \lambda_k$ for each k. Thus $\pi_{\alpha_k}(x_\lambda) \in U_{\alpha_k}$ for each k, that is, $x_\lambda \in \bigcap_{k=1}^n \pi_{\alpha_k}^{-1}(U_{\alpha_k}) = B$. Hence $x_\lambda \to y$. ∎

There is another convergence structure adequate to describe closure. This structure is called a filter, and the introduction of the reader to filters will consume the balance of this chapter. We will see that every net generates a filter in a natural way, and every filter generates a net in a natural way. The advantage of nets is that they look very much like sequences. The advantage of filters is that they do NOT look like sequences.

The similarity in the notation of nets to sequences brings a feeling of comfort and familiarity. The danger is that this comfort can lead to mistakes. One of the most common mistakes is to begin to think that a directed set is ordered better than it is (linearly, for example). As a graduate student, I once proved a remarkable result that settled an open question, and the proof was only moderately complicated. I was surprised that none of the other people who had worked on the problem had discovered this argument. After several days of thinking I had this big result, I found that I had cleverly disguised from myself the assumption that a directed set in my construction was actually linearly ordered. Down in flames!

Filters have the advantage (at least some people think it is an advantage) that they live entirely in the space. The directed set in the definition of a net is an object which lies outside the space, while a filter is defined entirely in terms of the space and subsets of the space.

Definition 16.18 *A **filter** on a set X is a nonempty collection $\mathcal{F}$ of nonempty subsets of X such that the following are true:*

1. *if $F_1, F_2 \in \mathcal{F}$, then $F_1 \cap F_2 \in \mathcal{F}$,*

2. *if $F \in \mathcal{F}$ and $F \subset G$, then $G \in \mathcal{F}$.*

So, a filter is a nonempty collection of nonempty sets that is closed under finite intersections and closed under supersets.

A word of caution: some authors do not exclude the empty set from membership in a filter as we have done, and they use the phrase "proper filter" to refer to what we call a filter.

Example 16.19 *If X is a topological space and $x \in X$, then $\mathcal{U}_x$, the complete neighborhood system at x, is a filter.*

Indeed, this is probably the prototype for the definition of a filter. The filter $\mathcal{U}_x$ is often called the *neighborhood filter* at x.

It is often more convenient to talk in terms of "bases" for filters.

Definition 16.20 *A **filterbase** on a set X is a nonempty collection $\mathcal{C}$ of nonempty subsets of X such that if $U_1, U_2 \in \mathcal{C}$, then there exists $U_3 \in \mathcal{C}$ such that $U_3 \subset U_1 \cap U_2$.*

Lemma 16.21 *If $\mathcal{C}$ is a filterbase on a set X, then*

$$\mathcal{F} = \{F \subset X : \exists U \in \mathcal{C} \text{ with } U \subset F\}$$

*is a filter, and $\mathcal{F}$ is called the **filter generated by the filterbase** $\mathcal{C}$.*

Proof. It is clear that $\mathcal{F}$ is closed under supersets. Suppose $F_1, F_2 \in \mathcal{F}$. Choose $U_1, U_2 \in \mathcal{C}$ with $U_1 \subset F_1$ and $U_2 \subset F_2$. Since $\mathcal{C}$ is a filterbase, there is an element $U_3 \in \mathcal{C}$ such that $U_3 \subset U_1 \cap U_2$. Now $U_3 \subset F_1 \cap F_2$, and thus $F_1 \cap F_2 \in \mathcal{F}$. Hence $\mathcal{F}$ is a filter. ∎

Continuing in the spirit of the example above, a neighborhood base at a point is a filterbase, and the filter that it generates is the neighborhood filter at the point.

Definition 16.22 *If $\mathcal{F}$ and $\mathcal{G}$ are filters on the same set X then we say $\mathcal{F}$ is* ***finer*** *than $\mathcal{G}$ provided $\mathcal{G} \subset \mathcal{F}$. We also describe this situation by saying that $\mathcal{G}$ is* ***coarser*** *than $\mathcal{F}$.*

To motivate our intuition here, because all filters are closed under supersets, all filters contain the "large" sets. When a filter is finer than another, it must contain more "small" sets.

Definition 16.23 *A filter $\mathcal{F}$ on a set X is called a* ***fixed*** *filter if there exists a point $x \in X$ with $x \in \cap \mathcal{F}$. A filter which is not fixed is called* ***free****, i.e., $\mathcal{F}$ is free provided $\cap \mathcal{F} = \varnothing$.*

Of course, neighborhood filters are always fixed. It takes a moment to convince yourself that free filters can even exist, but the following example is a useful one.

Example 16.24 *(Frechet filter) Let $\mathcal{C} = \{(a, \infty) : a \in \mathbb{R}\}$. It is obvious that $\mathcal{C}$ is a filterbase on $\mathbb{R}$. The filter $\mathcal{F}$ generated by this filterbase is called the Frechet filter. Notice that the Frechet filter is a free filter on the real numbers.*

Example 16.25 *Suppose $A \subset X$ and $A \neq \varnothing$. The collection*

$$\mathcal{F}_A = \{F \subset X : A \subset F\}$$

is a fixed filter on X.

Notice that $\mathcal{F}_A$ is finer than $\mathcal{F}_B$ if and only if $A \subset B$.

Definition 16.26 *Suppose $\mathcal{F}$ is a filter on a topological space X, and $x \in X$. We say $\mathcal{F}$* ***converges*** *to x, or x is a* ***limit point*** *of $\mathcal{F}$, if and only if $\mathcal{U}_x \subset \mathcal{F}$. In this case, we write $\mathcal{F} \to x$.*

Another way of saying this is that a filter converges to a point x if and only if every neighborhood of x contains an element of the filter.

Definition 16.27 *Suppose $\mathcal{F}$ is a filter on a topological space X and $x \in X$. We say $\mathcal{F}$* ***clusters*** *at x, or x is a* ***cluster point*** *of $\mathcal{F}$, if and only if $x \in \cap \{\overline{F} : F \in \mathcal{F}\}$.*

Lemma 16.28 *Suppose $\mathcal{F}$ is a filter on a topological space X, and $x \in X$. If $\mathcal{F} \to x$, then x is a cluster point of $\mathcal{F}$.*

Proof. Suppose $\mathcal{F} \to x$, $F \in \mathcal{F}$, and U is a neighborhood of x. Since $\mathcal{F} \to x$, $U \in \mathcal{F}$, and thus $U \cap F \in \mathcal{F}$. Hence $U \cap F \neq \varnothing$. Thus we have that $x \in \overline{F}$ for every $F \in \mathcal{F}$, i.e. x is a cluster point of $\mathcal{F}$. ∎

The converse may not be true, of course. In the real line, consider the filter $\mathcal{F}_{\mathbb{Q}}$ of all supersets of $\mathbb{Q}$. Each point of $\mathbb{R}$ is a cluster point of this filter, but no real number is a limit point.

Theorem 16.29 *Suppose $\mathcal{F}$ is a filter on a topological space X, and $x \in X$. If x is a cluster point of $\mathcal{F}$, then there is a finer filter $\mathcal{G} \supset \mathcal{F}$ such that $\mathcal{G} \to x$.*

Proof. Suppose that x is a cluster point of the filter $\mathcal{F}$. Consider

$$\mathcal{C} = \{U \cap F : U \in \mathcal{U}_x \text{ and } F \in \mathcal{F}\}.$$

Note that each element of $\mathcal{C}$ is nonempty. Suppose that $U_1 \cap F_1, U_2 \cap F_2 \in \mathcal{C}$. Now $U_1 \cap U_2$ is a neighborhood of x, and $F_1 \cap F_2 \in \mathcal{F}$. Hence $(U_1 \cap F_1) \cap (U_2 \cap F_2) = (U_1 \cap U_2) \cap (F_1 \cap F_2) \in \mathcal{C}$. Thus we see that $\mathcal{C}$ is a filterbase. Let $\mathcal{G}$ be the filter generated by $\mathcal{C}$. For each $U \in \mathcal{U}_x$, $U = U \cap X \in \mathcal{C}$, and thus $U \in \mathcal{G}$. So $\mathcal{G} \to x$. Also for each $F \in \mathcal{F}$, $F = X \cap F \in \mathcal{C}$, and thus $F \in \mathcal{G}$. So $\mathcal{G}$ is finer than $\mathcal{F}$. ∎

A filter for which there is no finer filter turns out to be a very interesting object.

Definition 16.30 *An **ultrafilter** is a filter that is contained in no finer filter. That is, a filter $\mathcal{F}$ is an ultrafilter if and only if whenever $\mathcal{G}$ is a filter and $\mathcal{F} \subset \mathcal{G}$, then $\mathcal{F} = \mathcal{G}$.*

It is natural to wonder whether such a filter could even exist, but consider the filter of all supersets of a single point. This filter is called the *principal ultrafilter* generated by the point. It would certainly have the ultrafilter property since the singleton of the point would be in the filter, and any additional sets would have to be disjoint from the singleton. A more difficult question is whether there are any nontrivial ultrafilters. This is answered in the affirmative by the following, assuming we are willing to accept the axiom of choice.

Theorem 16.31 *Every filter is contained in an ultrafilter.*

Proof. We will use Zorn's lemma to prove this result. Suppose $\mathcal{F}$ is a filter on a set X. Let $\mathbb{P} = \{\mathcal{G} : \mathcal{G}$ is a filter on X and $\mathcal{F} \subset \mathcal{G}\}$. Note that $\mathbb{P}$ is nonempty since $\mathcal{F} \in \mathbb{P}$. We order $\mathbb{P}$ by subset inclusion. Suppose that $\mathbb{K}$ is a nonempty linearly ordered subset of $\mathbb{P}$, and we will show that $\cup \mathbb{K} \in \mathbb{P}$. For each $\mathcal{G} \in \mathbb{K}$, $\mathcal{F} \subset \mathcal{G}$, and thus $\mathcal{F} \subset \cup \mathbb{K}$. It remains only to show that $\cup \mathbb{K}$ is a filter on X.

If $A, B \in \bigcup \mathbb{K}$, then we can choose $\mathcal{G}_1, \mathcal{G}_2 \in \mathbb{K}$ with $A \in \mathcal{G}_1$ and $B \in \mathcal{G}_2$. Now either $\mathcal{G}_1 \subset \mathcal{G}_2$ or $\mathcal{G}_2 \subset \mathcal{G}_1$, but in either case A and B are both elements of the same filter. Say it was $\mathcal{G}_1$; this means that $A \cap B \in \mathcal{G}_1$. Hence $A \cap B \in \bigcup \mathbb{K}$. Suppose $A \in \bigcup \mathbb{K}$ and $A \subset B$. Choose $\mathcal{G} \in \mathbb{K}$ with $A \in \mathcal{G}$. Since $\mathcal{G}$ is a filter, $B \in \mathcal{G}$, and thus $B \in \bigcup \mathbb{K}$. Therefore, $\bigcup \mathbb{K} \in \mathbb{P}$. Now, by Zorn's lemma, $\mathbb{P}$ has a maximal element, and that element will be an ultrafilter containing $\mathcal{F}$. ∎

Corollary 16.32 *There are free ultrafilters on* $\mathbb{R}$.

Proof. The ultrafilter containing the Frechet filter on the real numbers is a free ultrafilter. ∎

That ultrafilters are very strange objects is emphasized by the following result. This is often a convenient way to check that a particular filter is, or is not, an ultrafilter.

Theorem 16.33 *A filter* $\mathcal{F}$ *on a set* X *is an ultrafilter on* X *if and only if for every* $A \subset X$, *either* $A \in \mathcal{F}$ *or* $X \backslash A \in \mathcal{F}$.

Proof. Suppose $\mathcal{F}$ is an ultrafilter on X, and $A \subset X$. Suppose that there exists $F \in \mathcal{F}$ with $A \cap F = \varnothing$. We then have that $F \subset X \backslash A$, and thus $X \backslash A \in \mathcal{F}$. On the other hand, if no such F exists, then $\{A \cap F : F \in \mathcal{F}\}$ is a filterbase, and the filter $\mathcal{G}$ which it generates is finer than $\mathcal{F}$. Hence $\mathcal{F} = \mathcal{G}$, but $A = A \cap X \in \mathcal{G}$. Thus we have $A \in \mathcal{F}$, and we have the "only if" direction.

Now suppose that $\mathcal{F}$ is a filter, and for all $A \subset X$, either $A \in \mathcal{F}$ or $X \backslash A \in \mathcal{F}$. Suppose that $\mathcal{G}$ is a filter on X that is finer than $\mathcal{F}$. If $\mathcal{G} \neq \mathcal{F}$, then let $A \in \mathcal{G} \backslash \mathcal{F}$. Since $A \notin \mathcal{F}$, it must be the case that $X \backslash A \in \mathcal{F}$, but then $X \backslash A \in \mathcal{G}$ also, and $A \cap (X \backslash A) = \varnothing$, a contradiction. Hence $\mathcal{F} = \mathcal{G}$. Thus $\mathcal{F}$ is an ultrafilter. ∎

We now turn to the problem of describing topological properties with filters.

Theorem 16.34 *Suppose* X *is a space,* $x \in X$, *and* $A \subset X$. *The point* $x \in \overline{A}$ *if and only if there is a filter* $\mathcal{F}$ *on* X *with* $A \in \mathcal{F}$ *and* $\mathcal{F} \to x$.

Proof. Suppose that $\mathcal{F}$ is a filter which contains the set A and converges to x. We want to show that $x \in \overline{A}$. For any neighborhood U of x, $U \in \mathcal{F}$, and thus $U \cap A \neq \varnothing$. Hence $x \in \overline{A}$.

For the converse, suppose that $x \in \overline{A}$. Now the collection $\{A \cap U : U \in \mathcal{U}_x\}$ is a filterbase. Clearly, $\mathcal{U}_x$ is contained in the filter generated by this filterbase, by the superset property, and A is an element of this filterbase since $X \in \mathcal{U}_x$. ∎

Definition 16.35 *Suppose that $f : X \to Y$, and $\mathcal{F}$ is a filter on X. The collection $\{f(F) : F \in \mathcal{F}\}$ is a filterbase on Y since $f(F_1 \cap F_2) \subset f(F_1) \cap f(F_2)$. The filter on Y generated by this filterbase is denoted $f(\mathcal{F})$.*

Theorem 16.36 *Suppose X and Y are spaces, $f : X \to Y$, and $x \in X$. The function f is continuous at x if and only if whenever a filter $\mathcal{F}$ converges to x in X, we have that the filter $f(\mathcal{F})$ converges to $f(x)$ in Y.*

Proof. Suppose that f is continuous at x, and $\mathcal{F} \to x$. Let V be a neighborhood of $f(x)$ in Y. By continuity, choose a neighborhood U of x such that $f(U) \subset V$. By the superset property, we know that $V \in f(\mathcal{F})$. Hence $f(\mathcal{F}) \to f(x)$.

For the converse, suppose we have the filter convergence property, and let V be a neighborhood of $f(x)$. Since $\mathcal{U}_x$ is a filter on X, and clearly $\mathcal{U}_x \to x$, we know that $f(\mathcal{U}_x) \to f(x)$. Hence $V \in f(\mathcal{U}_x)$, that is, there exists $U \in \mathcal{U}_x$ such that $f(U) \subset V$. This is the definition of continuity at a point. ∎

Theorem 16.37 *Suppose $\mathcal{F}$ is a filter in the product space $\prod_{\alpha \in A} X_\alpha$. The filter $\mathcal{F}$ converges in $\prod_{\alpha \in A} X_\alpha$ if and only if the filter $\pi_\alpha(\mathcal{F})$ converges in the space X_α for every $\alpha \in A$.*

Proof. The "only if" direction follows from the previous result and the fact that projections are continuous. Suppose $\mathcal{F}$ is a filter in the product space $\prod_{\alpha \in A} X_\alpha$, and $\pi_\alpha(\mathcal{F}) \to x_\alpha$ in the space X_α for every $\alpha \in A$. Consider the point $x = (x_\alpha : \alpha \in A)$ in the product space. We wish to show that $\mathcal{F} \to x$. Let U be a basic open set in the product which contains x. We can describe $U = \bigcap_{k=1}^{n} \pi_{\alpha_k}^{-1}(U_{\alpha_k})$ where each U_{α_k} is an open set in X_{α_k} containing x_{α_k}. Since $\pi_{\alpha_k}(\mathcal{F}) \to x_{\alpha_k}$, we choose $F_k \in \mathcal{F}$ such that $\pi_{\alpha_k}(F_k) \subset U_{\alpha_k}$ for each $k = 1, 2, ..., n$. For each k, $F_k \subset \pi_{\alpha_k}^{-1}(U_{\alpha_k})$, and by the intersection property and induction, $\bigcap_{k=1}^{n} F_k \in \mathcal{F}$. Now U contains $\bigcap_{k=1}^{n} F_k$, and thus $U \in \mathcal{F}$. Hence $\mathcal{F} \to x$, and the proof is complete. ∎

We now mention a result about images of ultrafilters, which will be useful to us later.

Theorem 16.38 *If $\mathcal{F}$ is an ultrafilter on X, and $f : X \to Y$ is onto Y, then $f(\mathcal{F})$ is an ultrafilter on Y.*

Proof. Suppose $A \subset Y$, and consider $f^{-1}(A)$. If $f^{-1}(A) \in \mathcal{F}$, then, since f is onto, $A = f(f^{-1}(A)) \in f(\mathcal{F})$. Also, if $X \backslash f^{-1}(A) = f^{-1}(Y \backslash A) \in \mathcal{F}$, then $Y \backslash A = f(f^{-1}(Y \backslash A)) \in f(\mathcal{F})$. Since $\mathcal{F}$ is an ultrafilter, by theorem 16.33, one of these must be true, and thus $f(\mathcal{F})$ also satisfies the criterion in theorem 16.33. ∎

We have seen above that both nets and filters give us notions of convergence that demonstrate the same theorems for describing closure and for describing continuity. We now develop the relationship between the two structures and show that they are indeed equivalent.

Definition 16.39 *Suppose that $(x_\lambda : \lambda \in \Lambda)$ is a net in a set X. For each $\mu \in \Lambda$, we define the "tail following μ" to be the set*

$$T_\mu = \{x_\lambda : \lambda \geq \mu\}.$$

*Since Λ is a directed set, $\{T_\lambda : \lambda \in \Lambda\}$ is a filterbase on X. The filter that is generated by this filterbase is called the **filter generated by the net** $(x_\lambda : \lambda \in \Lambda)$.*

As we write this out, it seems complicated, but it is not. The filter generated by a net is just the filter having the tails of the net as a filterbase. A set will be in this filter if and only if it contains a tail of the net.

Definition 16.40 *Suppose $\mathcal{F}$ is a filter on a set X. We define*

$$\Lambda_{\mathcal{F}} = \{(y, F) : y \in F \in \mathcal{F}\}.$$

*We direct $\Lambda_{\mathcal{F}}$ by reverse inclusion on the second term, i.e., $(y, F) \leq (z, G)$ if and only if $G \subset F$. It is easy to check that $\Lambda_{\mathcal{F}}$ is a directed set with this direction. We define $x_{(y,F)} = y$ for each $(y, F) \in \Lambda_{\mathcal{F}}$. The net $(x_{(y,F)} : (y, F) \in \Lambda_{\mathcal{F}})$ is the **net generated by the filter** $\mathcal{F}$.*

We will see in the next two theorems that this correspondence between nets and filters generates the same idea of limit points.

Theorem 16.41 *Suppose X is a space and $x \in X$. A net $(x_\lambda : \lambda \in \Lambda)$ in X converges to x if and only if the filter generated by the net $(x_\lambda : \lambda \in \Lambda)$ converges to x.*

Proof. Suppose $x_\lambda \to x$, and suppose that U is a neighborhood of x. By the definition of convergence, there is λ_0 such that $T_{\lambda_0} \subset U$. Hence U is an element of the filter generated by the net $(x_\lambda : \lambda \in \Lambda)$. Since this is true for every neighborhood of x, the filter generated by $(x_\lambda : \lambda \in \Lambda)$ converges to x.

For the converse, suppose it is true that the filter generated by $(x_\lambda : \lambda \in \Lambda)$ converges to x. For any neighborhood V of x, V must be an element of this filter. Thus V must contain a tail of the net. Suppose $T_{\lambda_0} \subset V$. We then know that if $\lambda \geq \lambda_0$, then $x_\lambda \in T_{\lambda_0} \subset V$. Hence $x_\lambda \to x$. ∎

Theorem 16.42 *Suppose X is a space and $x \in X$. A filter $\mathcal{F}$ on the space X converges to x if and only if the net generated by the filter $\mathcal{F}$ converges to x.*

Proof. Suppose $\mathcal{F}$ is a filter on X and $\mathcal{F} \to x$. Let $U \in \mathcal{U}_x$. Since $U \in \mathcal{F}$, $(x, U) \in \Lambda_{\mathcal{F}}$. Suppose $(y, F) \in \Lambda_{\mathcal{F}}$, and $(y, F) \geq (x, U)$. By the definition of the direction, $F \subset U$, and thus $x_{(y,F)} = y \in U$. Hence $x_{(y,F)} \to x$.

For the converse, suppose that the net generated by the filter $\mathcal{F}$ converges to x, and suppose V is a neighborhood of x. Choose $(y, F) \in \Lambda_{\mathcal{F}}$ such that if $(z, G) \geq (y, F)$, then $x_{(z,G)} = z \in V$. For each point $z \in F$, we have $(z, F) \geq (y, F)$, and so it follows that $z \in V$. Hence $F \subset V$, and thus $V \in \mathcal{F}$. Since this is true for every neighborhood of x, $\mathcal{F} \to x$. ∎

Exercises

1. Which filters converge in a discrete space? Which filters converge in a space with the trivial topology? If X is an infinite space with the cofinite topology, which filters converge in X? Find a filter in a topological space that clusters but does not converge.

2. Mimic the proof in example 16.4 to show that any uncountable product of copies of a two-point discrete space fails to be first countable.

3. Prove that a net has a cluster point at x if and only if the filter generated by that net also has a cluster point at x. Prove that a filter has a cluster point at x if and only if the net generated by that filter also has a cluster point at x.

4. Prove that the filter generated by a subnet is finer than the filter generated by the original net.

5. Decide what should be meant by the word "ultranet," and show that the filter generated by an ultranet is an ultrafilter, and the net generated by an ultrafilter is an ultranet.

6. Is the translation process symmetric, that is, if you begin with a net, then look at the filter that it generates, and then look at the net it generates, do you have the net you began with? What if you start with a filter? Do you return to the filter you began with?

7. The set $\{(0, \varepsilon) : \varepsilon > 0\}$ is a filterbase on $\mathbb{R}$. Let $\mathcal{F}$ be the filter on $\mathbb{R}$ that is generated by this filterbase. Show that $\mathcal{F} \to 0$, even though there are sets

in the filter that do not contain the point 0. In particular, is it true that if a filter converges to a point, then each element of the filter must contain the point?

8. Prove that if an ultrafilter has a cluster point, then it must converge to that cluster point.

9. Prove that if a filter $\mathcal{F}$ on a set X is contained in only one ultrafilter on X, then $\mathcal{F}$ is an ultrafilter.

10. A space X is called a *Frechet-Urysohn space* provided that for every subset $A \subset X$, a point is in the closure of A if and only if there is a sequence in A that converges to the point. Prove that every first-countable space is a Frechet-Urysohn space. Prove that the quotient space obtained by identifying the closed subset $\mathbb{N}$ of the real line to a point is a Frechet-Urysohn space that is not first countable.

11. A space X is called a *sequential space* provided that a subset $A \subset X$ is closed if and only if A contains the limit of every convergent sequence that can be drawn from A. Prove that every Frechet-Urysohn space is sequential. Prove that the quotient space obtained from $\mathbb{R}$ by identifying $\frac{1}{n}$ with n, for each $n \in \mathbb{N}$, is a sequential space that is not Frechet-Urysohn.

12. A space X has *countable tightness* provided that if $A \subset X$ and $x \in \overline{A}$, then there is a countable set $C \subset A$ with $x \in \overline{C}$. Observe that every countable space has countable tightness. Prove that every sequential space has countable tightness. Let $X = \mathbb{N} \cup \{\infty\}$ where ∞ is some point that is not an element of $\mathbb{N}$. To define a topology on X, let $\mathcal{F}$ be a free ultrafilter on $\mathbb{N}$, a neighborhood of ∞ will have form $F \cup \{\infty\}$ where $F \in \mathcal{F}$, and all points of $\mathbb{N}$ will be isolated. Prove that X has countable tightness, but X is not sequential.

Chapter 17

Separation Axioms II

Separation axioms are statements about the richness of a topology. They measure, in some sense, how many open sets a space has. Are there enough to tell points apart? How well? Are there enough to tell points from closed sets? How about closed sets from closed sets? These are the kind of questions that are addressed by separation axioms. We introduced several of the separation axioms in chapter 6, and we revisit them in this chapter. We will also go a bit further in the hierarchy. We will continue to use the T_i notation, and we will leave it as an exercise to show that $T_i \implies T_j$ if $i > j$.

Definition 17.1 *A space X is called a T_0-**space** provided that for any two points of X there is an open subset of X that contains one of the points but not the other. We often say X is T_0.*

We are not going to say too much about T_0-spaces. The trivial topology on a set consisting of more than one point is not T_0. Many spaces, metric for example, are T_0. All subspaces and products of T_0-spaces are T_0, while quotients need not be.

Definition 17.2 *A space X is called a T_1-**space** provided that for any two points x, y of X there are open sets U and V such that $x \in U$ while $y \notin U$ and $y \in V$ while $x \notin V$.*

That is, X is T_1 provided that for any two points, each is contained in an open set that excludes the other. We usually think of T_1-spaces as the spaces where singletons are closed.

Theorem 17.3 *A space X is a T_1-space if and only if for each $x \in X$, $\{x\}$ is a closed set in X.*

Proof. Suppose that X is a T_1-space and $x \in X$. For each $y \in X$ with $x \neq y$, choose an open set V_y with $y \in V_y$ and $x \notin V_y$. The set $X \backslash \{x\} = \bigcup_{y \neq x} V_y$, which is an open set. Thus $\{x\}$ is a closed set.

For the converse, if each singleton is closed and x, y are distinct points of X, then $U = X \backslash \{y\}$ and $V = X \backslash \{x\}$ provide the desired separation. ∎

Example 17.4 *The Sierpiński space is a T_0-space which is not T_1. Recall that this example is the two-point set $\{0, 1\}$ with the topology $\tau = \{\varnothing, \{0\}, \{0, 1\}\}$.*

We will not prove, but we encourage the reader to prove, that all subspaces and products of T_1-spaces are again T_1. Furthermore, quotients of T_1-spaces need not be T_1 (or even T_0), but closed continuous images of T_1-spaces are T_1.

The minimum separation which most topologists are willing to accept is the Hausdorff separation axiom.

Definition 17.5 *We say X is a **Hausdorff space**, or a **T_2-space**, provided that for any two points $x, y \in X$ there exist disjoint open subsets U and V of X such that $x \in U$ and $y \in V$. We often say X is T_2.*

There are many examples of T_2-spaces, and there many are examples of spaces that are not T_2-spaces.

Example 17.6 *Any infinite set with the cofinite topology is a T_1-space that is not T_2. On the other hand, all metric spaces, as well as ω_1, $\mathbb{S}$, $\mathbb{M}$, and Γ are T_2.*

One of the nicest things about Hausdorff spaces is that limits are unique.

Theorem 17.7 *If X is a T_2 space, $\mathcal{F}$ is a filter on X, $\mathcal{F} \to x$, and $\mathcal{F} \to y$, then $x = y$.*

Proof. Suppose X is a T_2 space, $\mathcal{F}$ is a filter on X, $\mathcal{F} \to x$ and $\mathcal{F} \to y$. Suppose $x \neq y$. We choose disjoint open $U, V \subset X$ with $x \in U$ and $y \in V$. Now $U \in \mathcal{U}_x \subset \mathcal{F}$, and $V \in \mathcal{U}_y \subset \mathcal{F}$. Since $\mathcal{F}$ is a filter, $U \cap V \neq \varnothing$, a contradiction. ∎

Obviously by the translation process, limits of nets, and consequently limits of sequences, are also unique in Hausdorff spaces. It is also worth pointing out that the converse is also true: if x, y were distinct points of X which could not be separated by disjoint open sets, then $\{U \cap V : U \in \mathcal{U}_x \text{ and } V \in \mathcal{U}_y\}$ would be a filterbase for a filter that converges to both points.

Theorem 17.8 *A topological space X is Hausdorff if and only if the diagonal $\Delta = \{(x,x) : x \in X\}$ is a closed subset of $X \times X$.*

Proof. We leave most of the details as an exercise. As a starting point, show that if $U, V \subset X$, then $U \cap V = \varnothing$ if and only if $(U \times V) \cap \Delta = \varnothing$. ∎

We now present the preservation theorem for the Hausdorff property.

Theorem 17.9 *(Preservation theorem for T_2-spaces)*

1. *Every subspace of a T_2-space is also a T_2-space.*

2. *A nonempty product space $\prod_{\alpha \in I} X_a$ is T_2 if and only if X_α is T_2 for each $\alpha \in I$.*

3. *Quotients of T_2-spaces need not be T_2.*

Proof. Suppose $Y \subset X$, and X is a T_2-space. For any two distinct points $x, y \in Y$, choose disjoint open sets $U, V \subset X$ with $x \in U$ and $y \in V$. Now $U \cap Y$ and $V \cap Y$ are disjoint open sets in Y, and $x \in U \cap Y$, $y \in V \cap Y$. Thus Y is a T_2-space.

For part 2, since each factor space X_α is homeomorphic to a subspace of the product, the "only if" follows from part 1. Now suppose each factor space is T_2, and let $x = (x_\alpha : \alpha \in I)$ and $y = (y_\alpha : \alpha \in I)$ be distinct points in the product space $\prod_{\alpha \in I} X_a$. Since $x \neq y$, choose $\beta \in I$ such that $x_\beta \neq y_\beta$. Since X_β is a T_2-space, there exist disjoint open sets U, V in the space X_β with $x_\beta \in U$ and $y_\beta \in V$. Now $x \in \pi_\beta^{-1}(U)$, $y \in \pi_\beta^{-1}(V)$, and $\pi_\beta^{-1}(U) \cap \pi_\beta^{-1}(V) = \varnothing$. Hence the product space $\prod_{\alpha \in I} X_a$ is T_2.

For part 3., see the examples below. ∎

Example 17.10 *The following examples show that continuous and quotient images of T_2-spaces may fail to be T_2.*

1. *If X is any set with at least two points and the discrete topology, and X^t is the same set with the trivial topology, then the identity mapping is a continuous one-to-one function from the T_2-space X onto the non-T_2-space X^t.*

2. *Let X be the real line with the usual topology except that basic neighborhoods of 0 will have the form $(-\varepsilon, \varepsilon) \backslash A$ where $A = \{\frac{1}{n} : n \in \mathbb{N}\}$. It is easy to see that X is a T_2-space. Since A is a closed subset of X, the natural projection $\pi : X \to X/A$ is a closed continuous quotient mapping, but $[0]$ and A cannot be separated in the decomposition space. Hence X/A is not Hausdorff.*

3. *(Line with two origins) Let X be the subspace* $\mathbb{R} \times \{0,1\}$ *of* $\mathbb{R}^2$*, and let* Y
   be the space obtained from X *by identifying* $(x,0)$ *and* $(x,1)$ *for* $x \neq 0$*. In*
   Y*, the points* $(0,0)$ *and* $(0,1)$ *cannot be separated. The natural projection*
   $\pi : X \to Y$ *is an open continuous quotient mapping from the metric space*
   X *onto the non-*T_2*-space* Y*.*

We actually see, from part 3 above, that we will not be able to preserve any
of the separation axioms that are possessed by metric spaces (and that is all of
them) under quotient mappings, or even open continuous mappings.

From part one, we recall theorem 6.9 which says that continuous functions are
determined by their values on dense subsets of the domain if the range is T_2.

Theorem 17.11 *If* X *is a topological space,* Y *is a Hausdorff space,* $f : X \to Y$
is continuous, $g : X \to Y$ *is continuous, and* $\{x \in X : f(x) = g(x)\}$ *is dense in*
X*, then* $f = g$*.*

Our last result before going on to higher separation axioms is an easy remark,
but it is useful when we are constructing new examples. The fact that so many
natural conjectures have turned out to be false has resulted in the phenomenon
that it is probably more important in general topology than in any other discipline
in mathematics to be skillful in constructing new examples.

Theorem 17.12 *If* (X, τ) *is a* T_2*-space, and* τ_1 *is a finer topology on* X*, then*
(X, τ_1) *is also a* T_2*-space.*

Proof. Left to the reader. ∎

Definition 17.13 *A space* X *is called* **regular** *provided that for any closed set*
$F \subset X$ *and any point* $x \in X \backslash F$*, there exist disjoint open sets* $U, V \subset X$ *with*
$x \in U$ *and* $F \subset V$*. A space that is regular and* T_1 *is a* **T_3-space**.

We have defined T_3 to be regular and T_1 since it is then obvious that $T_3 \implies T_2$.
However, it is a fact, and we leave it as an exercise, that regular and T_0 also implies
T_2 (and thus T_3).

Example 17.14 *We describe a few* T_2*-spaces that fail to be regular.*

1. *Part 2 of example 17.10 is a* T_2*-space, and it fails to be regular for exactly*
   the same reason that the quotient fails to be T_2*.*

2. *(Radial plane) Let $X = \mathbb{R} \times \mathbb{R}$, and define a set $U \subset X$ to be open if and only if for each point $x \in U$, U contains a straight line segment of positive length emanating from x in every possible direction. It is easy to verify that this definition of open gives us a topology. The space is T_2 since the topology is finer than Euclidean. It is a good exercise to show that this space is not regular.*

3. *(Archimedean plane) This space is also sometimes called the "iron cross space." Let $x = (x_1, x_2) \in \mathbb{R}^2$. The cross of radius ε centered at x is the set*

$$C(x, \varepsilon) = ((x_1 - \varepsilon, x_1 + \varepsilon) \times \{x_2\}) \cup (\{x_1\} \times (x_2 - \varepsilon, x_2 + \varepsilon)).$$

This space, which we denote $\mathbb{A}^2$, has $\mathbb{R}^2$ as underlying set, and a set $U \subset \mathbb{A}^2$ is open if and only if for each $x \in U$ there exists $\varepsilon_x > 0$ such that $C(x, \varepsilon_x) \subset U$. The verification that this generates a topology is just like the corresponding result for metric spaces, and this topology is finer than the usual, so it is T_2. The odd thing is that, far from being metric, $\mathbb{A}^2$ is not regular, not first countable, and even not Frechet-Urysohn (see chapter 16, exercise 10).

Before proceeding to the preservation theorem, we give a characterization of regularity that is very useful. It states that points in open sets in regular spaces are really deep inside; that is, they are far enough inside the set that it is possible to fit another open set about the point whose closure still stays inside the original open set.

Theorem 17.15 *For any space X, the following are equivalent:*

1. *X is regular.*

2. *Each point of X has a neighborhood base consisting of closed sets.*

3. *If $x \in U \subset X$, and U is open, then there exists an open set $V \subset X$ such that $x \in V \subset \overline{V} \subset U$.*

Proof. First we prove $1 \implies 2$. Suppose X is regular, and $x \in X$. Let $\mathcal{B}_x = \{\overline{U} : U$ is open and $x \in U\}$. We will show that $\mathcal{B}_x$ is a neighborhood base at x. Suppose G is an open set, and $x \in G$. Since $X \backslash G$ is closed, by regularity we choose disjoint open sets $U, V \subset X$ with $x \in U$ and $X \backslash G \subset V$. Now $U \subset X \backslash V \subset X \backslash (X \backslash G) = G$, and thus $\overline{U} \subset X \backslash V \subset G$. Hence $\overline{U} \in \mathcal{B}_x$, and $x \in \overline{U} \subset G$. Thus $\mathcal{B}_x$ is a neighborhood base at x.

Now we prove 2 $\implies$ 3. Suppose $x \in U \subset X$, and U is open. Let $\mathcal{B}_x$ be a neighborhood base at x consisting of closed sets. Choose $B \in \mathcal{B}_x$ with $x \in B \subset U$. Let $V = IntB$. Since B is a neighborhood of x, $x \in V$, and since B is closed, $\overline{V} \subset B$. Hence we have that V is open, and $x \in V \subset \overline{V} \subset U$, as desired.

Finally, we show that 3 $\implies$ 1. Suppose that 3 is true, and $F \subset X$ is closed with $x \in X \backslash F$. Since $X \backslash F$ is open, by 3, we can choose an open set V with $x \in V \subset \overline{V} \subset X \backslash F$. Let $W = X \backslash \overline{V}$, and note that W is open. Further, $F \subset X \backslash \overline{V} = W$, and $V \cap W = \varnothing$. Hence X is regular. ∎

We now turn to the preservation theorem. We find that regularity is preserved by subspaces and products, but not by quotients. This is typical of separation properties.

Theorem 17.16 *(Preservation theorem for T_3-spaces)*

1. *Every subspace of a regular space (T_3-space) is also a regular space (T_3-space).*

2. *A nonempty product space $\prod_{\alpha \in I} X_\alpha$ is regular (T_3) space if and only if X_α is regular (T_3) for each $\alpha \in I$.*

3. *Quotients of T_3-spaces need not be regular.*

Proof. For 1 and 2, we need only prove regularity since we already know the preservation results for T_1. Suppose that $A \subset X$ where X is regular, and that F is a closed subset of A with $x \in A \backslash F$. Choose a closed set $H \subset X$ such that $H \cap A = F$. Now $x \in X \backslash H$, and X is regular, so we choose disjoint open subsets $U, V \subset X$ with $x \in U$ and $H \subset V$. We see that $U \cap A$ and $V \cap A$ are open subsets of A, that $x \in U \cap A$, and $F \subset V \cap A$. Hence A is regular.

The "only if" part of 2 follows from 1 since each factor is homeomorphic to a subspace of the product. For the "if" part, suppose that X_α is regular for each $\alpha \in I$. Suppose that $U \subset \prod_{\alpha \in I} X_\alpha$ is open and $x \in U$. Choose a basic open set $W = \bigcap_{k=1}^{n} \pi_{\alpha_k}^{-1}(W_{\alpha_k})$ with $x \in W \subset U$. Notice that $W = \prod_{\alpha \in I} W_\alpha$ where $W_\alpha = X_\alpha$ for $\alpha \notin \{\alpha_1, \alpha_2, ..., \alpha_n\}$. For each k, $k = 1, 2, ..., n$, X_{α_k} is regular, and $x_{\alpha_k} \in W_{\alpha_k}$. Thus we choose an open set V_{α_k} with $x_{\alpha_k} \in V_{\alpha_k} \subset \overline{V_{\alpha_k}} \subset W_{\alpha_k}$ for $k = 1, 2, ..., n$. For all other α, we let $V_\alpha = X_\alpha$. Let $V = \prod_{\alpha \in I} V_\alpha$, which is a basic open set, and we have $\overline{V} = \prod_{\alpha \in I} \overline{V_\alpha}$. Thus $x \in V \subset \overline{V} \subset W \subset U$. Hence $\prod_{\alpha \in I} X_\alpha$ is regular.

For part 3, see the examples below. ∎

Example 17.17 *Here we give several examples of losing regularity in quotients and continuous images.*

1. *If X is any space, then the identity from X to the same set with the trivial topology is continuous. So continuous maps can destroy any separation in a trivial way.*

2. *We showed above in example 17.10 that there is an open continuous mapping from a metric space onto a non-T_2-space.*

3. *The Sorgenfrey line $\mathbb{S}$ is a T_3-space since basic open sets of the form $[a, b)$ are both open and closed. By the preservation theorem it follows that $\mathbb{S} \times \mathbb{S}$ is also T_3. Let $Q = \{(x, -x) : x \in \mathbb{Q}\}$ and $P = \{(x, -x) : x \in \mathbb{R}\backslash\mathbb{Q}\}$. It was shown in exercise 14 in chapter 6 that there do not exist disjoint open sets in $\mathbb{S} \times \mathbb{S}$ that separate P and Q. Note that both P and Q are closed sets in $\mathbb{S} \times \mathbb{S}$. The decomposition space obtained by identifying Q to a point is a closed continuous image of $\mathbb{S} \times \mathbb{S}$, which is not regular even though it is T_2. The decomposition space obtained from $\mathbb{S} \times \mathbb{S}$ by identifying Q to a point and P to another point is a closed continuous image of $\mathbb{S} \times \mathbb{S}$ that is not T_2. Thus, closed continuous images can destroy regularity whether or not the image is Hausdorff.*

There is a small consolation prize in the preservation of separation by closed continuous quotients.

Theorem 17.18 *If X is a T_3-space, and A is a closed subspace of X, then the decomposition X/A, obtained from X by identifying A to a point, is a T_2-space.*

Proof. We leave this proof as an exercise. ∎

In order to present the separation axioms in order of increasing strength, we will now discuss the completely regular spaces. The completely regular spaces were not in the original T-*subscript* scheme, and because they are between (in strength) the T_3-spaces and the T_4-spaces (to be defined soon), the completely regular T_1-spaces are sometimes called $T_{3.5}$-spaces or $T_{3\frac{1}{2}}$-spaces. They are also called Tychonoff spaces.

Definition 17.19 *A space X is called **completely regular** provided that for any closed set $F \subset X$ and any point $x \in X\backslash F$ there is a continuous function $f : X \to [0, 1]$ such that $f(x) = 0$ and $F \subset f^{-1}(\{1\})$. A completely regular T_1-space is called a **Tychonoff space**, or sometimes a $\mathbf{T_{3.5}}$-space.*

Notice that if the closed set F in the definition above is nonempty, then we could state the condition $F \subset f^{-1}(\{1\})$ as $f(F) = 1$ which is a little more aesthetically pleasing. (Actually, we should say $f(F) = \{1\}$ rather than using the customary abuse of the notation.)

Lemma 17.20 *Every completely regular space is regular.*

Proof. Suppose that X is completely regular, F is a closed subspace of X, and $x \in X \backslash F$. Choose a continuous function $f : X \to [0,1]$ such that $f(x) = 0$ and $F \subset f^{-1}(1)$ by complete regularity. Let $U = f^{-1}([0, \frac{1}{2}))$ and $V = f^{-1}((\frac{1}{2}, 1])$, and note that these are open sets by the continuity of f. Clearly, $x \in U$ and $F \subset V$. Hence X is regular. ∎

It is worth noting that the range of the function in the definition of completely regular spaces does not really have to be $[0, 1]$. If we stated the definition with the interval $[a, b]$ and $f(x) = a$, and $F \subset f^{-1}(b)$, the same spaces would satisfy the definition since we could compose with the homeomorphism taking $[a, b]$ onto $[0, 1]$. Also if we stated the definition with the range being the entire real line and the point and the closed set mapped to two distinct points, then again the same spaces would satisfy the definition since we could follow the given function with the quotient mapping collapsing $[b, \infty)$ to $\{b\}$ and $(-\infty, a]$ to $\{a\}$.

Example 17.21 *We list some examples of completely regular spaces.*

1. *All metric spaces are completely regular.*

2. *All zero-dimensional spaces are completely regular. Recall that a space is called zero-dimensional if each point has a neighborhood base consisting of sets which are both open and closed. In particular, the Michael line $\mathbb{M}$, the Sorgenfrey line $\mathbb{S}$, and the countable ordinals ω_1 are all completely regular spaces.*

3. *The Moore plane, Γ, is completely regular.*

Proof. We leave these proofs as exercises. Here are some hints. For the metric case, think about the function $f(x) = d(x, F) = \inf \{d(x, y) : y \in F\}$. For the zero-dimensional case, note that the characteristic function of a set that is both open and closed is continuous. For a tangent ball, think about the Euclidean distance from the point of tangency along a straight line to the boundary. ∎

There are examples of regular spaces that are not completely regular. They are not really plentiful. One rather complicated example is the Tychonoff corkscrew, which we will develop in an exercise. We give here a somewhat easier example (which was discovered by Adam Mysior in 1980).

Example 17.22 *Let X be the upper-half plane with one additional point, i.e.* $X = (\mathbb{R} \times [0,\infty)) \cup \{(0,-1)\}$. *We let L denote the x-axis, $\mathbb{R} \times \{0\}$. For each point $z = (x,0) \in L$, we define $A_1(z) = \{x\} \times [0,2]$, the vertical line segment above z with height 2, and we define $A_2(z) = \{(x+y,y) : 0 \leq y \leq 2\}$, the line segment with slope 1 which extends from z to a height of 2.*

We are now ready to define neighborhoods. If $z = (x,y) \in X$, and $y > 0$, then we let $\mathcal{B}_z = \{\{z\}\}$. If $z \in L$, then we let $\mathcal{B}_z = \{(A_1(z) \cup A_2(z))\backslash F : F$ is finite and $z \notin F\}$. For $n \in \mathbb{N}$, we let $G_n = \{(0,-1)\} \cup ([n,\infty) \times [0,\infty))$, the extra point together with the points of the upper-half plane to the right of $x = n$. We let $\mathcal{B}_{(0,-1)} = \{G_n : n \in \mathbb{N}\}$. We show that this space is T_3, and not completely regular.

Proof. First we must show that the definition we have made is a valid assignment of neighborhood bases (see theorem 14.27). It is clear that $z \in B$ for all $z \in X$ and $B \in \mathcal{B}_z$. It is also clear that for any $z \in X$, and any $B_1, B_2 \in \mathcal{B}_z$, we have $B_1 \cap B_2 \in \mathcal{B}_z$. Finally, suppose $V \in \mathcal{B}_z$ and $t \in V$. We must find $W \in \mathcal{B}_t$ with $W \subset V$. Let let $z = (x,y)$ and $t = (t_1,t_2)$. If $y > 0$, then $V = \{z\}$, and thus $t = z$, and we let $W = V$, satisfying condition 3 of theorem 14.27.

Suppose that $y = 0$. Note that if $t_2 > 0$, then $W = \{t\} \subset V$. Now if $t_2 = 0$, then $t = z$, and $W = V \subset V$. Finally, suppose $y = -1$, i.e., $z = (0,-1)$. If $t_2 = -1$, then we again take $W = V$. If $t_2 = 0$, then let $W = A_1(t) \cup A_2(t)$, and we have $W \in \mathcal{B}_t$ and $W \subset V$. If $t_2 > 0$, then we take $W = \{t\}$, and $W \subset V$. Hence we have shown that if $V \in \mathcal{B}_z$, then for any $t \in V$, there is a $W \in \mathcal{B}_t$ with $W \subset V$. So we have a valid assignment of neighborhood bases.

It is easy to see that X is regular, since the elements of $\mathcal{B}_z$ are both open and closed if $z \neq (0,-1)$, and $\overline{G_{n+2}} \subset G_n$ for each $n \in \mathbb{N}$. That the complement of each singleton is open is clear, and thus we have that X is a T_3-space.

For each $i \in \mathbb{N}$, we let $L_i = [i-1,i] \times \{0\} \subset L$. Suppose that $f : X \to [0,1]$ is continuous, and $f(L_1) = 0$. Note that L_1 is a closed set in X. We will show that $f((0,-1))$ must also be 0, and thus X cannot be completely regular.

For each $n \in \mathbb{N}$, let $K_n = \{z \in L_n : f(z) = 0\}$. We will show by induction that K_n is infinite for each n. Obviously, K_1 is infinite. Suppose now that K_i is infinite, and we will show that K_{i+1} is infinite.

Suppose that $(c,0) \in K_i$. For each $m \in \mathbb{N}$, $(c,0) \in f^{-1}([0,\frac{1}{m}))$, and thus we choose a finite set F_m such that $[A_1((c,0)) \cup A_2((c,0))]\backslash F_m \subset f^{-1}([0,\frac{1}{m}))$. Now we see that $F_c = \bigcup_{m=1}^{\infty} F_m$ is countable, and $A_2((c,0))\backslash F_c \subset f^{-1}(0)$. Suppose that $\{(c_k, 0) : k \in \omega\}$ is a countably infinite subset of K_i. The set $F = \bigcup_{k \in \omega} F_{c_k}$ is therefore countable, and thus $A = \{x \in [i, i+1] : \exists y (x, y) \in F\}$ is countable. Suppose $z \in L_{i+1} \backslash (A \times \{0\})$. Then $A_1(z) \cap [A_2((c_k, 0))] \backslash F_{c_k}) \neq \varnothing$ for every $k \in \omega$. Hence $A_1(z) \cap f^{-1}(0)$ is infinite, so every neighborhood of z intersects $f^{-1}(0)$. By continuity, $f^{-1}(0)$ is closed. Hence $z \in f^{-1}(0)$, and thus $z \in K_{i+1}$. Since A is countable, there are infinitely many (in fact, continuum many) such z's. Hence K_{i+1} is infinite. Hence, by the principle of mathematical induction, K_n is infinite for all n.

Since every basic neighborhood of $(0, -1)$ contains all but finitely many of the K_n's, every neighborhood of $(0, -1)$ intersects $f^{-1}(0)$, and thus $(0, -1) \in \overline{f^{-1}(0)}$. Since $f^{-1}(0)$ is closed, $(0, -1) \in f^{-1}(0)$. Hence X cannot be completely regular. ∎

Now we do the preservation theorem.

Theorem 17.23 *(Preservation theorem for Tychonoff spaces)*

1. *Every subspace of a completely regular space ($T_{3.5}$-space) is also a completely regular space ($T_{3.5}$-space).*

2. *A nonempty product space $X = \prod_{\alpha \in I} X_a$ is completely regular (resp., $T_{3.5}$) if and only if X_α is completely regular (resp., $T_{3.5}$) for each $\alpha \in I$.*

3. *Quotients of $T_{3.5}$-spaces need not be completely regular.*

Proof. For 1 and 2, we need only prove complete regularity since we already know the preservation results for T_1. Suppose that $A \subset X$, where X is completely regular, and that F is a closed subset of A with $x \in A \backslash F$. Choose a closed set $H \subset X$ such that $H \cap A = F$. Since X is completely regular, choose a continuous function $f : X \to [0, 1]$ such that $f(x) = 0$ and $H \subset f^{-1}(1)$. Now the restriction $f|A : A \to [0, 1]$ is also continuous and separates x and F in the subspace A.

Since each factor is homeomorphic to a subspace of the product, the "only if" direction of part 2 follows. Suppose that each factor space X_α is completely regular. We will show that if $x \in X = \prod_{\alpha \in I} X_a$ and U is a basic open set containing x, then there is a continuous function $f : X \to [0, 1]$ such that $f(x) = 1$ and $X \backslash U \subset f^{-1}(0)$. Clearly, this suffices to show that X is completely regular.

Suppose $U = \bigcap_{k=1}^{n} \pi_{\alpha_k}^{-1}(U_{\alpha_k})$ where U_{α_k} is open in X_{α_k} for $k = 1, 2, ..., n$, and $x \in U$. For each point k, $x_{\alpha_k} \in U_{\alpha_k}$ and the space X_{α_k} is completely regular. So we choose a continuous function $f_k : X_{\alpha_k} \to [0, 1]$ such that $f_k(x_{\alpha_k}) = 1$ and $X_{\alpha_k} \backslash U_{\alpha_k} \subset f_k^{-1}(0)$. Define $f : X \to [0, 1]$ by $f = \min\{f_1 \circ \pi_{\alpha_1}, f_2 \circ \pi_{\alpha_2}, ..., f_n \circ$

π_{α_n}}. It is an easy exercise to show that the minimum of finitely many continuous functions is continuous. Furthermore, $f(x) = 1$, and if $y \notin U$, then there exists some k such that $y_{\alpha_k} \notin U_{\alpha_k}$. Hence $f(y) = f_k(y_{\alpha_k}) = 0$. This completes the proof of part 2.

For part 3, since metric spaces are completely regular, and $\mathbb{S} \times \mathbb{S}$ is completely regular, we can use the same examples as in example 17.17. ∎

If X is a topological space, the set of all continuous real-valued functions on X is one of the more interesting examples that can be constructed from X.

Definition 17.24 *Suppose X is a space. $C(X)$ is the set of all **continuous real-valued functions** with domain X. $C^*(X)$ is the set of all **bounded** continuous real valued functions on X.*

One of the things that makes these sets interesting is that the real numbers induce algebraic structure on the sets by doing the operations pointwise, i.e., $(f + g)(x) = f(x) + g(x)$, $(fg)(x) = f(x)g(x)$, and $(rf)(x) = rf(x)$. So $C(X)$ and $C^*(X)$ are rings and vector spaces. If X is compact, or even countably compact, and $f \in C(X)$, then $\{f^{-1}(-n, n) : n \in X\}$ must have a finite subcover. Hence, $C(X) = C^*(X)$, and we can define $||f|| = \sup\{|f(x)| : x \in X\}$ for $f \in C^*(X)$, making $C^*(X)$ into a normed linear space, and thus making it into a topological space. One can also get a topology for these spaces by looking at them as subspaces of the product space $\mathbb{R}^X$. The interplay among the algebraic structure, the normed linear space structure, and the product space structure is a very rich area of study.

It is clear from the definition of complete regularity that the following is true.

Lemma 17.25 *If X is completely regular, then $C^*(X)$ separates points from closed sets in X.*

Consequently, we have the following.

Lemma 17.26 *If X is completely regular, then X has the weak topology generated by $C^*(X)$.*

If we also know that X is T_1, then $C^*(X)$ separates points in X. For each $f \in C^*(X)$, let I_f be the closed bounded interval $[\inf f(X), \sup f(X)]$. From the embedding lemma, we get the following.

Theorem 17.27 *If X is a completely regular T_1-space, then the evaluation mapping $e : X \to \prod_{f \in C^*(X)} I_f$ is an embedding.*

A product space where each factor space is a closed bounded interval in the real line is called a *cube*. So from the theorem above, we see that every Tychonoff space is homeomorphic to a subspace of a cube. On the other hand, since any subspace of the real line is a metric space, and thus completely regular, any cube is a Tychonoff space. Since every subspace of a Tychonoff space is again Tychonoff, we have now the following result.

Theorem 17.28 *A space X is a completely regular T_1-space if and only if X is homeomorphic to a subspace of a cube.*

Proof. We have sketched the proof in the preceding discussion. ∎

We now define the normal spaces.

Definition 17.29 *A space X is called **normal** if and only if for each pair H, K of disjoint closed subsets of X, there exists a pair U, V of disjoint open subsets of X such that $H \subset U$ and $K \subset V$. A normal T_1-space is called a T_4-**space**.*

First we state a characterization which is used nearly as much as the definition.

Lemma 17.30 *A space X is normal if and only if whenever A is a closed subset of X, U is an open subset of X, and $A \subset U$, there exists an open set V with $A \subset V \subset \overline{V} \subset U$.*

Proof. We leave this as an exercise. ∎

We will see that normality is quite a strong condition for a topological space to satisfy. Indeed, the questions surrounding this property have been among the most interesting, and difficult, questions in the field of general topology. The rise of set theory in topology is largely due to the investigations of normality. From the very early days, set theory, or at least cardinal arithmetic, was interwoven with normality. The next result was used by Jones to show that if the continuum hypothesis is true, then all separable normal Moore spaces are metrizable. In the paper containing this result, the normal Moore space problem was first posed. We will have much more to say about this in chapter 21.

Theorem 17.31 *(Jones' lemma) If X is normal, D is a dense subset of X, and S is a closed discrete subspace of X, then $2^{|S|} \leq 2^{|D|}$.*

Proof. To prove the result, we need to find a one-to-one function from $\mathcal{P}(S)$, the power set of S, into $\mathcal{P}(D)$. For each $A \subset S$, A and $S \setminus A$ are disjoint closed sets in

X. By normality, we choose U_A, V_A
$S \backslash A \subset V_A$. Define $\phi : \mathcal{P}(S) \to \mathcal{P}(D)$ by
is one-to-one. Suppose $A, B \subset S$ and $A \neq$ ... open sets with $A \subset U_A$ and
    We suppose $x \in A \backslash B$ (the other case is ana... U_A. We need to show that ϕ
we know that $x \in U_A \cap V_B$. Since D is dense, $B \neq \varnothing$ or $B \backslash A \neq \varnothing$.
and since $U_B \cap V_B = \varnothing$, $z \notin U_B$. Thus $z \in D \cap U_A$ ...ce $x \in A$ and $x \in S \backslash B$,
$\phi(A) \neq \phi(B)$. So we have shown that ϕ is a one-to-one $z \in D \cap (U_A \cap V_B)$,
that $|\mathcal{P}(S)| \leq |\mathcal{P}(D)|$, as desired. ∎ $z \notin \phi(B)$. Hence
    , and it follows

Corollary 17.32 *If X is a separable space with a closed discrete* ...
dinality $\mathfrak{c}$, then X is not normal.   ...ce of car-

Proof. Since $\mathfrak{c} = 2^\omega$, and $2^{\mathfrak{c}} > \mathfrak{c}$, it is impossible that $2^{\mathfrak{c}} \leq 2^\omega$, and the result now follows from Jones' lemma. ∎

This corollary is a very easy way to show that certain spaces cannot be normal.

Example 17.33 *We can use Jones' lemma to show the following:*

1. *The Sorgenfrey plane $\mathbb{S} \times \mathbb{S}$ is not normal.*

2. *The Moore plane Γ is not normal.*

We leave it as an exercise to show that both of these spaces are separable and have closed discrete subspaces of size $\mathfrak{c}$. In part one, we outlined a proof (using the Baire category theorem) to show that $\mathbb{S} \times \mathbb{S}$ is not normal. A similar approach can be used to give a more constructive proof that Γ is not normal. One can use the Baire category theorem to show that the rational points and the irrational points on the x-axis cannot be separated by disjoint open sets in Γ.
    On the other hand, there are many normal spaces.

Example 17.34 *Every metrizable space is normal.*

Proof. Suppose that (X, d) is a metric space, and that H, K are disjoint closed subsets of X. For each $x \in H$, we know that $x \notin K$, so we choose $\varepsilon_x > 0$ such that $B(x, \varepsilon_x) \cap K = \varnothing$. Similarly, for each $y \in K$, we choose $\delta_y > 0$ such that $B(y, \delta_y) \cap H = \varnothing$. We define $U = \bigcup_{x \in H} B(x, \frac{\varepsilon_x}{2})$ and $V = \bigcup_{y \in K} B(y, \frac{\delta_y}{2})$. Clearly both U and V are open, $H \subset U$, and $K \subset V$. We show that $U \cap V = \varnothing$. If not, then let $z \in U \cap V$, and choose $x \in H, y \in K$ with $z \in B(x, \frac{\varepsilon_x}{2}) \cap B(y, \frac{\delta_y}{2})$. Suppose that $\varepsilon_x \geq \delta_y$. Now $d(x, y) \leq d(x, z) + d(z, y) < \frac{\varepsilon_x}{2} + \frac{\delta_y}{2} \leq \varepsilon_x$. Hence $y \in K \cap B(x, \varepsilon_x)$, a contradiction. In case $\varepsilon_x \leq \delta_y$, then $x \in H \cap B(y, \delta_y)$, also a contradiction. Hence no such z can exist, and the proof is complete. ∎

pace is normal.

Example 17.35 *Every* .n 7.8. ■

Proof. This is prove *enfrey line is normal.*

Example 17.36 a challenging exercise to show that $\mathbb{S}$ is hereditarily Lindelöf.

Proof. We lea rem 6.25 that regular Lindelöf spaces are normal. ■
It is shown j.

Exampl .37 *The Michael line is normal.*

rro It is also the case that $\mathbb{M}$ satisfies stronger properties that will force it to be normal (namely paracompactness), but again, we leave it as a very good exercise to show directly that $\mathbb{M}$ is normal. ■

Example 17.38 *The space of countable ordinals ω_1 is normal.*

Proof. This is exercise 8 of the project in chapter 7. ■

We now do the preservation theorem. We see that the behavior of normality is not like the other separation axioms at all.

Theorem 17.39 *(Preservation theorem for T_4-spaces)*

1. *Subspaces of T_4-spaces need not be normal, but closed subspaces of normal spaces are normal.*

2. *Products of T_4-spaces need not be normal.*

3. *Quotients of T_4-spaces need not be normal, but closed continuous images of normal (T_4) spaces are normal (T_4). In particular, upper semi-continuous decompositions of normal spaces are normal.*

Proof. We will prove the positive results and give examples below to illustrate the negative results. Suppose X is normal and A is a closed subspace of X. We wish to show that A is normal. Suppose H and K are disjoint closed subspaces of A. Since A is closed, H and K are closed in X. So we choose the disjoint open sets U and V in X with $H \subset U$ and $K \subset V$ by the normality of X. Now $A \cap U$ and $A \cap V$ are disjoint open sets in A that provide the separation of H and K in the space A.

For part 3, suppose X is normal and $f : X \rightarrow Y$ is a closed continuous surjection. Suppose that H, K are disjoint closed subsets in Y. By the continuity, $f^{-1}(H)$ and $f^{-1}(K)$ are disjoints closed subsets in X. By normality of X, choose disjoint open sets U, V such that $f^{-1}(H) \subset U$ and $f^{-1}(K) \subset V$. Let $U_1 = Y \backslash f(X \backslash U)$ and $V_1 = Y \backslash f(X \backslash V)$. Since f is a closed mapping, it is clear that U_1 and V_1 are open sets in Y. Since U and V are disjoint, $X \backslash U$ and $X \backslash V$ cover X, and thus if $y \in U_1 \cap V_1$ and $y = f(x)$, we would have $x \notin (X \backslash U) \cup (X \backslash V)$, a contradiction. Hence $U_1 \cap V_1 = \varnothing$. Suppose $y \in H$. We claim that $y \in U_1$. If that were not the case, then we could choose $x \in X \backslash U$ with $y = f(x)$, but then we would have $x \in f^{-1}(H)$, and thus $x \in U$, a contradiction. Hence $H \subset U_1$. It is similarly true that $K \subset V_1$, and the proof is complete. ∎

Example 17.40 *Here we give examples to illustrate the negative results in the preservation theorem for T_4-spaces.*

1. *We showed that any $T_{3.5}$-space can be embedded in a cube. We will see from the Tychonoff theorem (to be proved soon) that any cube is a compact T_2-space, and thus any cube is a T_4-space. Hence the preservation to subspaces fails as badly as it possibly could; every Tychonoff space can be embedded into a normal space. To give a concrete example, the Moore plane Γ is a non-normal completely regular T_1-space, and hence it is a non-normal subspace of some cube which is a T_4-space.*

2. *The Sorgenfrey line $\mathbb{S}$ is a T_4-space, but $\mathbb{S} \times \mathbb{S}$ is not normal.*

3. *We saw earlier an example of an open continuous image of a metric space that was T_1, but not T_2. Obviously, this shows that quotients can destroy normality.*

The failure of the preservation theorem for normality has lead to much investigation in set-theoretic topology. More precisely, the question of what it takes to make the product of two spaces be normal has long been the object of study. Obviously, if both X and Y are metrizable, then $X \times Y$ is normal (actually, metrizable), and if both are compact Hausdorff, then the product is normal (actually compact Hausdorff). These results are unsatisfying because the normality is an accidental by-product of a another very strong property that is possessed by the product.

On the other hand, $\mathbb{S}$ is a very nice space, and not only $\mathbb{S}$ itself but every subspace of $\mathbb{S}$ is normal, and yet $\mathbb{S} \times \mathbb{S}$ is not normal. So we cannot prove that the square of even a very nice space is normal, but perhaps we can make $X \times Y$ be normal for any normal X if Y is nice enough. This line of thinking is given a setback by the Michael line. The Michael line is again a very nice space; it is normal, and we will see later that it is even paracompact. However the product of $\mathbb{M}$ with the irrationals is not normal (exercise 13, chapter 6). Since the space of irrationals has a compatible complete metric, the product theorem such as mentioned above fails even if Y is a complete metric space.

What if Y is a compact metric space? If X is normal, surely one should be able to show that $X \times [0,1]$ is normal. This was Dowker's Conjecture, and a counterexample to Dowker's conjecture has come to be called a Dowker space. Mary Ellen Rudin, in 1971, constructed the first example in ZFC of a Dowker space. While many other examples of Dowker spaces were constructed in various models of set theory, this was the only known "real" example for over twenty years. Indeed, there was hope that it could be shown to be consistent that Dowker spaces that were "small" could not exist. The Rudin example was very large, and all its cardinal functions (like cardinality of bases, cardinality of local bases, cardinality of dense subspaces, etc.) were also very large. In the early 1990's, Zoltan Balogh created a remarkable technique for constructing Dowker spaces of various types, which were small in the sense of both size and cardinal functions, and they could be designed to have many other properties, all within ZFC.

We now turn to some characterizations of normality. In part one, we proved Urysohn's lemma from scratch and then proved Tietze's extension theorem from that. In this portion of the book, we will do the opposite. We will prove the Tietze extension theorem from scratch, and we will then use that to prove Urysohn's lemma.

In some ways reminiscent of the Hahn-Banach theorem from analysis, the Tietze extension theorem characterizes when we can expect to be able to extend a continuous real-valued function from a closed subspace to a continuous function on the entire space while maintaining the same bounds on the range. Purely for notational convenience, we will assume that the range in this theorem is the interval $[0,1]$; any other closed bounded interval or $\mathbb{R}$ itself would work equally as well.

The idea of the proof of the "only if" portion of the Tietze extension theorem is not actually as hard as it looks below. We have a closed subspace that has a scale from 0 to 1 on it. We want to use the normality to "spread" that scale across the space. This has to be done a little carefully since the sets $[0,r)$ and $(r,1]$ are not closed, and so we will have to back off from the point r a bit in each direction to get disjoint closed sets to use the normality, but we don't want our open sets to intersect the subspace in the gap where we backed away from r. We will do this construction inductively using the countability of the rationals, and at each stage we will also be careful not to disturb the construction done earlier.

Theorem 17.41 *(Tietze extension theorem) A space X is normal if and only if* *whenever A is a closed subset of X and f : A → [0, 1] is continuous, there is a continuous $\widehat{f} : X \to [0, 1]$ such that $\widehat{f}(x) = f(x)$ for all $x \in A$.*

Proof. For the easy direction, if H and K are disjoint closed subsets of X, and X has the extension property, consider the function $f : H \cup K \to [0, 1]$ defined by

$$f(x) = \begin{cases} 0, & \text{if } x \in H \\ 1, & \text{if } x \in K. \end{cases}$$

Since $f|H$ and $f|K$ are both constant functions and $H \cap K = \varnothing$, f is a continuous function on the closed subspace $H \cup K$. Let $\widehat{f} : X \to [0, 1]$ be the extension, and note the $H \subset \widehat{f}^{-1}([0, \frac{1}{2}))$ and $K \subset \widehat{f}^{-1}((\frac{1}{2}, 1])$, By continuity, this provides the open sets that establish the normality of X.

Let us proceed to the converse. Suppose X is normal, A is a closed subspace of X, and $f : A \to [0, 1]$ is continuous. Notice that since A is closed, any closed subspace of A is closed in X. Let $Q = \mathbb{Q} \cap [0, 1]$, and by induction we will construct, for each $r \in Q$, an open set $W_r \subset X$ such that

$$f^{-1}((r, 1]) \subset W_r,$$

$$f^{-1}([0, r)) \cap \overline{W_r} = \varnothing,$$

and

$$\text{if } r < s, \text{ then } W_r \supset \overline{W_s}$$

To do this construction, we first list $Q = \{r_n : n \in \mathbb{N}\}$. The initial step of the construction is r_1. For each $m \in \mathbb{N}$, use the normality to find three disjoint pairs of open sets. Choose U_m^1, V_m^1 disjoint open such that $f^{-1}([0, r_1 - \frac{1}{m}]) \subset U_m^1$ and $f^{-1}([r_1 + \frac{1}{m}, 1]) \subset V_m^1$. Choose U_m^2, V_m^2 disjoint open such that $f^{-1}([0, r_1 - \frac{1}{m}]) \subset U_m^2$ and $f^{-1}([r_1, 1]) \subset V_m^2$. Choose U_m^3, V_m^3 disjoint open such that $f^{-1}([0, r_1]) \subset U_m^3$ and $f^{-1}([r_1 + \frac{1}{m}, 1]) \subset V_m^3$. Let $U_m = U_m^1 \cap U_m^2 \cap U_m^3$ and $V_m = V_m^1 \cap V_m^2 \cap V_m^3$. Let $S_1 = U_1$ and $T_1 = V_1$, and for $m > 1$ let

$$\begin{aligned} S_m &= U_m \backslash \overline{(T_1 \cup T_2 \cup ... \cup T_{m-1})} \\ T_m &= V_m \backslash \overline{(S_1 \cup S_2 \cup ... \cup S_{m-1})}. \end{aligned}$$

Notice that $\bigcup_{m=1}^{\infty} S_m$ and $\bigcup_{m=1}^{\infty} T_m$ are disjoint open sets containing $f^{-1}([0, r_1))$ and $f^{-1}((r_1, 1])$ respectively. Let $W_{r_1} = \bigcup_{m=1}^{\infty} T_m$.

Suppose that $n > 1$, and we have the construction completed as claimed for $j < n$. Choose $j, k < n$ such that $r_k = \max \{r_i : i < n, r_i < r_n\}$ and $r_j =$

$\min \{r_i : i < n, r_i > r_n\}$. Notice that by the inductive hypothesis,

$$W_{r_k} = \bigcap \{W_{r_i} : i < n, r_i < r_n\},$$
$$\overline{W_{r_j}} = \bigcup \{\overline{W_{r_i}} : i < n, r_i > r_n\},$$

and

$$\overline{W_{r_j}} \subset W_{r_k}.$$

It could be the case, of course, that one or the other of r_k, r_j does not exist. If that is the case, we take $W_{r_j} = \varnothing$ or $W_{r_k} = X$ as needed and proceed with the construction as indicated below. For each $m \in \mathbb{N}$, choose three disjoint pairs of open sets $U_m^1, V_m^1,$ $U_m^2, V_m^2,$ and U_m^3, V_m^3 such that the following are true:

1. $f^{-1}([0, r_n - \frac{1}{m}]) \cup (X \backslash W_{r_k}) \subset U_m^1$ and $f^{-1}([r_n + \frac{1}{m}, 1]) \cup \overline{W_{r_j}} \subset V_m^1$

2. $f^{-1}([0, r_n - \frac{1}{m}]) \cup (X \backslash W_{r_k}) \subset U_m^2$ and $f^{-1}([r_n, 1]) \cup \overline{W_{r_j}} \subset V_m^2$

3. $f^{-1}([0, r_n]) \cup (X \backslash W_{r_k}) \subset U_m^3$ and $f^{-1}([r_n + \frac{1}{m}, 1]) \cup \overline{W_{r_j}} \subset V_m^3$.

Let $U_m = U_m^1 \cap U_m^2 \cap U_m^3$ and $V_m = V_m^1 \cap V_m^2 \cap V_m^3$ for each m. Let

$$S_1 = U_1, T_1 = V_1,$$

and for $m > 1$ let

$$S_m = U_m \backslash \overline{T_1 \cup T_2 \cup ... \cup T_{m-1}}$$

and

$$T_m = V_m \backslash \overline{(S_1 \cup S_2 \cup ... \cup S_{m-1})}.$$

Now we can easily see that $\bigcup_{m=1}^{\infty} S_m$ and $\bigcup_{m=1}^{\infty} T_m$ are disjoint open sets containing $f^{-1}([0, r_n)) \cup (X \backslash W_{r_k})$ and $f^{-1}((r_n, 1]) \cup \overline{W_{r_j}}$ respectively. Let $W_{r_n} = \bigcup_{m=1}^{\infty} T_m$, and it is routine to check that the inductive hypothesis is satisfied. Thus, by induction, the construction is complete.

We define $\widehat{f} : X \to [0, 1]$ by $\widehat{f}(x) = \inf \{r \in Q : x \notin \overline{W_r}\}$. We claim that this is the function we seek.

Suppose $x \in A$. If $r \in Q$ and $x \notin \overline{W_r}$, then $x \notin f^{-1}((r, 1])$, and thus $f(x) \leq r$. Hence $f(x) \leq \widehat{f}(x)$. If it were true that $f(x) < \widehat{f}(x)$, then we could choose $r \in Q$ with $f(x) < r < \widehat{f}(x)$. This gives us $x \in f^{-1}([0, r))$ and thus $x \notin \overline{W_r}$, so $r \geq \widehat{f}(x)$, a contradiction. Hence $\widehat{f}$ is indeed an extension of f.

It remains only to show that $\widehat{f}$ is continuous, and we do this by considering subbasic open sets in $[0, 1]$. Suppose that α is an irrational in $[0, 1]$, and notice that $\widehat{f}^{-1}((\alpha, 1]) = \bigcup_{r>\alpha} W_r$ and $\widehat{f}^{-1}([0, \alpha)) = \bigcup_{r<\alpha}(X \backslash \overline{W_r})$. Since the sets of form $[0, \alpha)$ and of form $(\alpha, 1]$, where α is irrational, give us a subbase for $[0, 1]$, this shows that $\widehat{f}$ is continuous on X, and the proof is complete. ∎

We now get Urysohn's lemma as a corollary.

Theorem 17.42 *(Urysohn's lemma) A space X is normal if and only if whenever H and K are disjoint closed subsets of X there exists a continuous function f : $X \to [0, 1]$ such that $H \subset f^{-1}(0)$ and $K \subset f^{-1}(1)$.*

Proof. If there were such a function, then $f^{-1}([0, \frac{1}{2}))$ and $f^{-1}((\frac{1}{2}, 1])$ would be disjoint open sets containing H and K respectively, so the "if" direction is clear. Now suppose that X is normal and H, K are disjoint closed sets. Now $H \cup K$ is closed, and if we define $f : H \cup K \to [0, 1]$ by

$$ f(x) = \begin{cases} 0, & \text{if } x \in H \\ 1, & \text{if } x \in K, \end{cases} $$

then f is continuous on $H \cup K$, and the extension given us by the Tietze extension theorem is the function we seek. ∎

From Urysohn's lemma it is clear that normal implies completely regular if points are closed sets, and thus $T_4 \implies T_{3.5}$. Strangely enough, normal does not even imply regular without knowing that points are closed sets.

Example 17.43 *The Sierpiński space $X = \{0, 1\}$ with $\tau = \{\varnothing, X, \{0\}\}$ is normal because if H, K are disjoint closed sets, then one of them must be empty. However this space is not regular since the point 0 cannot be separated from the closed set $\{1\}$.*

We include the definitions of T_5 and T_6 without much comment. These are certainly important ideas, but we will leave those for the next course in general topology.

Definition 17.44 *A space X is called **hereditarily normal** if and only if each subspace of X is normal. This notion is also sometimes called **completely normal**. A $\mathbf{T_5}$-space is a T_1-space which is hereditarily normal.*

Definition 17.45 *A space X is called **perfectly normal** if and only if it has exact Urysohn functions, i.e. if H and K are disjoint nonempty closed subsets of X, then there exists a continuous function $f : X \to [0, 1]$ such that $H = f^{-1}(0)$ and $K = f^{-1}(1)$. A $\mathbf{T_6}$-space is a T_1-space that is perfectly normal.*

It is not trivial to show that perfectly normal spaces are hereditarily normal, but it is true, as we shall see in the exercises. The following lemma is often given as the definition of perfect normality. Recall that a G_δ-set is a set which is the intersection of some countable collection of open sets.

Lemma 17.46 *A space X is perfectly normal if and only if X is normal and each closed subset of X is a G_δ-set.*

Proof. We leave this as an exercise. ∎

Because of this result, spaces in which all closed sets are G_δ-sets are often called *perfect* spaces. So that a perfectly normal space is a space that is both perfect and normal. A word of caution is needed here though. When analysts use the word *perfect*, they usually mean that every point of the space is a point of accumulation of the space. Indeed, when you read the early works in set theory and topology (Cantor's work, for example) the analysis usage is much more common than the one we have just defined.

The preservation theorem for normality is really quite different from the preservation theorems for the other separation axioms. We will see in the next chapters that the behavior of normality has characteristics of covering properties. The covering property nature of normality is made somewhat more clear by the following characterization of normality. Before stating this result, we will need several more definitions. We list the basic definitions dealing with covers together.

Definition 17.47 *Suppose X is a space.*

1. *A collection $\mathcal{U}$ of subsets of X is a **cover**, or a covering, of X if and only if $\cup \mathcal{U} = X$.*

2. *If the elements of a cover $\mathcal{U}$ are all open sets, then we say $\mathcal{U}$ is an **open cover** of X.*

3. *If the elements of a cover $\mathcal{U}$ are all closed sets, then we say $\mathcal{U}$ is a **closed cover** of X.*

4. *If a cover $\mathcal{U}$ has the property that for each point $x \in X$ the set*

$$\{U \in \mathcal{U} : x \in U\}$$

*is a finite set, then we say that $\mathcal{U}$ is a **point finite cover** of X. Similarly, we define point finite collections of subsets even if they do not cover.*

5. *If a cover $\mathcal{U}$ has the property that for each point $x \in X$ there is a neighborhood G of x such that the set*

$$\{U \in \mathcal{U} : U \cap G \neq \varnothing\}$$

*is a finite set, then we say that $\mathcal{U}$ is a **locally finite cover** of X. Similarly, we define locally finite collections of subsets even if they do not cover.*

6. *If a cover $\mathcal{U}$ has the property that for each point $x \in X$ there is a neighborhood G of x such that*
$$|\{U \in \mathcal{U} : U \cap G \neq \varnothing\}| \leq 1,$$

*then we say that $\mathcal{U}$ is a **discrete cover** of X. Similarly, we define discrete collections of subsets even if they do not cover.*

It is clear that any discrete collection of subsets is locally finite, and that any locally finite collection is point finite.

Definition 17.48 *If $\mathcal{U} = \{U_\alpha : \alpha \in I\}$ is an open cover of X, then a **shrinking** of $\mathcal{U}$ is an open cover $\mathcal{V} = \{V_\alpha : \alpha \in I\}$ such that for each $\alpha \in I$ it is true that $\overline{V_\alpha} \subset U_\alpha$. If an open cover of X has a shrinking, then we say the cover is shrinkable.*

Theorem 17.49 *A space X is normal if and only if every point-finite open cover of X is shrinkable.*

Proof. Suppose that every point-finite open cover of X is shrinkable, and let H and K be disjoint closed subsets of X. Notice that because H and K are disjoint, $\{X\backslash H, X\backslash K\}$ is a point finite open cover of X. Choose a shrinking $\{U, V\}$ such that $\overline{U} \subset X\backslash H$ and $\overline{V} \subset X\backslash K$. Since $U \cup V = X$, and thus $\overline{U} \cup \overline{V} = X$, it follows that $(X\backslash\overline{U}) \cap (X\backslash\overline{V}) = \varnothing$. Moreover, since $\overline{U} \subset X\backslash H$, it follows that $H \subset X\backslash\overline{U}$, and similarly $K \subset X\backslash\overline{V}$. Hence X is normal.

For the converse, suppose that X is normal, and $\mathcal{U} = \{U_\alpha : \alpha \in I\}$ is a point finite open cover of X, and we assume that the indexing is *faithful* in the sense that any set appears on the list only once. We wish to construct a shrinking of $\mathcal{U}$. Using the axiom of choice, we well-order I, and so we assume that $I = \kappa$ for some ordinal κ. We construct the shrinking by recursion on κ. Let $F_0 = U_0\backslash\bigcup_{\gamma>0} U_\gamma = X\backslash\bigcup_{\gamma>0} U_\gamma$, and notice that F_0 is a closed set.

By normality, choose an open set V_0 such that $F_0 \subset V_0 \subset \overline{V_0} \subset U_0$. Suppose $\beta \in I$, and for each $\alpha < \beta$, we have constructed F_α, V_α such that $F_\alpha = U_\alpha\backslash(\bigcup_{\gamma<\alpha} V_\gamma \cup \bigcup_{\gamma>\alpha} U_\gamma)$ is a closed set, and $F_\alpha \subset V_\alpha \subset \overline{V_\alpha} \subset U_\alpha$. Let $F_\beta = X\backslash(\bigcup_{\gamma<\beta} V_\gamma \cup \bigcup_{\gamma>\beta} U_\gamma)$, and notice that F_β is a closed set.

We claim that $F_\beta \subset U_\beta$. To see this, suppose that $x \in F_\beta$. By the point finiteness, we choose $\alpha_0 \in I$ where $\alpha_0 = \max\{\gamma \in I : x \in U_\gamma\}$. Since $x \notin \bigcup_{\gamma > \beta} U_\gamma$, we know that $\alpha_0 \leq \beta$. If $\alpha_0 < \beta$, then either $x \in F_{\alpha_0}$ or $x \in V_\gamma$ for some $\gamma < \alpha_0$. If $x \in F_{\alpha_0}$, then $x \in V_{\alpha_0}$, and thus $x \notin F_\beta$, a contradiction. If $x \in V_\gamma$ for some $\gamma < \alpha_0 < \beta$, we again have $x \notin F_\beta$, a contradiction. Hence $\alpha_0 = \beta$, and we have that $x \in U_\beta$.

By normality, we choose an open set V_β such that $F_\beta \subset V_\beta \subset \overline{V_\beta} \subset U_\beta$. Notice that since we showed that $F_\beta \subset U_\beta$, it follows that $F_\beta = U_\beta \backslash (\bigcup_{\gamma < \beta} V_\gamma \cup \bigcup_{\gamma > \beta} U_\gamma)$, and the inductive hypothesis is satisfied.

To see that $\{V_\alpha : \alpha \in I\}$ is a shrinking of $\mathcal{U}$, all that is not obvious is that $\{V_\alpha : \alpha \in I\}$ covers X. Suppose that $x \in X$, and let $\alpha_0 = \max\{\alpha \in I : x \in U_\alpha\}$, and recall that this is possible since $\mathcal{U}$ is point finite. If there is a $\gamma < \alpha_0$ with $x \in V_\gamma$, then we have x covered. Otherwise, $x \in F_{\alpha_0} \subset V_{\alpha_0}$, and again x is covered. Hence the proof is complete. ∎

It is easy to see that we can extend normality to separate any finite pairwise disjoint collection of closed sets by pairwise disjoint open sets. It is also true, though not as easy to prove, that we can extend this to countable pairwise disjoint collections of closed sets, provided that the union of any subcollection of the closed sets remains closed. Such collections are called closure preserving.

Definition 17.50 *If X is a space and $\mathcal{C}$ is a collection of subsets of X, then we call $\mathcal{C}$ a **closure preserving** collection provided that for any subcollection $\mathcal{A} \subset \mathcal{C}$, it is true that $\overline{\cup \mathcal{A}} = \cup \{\overline{A} : A \in \mathcal{A}\}$.*

Lemma 17.51 *Every locally finite collection is closure preserving. In particular, discrete collections are closure preserving.*

Proof. Left as an exercise. ∎

Lemma 17.52 *Every pairwise-disjoint, closure-preserving collection of closed sets is a discrete collection.*

Proof. Suppose $\mathcal{C}$ is a pairwise-disjoint, closure-preserving collection of closed subsets of a space X. Suppose $x \in X$. We want a neighborhood of x that intersects at most one element of $\mathcal{C}$. First, suppose $x \notin \cup \mathcal{C}$. Since $X \backslash \cup \mathcal{C}$ is an open set, it is a neighborhood of x that intersects no element of $\mathcal{C}$. If $x \in \cup \mathcal{C}$, then we choose the unique element $C_0 \in \mathcal{C}$ such that $x \in C_0$. Now $X \backslash \cup (\mathcal{C} \backslash \{C_0\})$ is an open set, and thus it is a neighborhood of x which intersects only one (namely C_0) element of $\mathcal{C}$. ∎

Since discrete collections are both pairwise disjoint and closure preserving, we can capture both of these ideas for collections of closed sets, without losing any generality, by using only discrete collections.

Definition 17.53 *A space X is called **collectionwise normal** (CWN) provided that for any faithfully indexed discrete collection $\{F_\alpha : \alpha \in I\}$ of nonempty closed subsets of X there is a faithfully indexed pairwise disjoint collection $\{U_\alpha : \alpha \in I\}$ of open subsets of X such that $F_\alpha \subset U_\alpha$ for each $\alpha \in I$.*

We have used the phrase "faithfully indexed" as we did in the proof above, that is, a collection $\{F_\alpha : \alpha \in I\}$ is faithfully indexed if and only if $\alpha \neq \beta \implies F_\alpha \neq F_\beta$. In particular, we have avoided the possibility of taking U_α to be the entire space X, for each α, in the definition of CWN, which would make all spaces CWN.

It is obvious that CWN spaces are normal. The proof that metric spaces are collectionwise normal is very similar to the proof that metric spaces are normal. (We leave that as an exercise.) The existence of normal spaces that are not CWN is at the heart of some of the most famous problems in set-theoretic topology. There are models of set theory in which there are subspaces of the Moore plane, Γ, that are normal and not CWN, but it was shown in the late 1970's by Peter Nyikos that if there are certain very large cardinal numbers, then it is consistent that all first-countable normal spaces are CWN.

Theorem 17.54 *A space X is collectionwise normal if and only if for any faithfully indexed discrete collection $\{F_\alpha : \alpha \in I\}$ of nonempty closed subsets of X there is a faithfully indexed discrete collection $\{U_\alpha : \alpha \in I\}$ of open subsets of X such that $F_\alpha \subset U_\alpha$ for each $\alpha \in I$.*

Proof. We leave the details as an exercise. Here is a place to start. If $\mathcal{F}$ is a discrete collection of closed subsets of a CWN space X and $\mathcal{U}$ is a pairwise disjoint collection of open sets that separates $\mathcal{F}$, then find an open set V such that $\cup\mathcal{F} \subset V \subset \overline{V} \subset \cup\mathcal{U}$. Now show that $\{U \cap V : U \in \mathcal{U}\}$ is the collection we seek. ∎

Exercises

1. (T_0- and T_1-spaces) Prove that every subspace of a T_0-space is a T_0-space, and that every subspace of a T_1-space is a T_1-space. Prove that a nonempty product space is T_0 if and only if each factor space is T_0. Prove that a nonempty product space is T_1 if and only if each factor space is T_1.

2. Prove theorem 17.8; a topological space X is Hausdorff if and only if the diagonal $\Delta = \{(x,x) : x \in X\}$ is a closed subset of $X \times X$. Prove that if (X, τ_1) is a T_2-space and τ_2 is finer than τ_1, then (X, τ_2) is also a Hausdorff space.

3. Prove theorem 17.18; the quotient space X/A obtained by identifying a single closed set A to a point is T_2 if the space X is T_3. What can you say if X is T_4? $..T_2$? $..T_1$?

4. Prove theorem 17.28; a space is a Tychonoff space if and only if it can be embedded in a cube.

5. Prove that the Moore plane, Γ, is completely regular. Prove that Γ is not normal by using Jones' lemma. Prove that the Sorgenfrey plane $\mathbb{S} \times \mathbb{S}$ is not normal by using Jones' lemma.

6. Prove that a space X is normal if and only if whenever F is a closed subset of X, U is an open subset of X, and $F \subset U$, there is an open subset V of X such that $F \subset V \subset \overline{V} \subset U$.

7. Prove that the Michael line, $\mathbb{M}$, is normal.

8. Prove that the Sorgenfrey line $\mathbb{S}$ is hereditarily Lindelöf.

9. (Tychonoff plank) The Tychonoff plank is the subspace $\mathbb{T}$ obtained from the product $(\omega_1 + 1) \times (\omega + 1)$ by removing the top right corner, i.e. $\mathbb{T} = [(\omega_1 + 1) \times (\omega+1)] \backslash \{(\omega_1, \omega)\}$. Let $A = \{(\alpha, \omega) : \alpha < \omega_1\}$ and $B = \{(\omega_1, n) : n < \omega\}$, that is, A is the top edge and B is the right edge of $\mathbb{T}$. Show that T is a completely regular T_1-space, but A and B are disjoint closed subsets of $\mathbb{T}$ that cannot be separated. Hence $\mathbb{T}$ is not normal.

10. (Tychonoff corkscrew) Let $\mathbb{T}$ be the Tychonoff plank and $\mathbb{Z}$ be the integers, positive, negative, and zero. We form a quotient X of $\mathbb{T} \times \mathbb{Z}$ in the following way: if n is even, then we identify the points (ω, α, n) and $(\omega, \alpha, n+1)$ for each $\alpha < \omega_1$, and if n is odd, then we identify the points (k, ω_1, n) and $(k, \omega_1, n+1)$ for each $k < \omega$. So, thinking of $\mathbb{T} \times \mathbb{Z}$ as $\mathbb{Z}$ many copies of $\mathbb{T}$, we glue the top edge of each even layer to the layer above, and we glue the right edge of each odd layer to the layer above. To this quotient space we attach two more points, a and b, to obtain our space $\mathbb{T}_c$. The quotient space X is open in the space $\mathbb{T}_c$, basic neighborhoods of a have form $U_n(a) = \{a\} \cup \bigcup_{k=n}^{\infty} \mathbb{T} \times \{k\}$, and basic neighborhoods of b will have form $U_n(b) = \{b\} \cup \bigcup_{k=n}^{\infty} \mathbb{T} \times \{-k\}$.

Prove that if $f : \mathbb{T}_c \to [0,1]$ is continuous, and $f(a) = 0$, then $f(b) = 0$. From this it follows that $\mathbb{T}_c$ is not completely regular, and now prove that $\mathbb{T}_c$ is regular.

11. (Ranges aren't critical) Prove the following analogues of the definition of completely regular and of the Tietze extension theorem.

   (a) A space X is completely regular if and only if whenever F is a closed subspace of X and $x \in X \backslash F$, there is a continuous function $f : X \to \mathbb{R}$ and $a \neq b$ such that $f(x) = a$ and $F \subset f^{-1}(b)$.

   (b) A space X is normal if and only if whenever A is a closed subspace of X and $f : A \to \mathbb{R}$ is continuous, there is a continuous function $\widehat{f} : X \to \mathbb{R}$ such that $\widehat{f}(x) = f(x)$ for all $x \in A$.

12. (Zero-sets) A subset Z of a space X is called a zero-set if and only if there is a continuous real-valued function $f : X \to \mathbb{R}$ with $Z = f^{-1}(0)$. Prove that a zero set in a space X is a closed G_δ-set in X. Prove that a space X is completely regular if and only if for each point $x \in X$ the neighborhoods of x that are zero sets form a neighborhood base at x.

13. (Perfectly normal spaces)

   (a) Prove lemma 17.46; a space X is perfectly normal if and only if X is normal and each closed subset of X is a G_δ-set in X. Hint: if $A = \bigcap_{n=1}^{\infty} G_n$ where A is closed and each G_n is open, then we can choose for each n a continuous $f_n : X \to [0,1]$ with $f_n(A) = 0$ and $f_n(X \backslash G_n) = 1$. Let $f_A = \sum_{n=1}^{\infty} \frac{f_n}{2^n}$. Now if A and B are disjoint closed sets which are also G_δ-sets, consider $f(x) = \frac{f_A(x)}{f_A(x) + f_B(x)}$.

   (b) Prove that metric spaces are perfectly normal.

   (c) Prove that every perfectly normal space is hereditarily normal.

   (d) Prove that every subspace of a perfectly normal space is perfectly normal.

14. Prove that locally finite collections are closure preserving.

15. Prove that all metric spaces are collectionwise normal.

16. Prove theorem 17.54; a space X is collectionwise normal if and only if for any faithfully indexed discrete collection $\{F_\alpha : \alpha \in I\}$ of nonempty closed

subsets of X there is a faithfully indexed discrete collection $\{U_\alpha : \alpha \in I\}$ of open subsets of X such that $F_\alpha \subset U_\alpha$ for each $\alpha \in I$.

17. (Alexandroff double) Let X be a space. The Alexandroff double of X is the space having $D(X) = X \times \{0, 1\}$ as underlying set. Basic neighborhoods of points $(x, 0)$ on the bottom copy of X are sets of form $(U \times \{0, 1\}) \backslash F$ where F is a finite subset of $X \times \{1\}$, U is an open subset of X, and $x \in U$; and points $(x, 1)$ on the top copy of X are isolated. Prove the basic neighborhoods as described do define a topology. Prove that X is homeomorphic to the closed subspace $X \times \{0\}$ of $D(X)$. Prove that if X is T_i, then so is $D(X)$ for $i = 1, 2, 3, 3.5, 4$. Find an example of a perfectly normal space X so that $D(X)$ is not perfect.

18. Prove that the following conditions on a space X are equivalent.

   (a) X is hereditarily normal.

   (b) Every open subspace of X is normal.

   (c) If A and B are subsets of X such that $A \cap \overline{B} = \varnothing = \overline{A} \cap B$, then there exist disjoint open sets $U, V \subset X$ with $A \subset U$ and $B \subset V$.

Chapter 18

Compactness and Countability Properties

In this chapter we will recall some properties that were treated fairly thoroughly in part one. We will recall, and sometimes extend, the theorems that we had for these. In particular, we will include here the preservation theorem for each of these properties. The properties that we will consider here are second countability, separability, the Lindelöf property, compactness, and local compactness.

Definition 18.1 *(Recall definition 4.25.) A space X is called **second countable** if and only if there is a countable collection of open subsets of X that forms a base for the topology on X.*

Example 18.2 *We list a few examples.*

1. *The space of real numbers $\mathbb{R}$ is second countable since*

$$\mathcal{B} = \{(r, s) : r < s \text{ and } r, s \in \mathbb{Q}\}$$

   is a countable base for the Euclidean topology on $\mathbb{R}$.

2. *Any separable metric space is second countable. Indeed, the Urysohn metrization theorem (given as a project in chapter 6) states that for T_3-spaces, being second countable is equivalent to being a separable metric space.*

3. *Any space with an uncountable set of isolated points is not second countable. So, any uncountable discrete space, the Michael line $\mathbb{M}$, and the space of countable ordinals ω_1 are not second countable.*

Now let us move on to the preservation theorem.

Theorem 18.3 *(Preservation theorem for second-countable spaces)*

1. *Every subspace of a second-countable space is second countable.*

2. *A nonempty product of Hausdorff spaces is second countable if and only if each factor is second countable and all but countably many factors consist of only one point.*

3. *Quotients generally need not preserve second countability, but every open continuous image of a second countable space is second countable.*

Proof. The proof of 1 is easy. If $\mathcal{B}$ is a countable base for the space X, and $A \subset X$, then $\{B \cap A : B \in \mathcal{B}\}$ is a countable base for A.

For 2, suppose $X = \prod_{\alpha \in I} X_\alpha$ is a nonempty product of T_2-spaces, and X is second countable. Since each factor is homeomorphic to a subspace of X, each factor, by part 1, must be second countable. If there were an uncountable family of factors with more than one point, then we could find a two-point discrete subspace inside each of those factors. Hence X would contain a subspace that is the uncountable product of two-point discrete spaces, and thus that subspace is not even first countable. Hence X would have a non-second-countable subspace, and thus X would not be second countable, a contradiction.

Conversely, suppose that $X = \prod_{\alpha \in I} X_\alpha$ is a product of second countable spaces and I is countable. For each $\alpha \in I$, let $\mathcal{B}_\alpha$ be a countable base for the topology on X_α. Let $\mathcal{F}$ be the set of all finite subsets of I. Note that $\mathcal{F}$ is countable. Suppose that $F = \{\alpha_1, \alpha_2, ..., \alpha_n\} \in \mathcal{F}$. Define $\mathcal{B}_F = \{\bigcap_{i=1}^{n} \pi_{\alpha_i}^{-1}(B_i) : B_i \in \mathcal{B}_{\alpha_i}\}$, and note that since $\prod_{\alpha \in F} \mathcal{B}_\alpha$ is a finite product of countable sets, $\mathcal{B}_F$ is a countable collection. Now $\mathcal{B} = \bigcup_{F \in \mathcal{F}} \mathcal{B}_F$ is a countable union of countable collections, and thus $\mathcal{B}$ is countable. Further, $\mathcal{B}$ is clearly a base for X, and hence X is second countable.

An example to show the lack of preservation by quotients will be given below. Let us prove the preservation by open continuous mappings. Suppose that $f : X \to Y$ is an open continuous mapping of X onto Y, and X is second countable. Choose a countable base for X, and call it $\mathcal{B}$. We will show that $\{f(B) : B \in \mathcal{B}\}$ is a base for Y. Suppose that U is an open subset of Y, and $y \in U$. Choose $x \in X$ with $f(x) = y$. By the continuity of f, $f^{-1}(U)$ is an open set in X, and clearly $x \in f^{-1}(U)$. Choose $B \in \mathcal{B}$ with $x \in B \subset f^{-1}(U)$. Now we see that $y \in f(B) \subset U$. Hence $\{f(B) : B \in \mathcal{B}\}$ is a base for Y, and since $\mathcal{B}$ is countable, so must be $\{f(B) : B \in \mathcal{B}\}$. Thus Y is second countable. ∎

We now give an example of a non-second countable closed continuous image of a second-countable space.

Example 18.4 *For each $n \in \omega$, let $I_n = [0,1]$. Let $X = \sum_{n=0}^{\infty} I_n$, the disjoint union of this countable set of copies of the unit interval. Let A be the closed subset of X consisting of all the left endpoints, i.e. $A = \{(0,n) : n \in \omega\}$. Now we collapse A to a point. The resulting quotient space X/A is not first countable at the distinguished point A, and hence X/A is not second countable. (We leave the proof of this as an exercise.) Since we are identifying a single closed set to a point, the natural quotient mapping is a closed continuous mapping.*

The following is a very useful theorem for deciding when a space is second countable. It says that if you have a base made up of sets with a nice form, and the space has a countable base, then there is a countable base made up of the sets with the nice form (recall theorem 4.36).

Theorem 18.5 *If a space X is second countable, and $\mathcal{B}$ is a base for X, then there is a countable $\mathcal{G} \subset \mathcal{B}$ such that $\mathcal{G}$ is a base for X.*

Proof. Since the proof is given in theorem 4.36, we will omit it here. ∎

We can use this theorem to show that the Sorgenfrey line is not second countable.

Example 18.6 *The Sorgenfrey line $\mathbb{S}$ is not second countable.*

Proof. If $\mathbb{S}$ were second countable, then there would have to be a base of form $\mathcal{B} = \{[a_n, b_n) : n \in \omega\}$. Choose a point $x \in \mathbb{S} \setminus \{a_n : n \in \omega\}$. If $x \in [a_n, b_n) \subset [x, x+1)$, then $x = a_n$, a contradiction. Hence $\mathcal{B}$ cannot be a base. ∎

The connection of second countability with covering properties is given by Lindelöf's theorem.

Theorem 18.7 *(Lindelöf's theorem, see theorem 4.31) If X is second countable and $\mathcal{U}$ is a collection of open subsets of X, then there is a countable subcollection $\mathcal{G} \subset \mathcal{U}$ such that $\cup \mathcal{G} = \cup \mathcal{U}$.*

Definition 18.8 *A space X is called a **Lindelöf space** if and only if every open cover of X has a countable subcover.*

Lemma 18.9 *If every open subspace of X is Lindelöf, then every subspace of X is Lindelöf.*

Proof. This is left as an exercise. ∎

Spaces in which every subspace is Lindelöf are called *hereditarily Lindelöf spaces*. From Lindelöf's theorem and the lemma above, it follows that second-countable spaces are hereditarily Lindelöf.

We saw in part one that regular Lindelöf spaces are normal. It follows from the next result that regular hereditarily Lindelöf spaces are perfectly normal. We will leave this result as well as the proof of the following result as exercises.

Theorem 18.10 *If X is a T_3-space which is hereditarily Lindelöf, then every open subset of X is an F_σ-set (i.e. the countable union of closed sets) in X.*

Corollary 18.11 *If X is a T_3 hereditarily Lindelöf space, then X is perfectly normal.*

Proof. Left to the reader. ∎

It was an exercise in chapter 4 to show that the Sorgenfrey line is Lindelöf. It is actually no more difficult to show that $\mathbb{S}$ is hereditarily Lindelöf. We leave this as an exercise. Thus we have an example of a T_3, hereditarily Lindelöf, and thus perfectly normal, space that is not second countable.

Before we move on to the preservation theorem for Lindelöf spaces, we recall a result (given as exercise 7 in chapter 6).

Theorem 18.12 *(Tube lemma) Suppose that $\mathcal{U}$ is a collection of open sets in $X \times Y$, where Y is compact, and $\mathcal{U}$ covers $\{x\} \times Y$, where $x \in X$. There exists a finite subcollection $\mathcal{U}_1 \subset \mathcal{U}$ and an open set $V \subset X$ with $x \in V$ such that $\mathcal{U}_1$ covers $V \times Y$.*

Theorem 18.13 *(Preservation theorem for Lindelöf spaces)*

1. *Subspaces of Lindelöf spaces need not be Lindelöf, but closed subspaces of Lindelöf spaces are Lindelöf.*

2. *Products of (even two) Lindelöf spaces need not be Lindelöf, but a product of form $X \times K$, where X is Lindelöf and K is compact, is Lindelöf.*

3. *If X is Lindelöf and $f : X \to Y$ is a continuous mapping of X onto Y, then Y is Lindelöf.*

Proof. For each part of this theorem, we have something positive to do. (We will leave the negative results for the examples below.) Suppose X is a Lindelöf space, and let A be a closed subspace of X. Suppose $\mathcal{U} = \{U_\alpha : \alpha \in I\}$ is an open cover of A. For each $\alpha \in I$, we choose an open set $V_\alpha \subset X$ such that $U_\alpha = A \cap V_\alpha$. Now $\{X \backslash A\} \cup \{V_\alpha : \alpha \in I\}$ is an open cover of X. Since X is Lindelöf, choose $\{\alpha_n : n \in \omega\}$ such that $X = (X \backslash A) \cup \bigcup_{n \in \omega} V_{\alpha_n}$ and it is easy to see that $A = \bigcup_{n \in \omega} U_{\alpha_n}$. Hence every open cover of A has a countable subcover.

For part 2, suppose that X is Lindelöf and K is compact, and let $\mathcal{U}$ be an open cover of $X \times K$. For each $x \in X$, by the tube lemma, we choose a finite subcollection $\mathcal{U}_x \subset \mathcal{U}$ and an open set $V_x \subset X$ such that $V_x \times K \subset \bigcup \mathcal{U}_x$. Now $\{V_x : x \in X\}$ is an open cover of X, so we choose a countable set $\{x_n : n \in \omega\}$ such that $X = \bigcup_{n \in \omega} V_{x_n}$. Being the countable union of finite collections, $\bigcup_{n \in \omega} \mathcal{U}_{x_n}$ is a countable subcollection of $\mathcal{U}$. It clearly covers $X \times K$. Hence $X \times K$ is Lindelöf.

Finally, for part 3, suppose $f : X \to Y$ is a continuous mapping of X onto Y, and suppose that X is Lindelöf. Let $\mathcal{U}$ be an open cover of Y. Consider $\mathcal{V} = \{f^{-1}(U) : U \in \mathcal{U}\}$. By continuity, $\mathcal{V}$ is an open collection, and since $\mathcal{U}$ covers the range of f, $\mathcal{V}$ is a cover of X. Since X is Lindelöf, we choose a countable subcollection $\{f^{-1}(U_n) : n \in \omega\} \subset \mathcal{V}$ such that $X = \bigcup_{n \in \omega} f^{-1}(U_n)$. If $y \in Y$, we can choose $x \in X$ with $f(x) = y$, and there must exist $n \in \omega$ with $x \in f^{-1}(U_n)$, and thus $y \in U_n$. Hence $\{U_n : n \in \omega\}$ is the desired countable subcover, and we see that Y is Lindelöf. ∎

Example 18.14 *Here is a list of some examples to show that the negative results above are true.*

1. *The ordinal space $\omega_1 + 1$ is a Lindelöf space, in fact a compact space, but the open subspace ω_1 is not Lindelöf since $\{[0, \alpha) : \alpha \in \omega_1\}$ can have no countable subcover.*

2. *The Sorgenfrey line $\mathbb{S}$ is hereditarily Lindelöf, but $\mathbb{S} \times \mathbb{S}$ is not even normal.*

We recall from part one that a space is called *separable* if and only if it has a countable dense subset. In part one of this book, we showed that second countable, Lindelöf and separable are equivalent for metric spaces (see theorem 4.35). We also showed in part one that all second-countable spaces are separable, whether metric or not, and since second countability in hereditary, it follows that every second-countable space is hereditarily separable. By Lindelöf's theorem, every second-countable space is also hereditarily Lindelöf.

We now explore separability and the relationship between separability and the Lindelöf property. First we will do the preservation theorem.

Theorem 18.15 *(Preservation theorem for separable spaces)*

1. *Subspaces of separable spaces need not be separable, but open subspaces of separable spaces are separable.*

2. *A product of T_2-spaces, where each factor has at least two points is separable if and only if each factor is separable and there are at most $\mathfrak{c}$ factors.*

3. *If X is separable and $f : X \to Y$ is a continuous mapping of X onto Y, then Y is separable.*

Proof. For the positive part of 1, suppose X is a space and D is a countable dense subset of X. If U is an open subspace of X, and V is an open subset of U, then choose an open set W in X with $W \cap U = V$. Now $W \cap U$ is open in X, and so we know that $(W \cap U) \cap D \neq \varnothing$. From this it follows that $\varnothing \neq (V \cap U) \cap D = V \cap (U \cap D)$. Hence $U \cap D$ is dense in U, and $U \cap D$ is clearly countable. Thus U is separable.

Let us go ahead with statement 3, since it too is easy. Suppose $f : X \to Y$ is continuous and onto, and suppose that D is a countable dense subset of X. By continuity, $Y = f(X) = f(\overline{D}) \subset \overline{f(D)}$, and clearly $f(D)$ is countable. Thus Y has a countable dense subset.

Now we prove statement 2. Suppose that $X = \prod_{\alpha \in I} X_\alpha$, and for the "only if" direction suppose that X is separable. Since each projection mapping is continuous and onto, it follows from 2 that each factor is separable. We show that I has cardinality less than or equal the continuum by constructing a one-to-one mapping from I into the power set of a countable dense subset. Suppose that D is a countable dense subset of X. For each $\alpha \in I$, choose a pair U_α, V_α of disjoint open subsets of X_α. Define $\phi : I \to \mathcal{P}(D)$ by $\phi(\alpha) = D \cap \pi_\alpha^{-1}(U_\alpha)$. Now if $\alpha \neq \beta$, then $\pi_\alpha^{-1}(U_\alpha) \cap \pi_\beta^{-1}(V_\beta) \neq \varnothing$, and since this is a basic open set, $D \cap \pi_\alpha^{-1}(U_\alpha) \cap \pi_\beta^{-1}(V_\beta) \neq \varnothing$. Choose a point $x \in D \cap \pi_\alpha^{-1}(U_\alpha) \cap \pi_\beta^{-1}(V_\beta)$. Now $x \in \phi(\alpha)$, but since $U_\beta \cap V_\beta = \varnothing$, $x \notin \phi(\beta)$. Hence $\phi(\alpha) \neq \phi(\beta)$, so ϕ is a one-to-one function. Thus $|I| \leq |\mathcal{P}(D)| \leq 2^\omega = \mathfrak{c}$.

Now for the converse, suppose that each X_α is separable, and let $D_\alpha = \{d_{\alpha,n} : n \in \mathbb{N}\}$ be a countable dense subset of X_α. Assuming that $|I| \leq \mathfrak{c}$, we can embed I as a set into the reals. So we assume that $I \subset \mathbb{R}$. Let $\mathcal{J}$ be the collection of all finite sets of pairwise disjoint open intervals with rational endpoints. Since $\mathbb{Q}$ is countable, we see that $\mathcal{J}$ is countable. If $J = \{J_1, J_2, ..., J_n\} \in \mathcal{J}$, and $k_1, k_2, ..., k_n$ are natural numbers, we define a point of the product space $p = p(J; k_1, k_2, ..., k_n)$ by

$$p_\alpha = \begin{cases} d_{\alpha, k_i}, & \text{if } \alpha \in J_i, \\ d_{\alpha, 1}, & \text{otherwise.} \end{cases}$$

Let

$$D = \{p(J; k_1, k_2, ..., k_n) : J \in \mathcal{J} \text{ and } (k_1, k_2, ..., k_n) \in \mathbb{N}^n \text{ where } n = |J|\}.$$

Since any finite product of countable sets is countable, and any countable union of countable sets is countable, it follows that D is countable. We will show that D is dense in X. Suppose that $B = \bigcap_{i=1}^n \pi_{\alpha_i}^{-1}(U_{\alpha_i})$ is a nonempty basic open set in X. Since $\{\alpha_1, \alpha_2, ..., \alpha_n\}$ is a finite set of points in $\mathbb{R}$, we can choose a finite pairwise disjoint set of open intervals with rational endpoints $J = \{J_1, J_2, ..., J_n\}$ with $\alpha_i \in J_i$ for $i = 1, 2, ..., n$. For each i, choose $k_i \in \mathbb{N}$ such that $d_{\alpha_i, k_i} \in U_{\alpha_i}$ by the density of D_{α_i}. Now we see that $p(J; k_1, k_2, ..., k_n) \in D \cap B$. Therefore D is a countable dense subset of X. ∎

To see the negative result in part 1 of the theorem above, we consider the following.

Example 18.16 *The Moore plane Γ is a separable space since the points not on the x-axis with both coordinates rational form a countable dense subspace. However, the x-axis is a discrete space, in the subspace topology, of cardinality $\mathfrak{c}$. Hence we have a nonseparable closed subspace of a separable space.*

One easily sees that there is no implication relation between separable and Lindelöf. The Moore plane is a separable space that is not Lindelöf, and the ordinal space $\omega_1 + 1$ is a Lindelöf space (even compact) that is not separable. The question becomes more interesting when we look at the hereditary versions of these properties.

This is the famous "S- and L-space problem." An S-space is a hereditarily separable T_3-space that is not Lindelöf. An L-space is a hereditarily Lindelöf T_3-space that is not separable. In the last thirty-five years, there have been many spaces of both types constructed in various models of set theory. Gödel's constructible universe, for example, contains both types of spaces. Stevo Todorčević has constructed a model that has no S-spaces. What that means is that the existence of S-spaces is undecidable by the usual axioms for set theory. The existence of such spaces causes no contradictions, and the nonexistence of such spaces also causes no contradictions. At this writing, it is not known whether there is a model of set theory that contains no L-spaces.

We have assumed regularity because if we take a subspace of the real line of size ω_1 and assume the set has been well ordered by the function witnessing the cardinality, then, if the usual subspace topology from $\mathbb{R}$ is augmented by declaring initial segments in the well order to be open, one obtains a T_2-space that is hereditarily separable and not Lindelöf (also not regular). On the other hand, by declaring initial segments to be closed, one obtains a hereditarily Lindelöf space that is not separable (and also not regular). These examples give us a clue to both the structure and the duality between these two types of spaces.

Definition 18.17 *A space X is called **left separated** if and only if X can be well ordered in such a way that initial segments are closed.*

Definition 18.18 *A space X is called **right separated** if and only if X can be well ordered in such a way that initial segments are open.*

Theorem 18.19 *A space X is hereditarily separable if and only if X has no uncountable left-separated subspace. A space X is hereditarily Lindelöf if and only if X has no uncountable right-separated subspace.*

Proof. A left separated subspace of order type ω_1 is not separable. A right separated subspace of order type ω_1 is not Lindelöf. For the converse, suppose that X is not hereditarily Lindelöf. Choose a collection $\mathcal{U}$ of open subsets of X such that for any countable subcollection $\mathcal{V} \subset \mathcal{U}$ we have $\cup \mathcal{U} \setminus \cup \mathcal{V} \neq \varnothing$. We will construct the right separated subspace by induction. Choose $U_0 \in \mathcal{U}$, and $x_0 \in U_0$. Suppose that $\alpha < \omega_1$, and for $\beta < \alpha$, we have $x_\beta \in U_\beta \in \mathcal{U}$ chosen. Choose $x_\alpha \in \cup \mathcal{U} \setminus (\bigcup_{\beta < \alpha} U_\beta)$, and choose $U_\alpha \in \mathcal{U}$ with $x_\alpha \in U_\alpha$. Continue by induction to construct $A = \{x_\alpha : \alpha < \omega_1\}$. It is easy to check that A is an uncountable right separated subspace. We leave the hereditarily separable case as an exercise. ∎

In particular, if we have an S-space, then it must contain an uncountable right-separated subspace, and that subspace is also an S-space. Also, if we have an L-space, then it must contain an uncountable left-separated subspace, and that subspace is also an L-space. So the structure of these spaces is well determined. There is an L-space if and only if there is an uncountable left-separated space with no uncountable right-separated subspace. There is an S-space if and only if there is an uncountable right-separated space with no uncountable left-separated subspace. This striking duality is part of what makes this problem so fascinating.

We now return to the compact spaces. Let us first recall the definition.

Definition 18.20 *(Recall definition 7.1) A space X is **compact** if and only if every open cover of X has a finite subcover.*

We now state a theorem that contains some interesting and useful characterizations of compactness. We say that a collection of sets has the *finite intersection property* provided that the intersection of any finite subcollection is nonempty.

Theorem 18.21 *If X is a topological space, then the following are equivalent:*

1. *X is compact.*

2. *If $\mathcal{F}$ is a collection of closed subsets of X with the finite intersection property, then $\cap \mathcal{F} \neq \varnothing$.*

3. *Every filter on X has a cluster point.*

4. *Every ultrafilter on X converges.*

Proof. First we show that $1 \implies 2$. Suppose that X is compact and $\mathcal{F}$ is a collection of closed subsets of X with the finite intersection property. Aiming toward a contradiction, suppose that $\cap \mathcal{F} = \varnothing$. Since the elements of F are closed sets, and $\cap \mathcal{F} = \varnothing$, $\{X \backslash F : F \in \mathcal{F}\}$ is an open cover of X. Since X is compact, choose a finite subcover $\{X \backslash F_k : k = 1, 2, ..., n\}$. Now we see that $X = \bigcup_{k=1}^{n}(X \backslash F_k)$, and thus $\varnothing = \bigcap_{k=1}^{n} F_k$, which contradicts the finite intersection property. Thus the result is established.

Now we show that $2 \implies 3$. Suppose that 2 is true and $\mathcal{G}$ is a filter on X. Since every filter must satisfy the finite intersection property itself, $\{\overline{F} : F \in \mathcal{G}\}$ is a collection of closed subsets of X that satisfies the finite intersection property. Hence by 2, $\cap \{\overline{F} : F \in \mathcal{G}\} \neq \varnothing$. Hence $\mathcal{G}$ clusters (see definition 16.27).

To see that $3 \iff 4$, recall from exercise 8 of chapter 16, that if an ultrafilter has a cluster point, then it converges to it. On the other hand, if every ultrafilter on X converges, and $\mathcal{F}$ is a filter on X, then we extend $\mathcal{F}$ to an ultrafilter $\mathcal{G}$. Choose $x \in X$ with $\mathcal{G} \to x$. We note that x is an element of the closure of each element of $\mathcal{G}$, and consequently x is an element of the closure of each element of $\mathcal{F}$, i.e., $\mathcal{F}$ has a cluster point.

Finally, to complete the proof, we show that $3 \implies 1$. Suppose every filter on X has a cluster point, and let $\mathcal{U}$ be an open cover of X with no finite subcover. Consider $\mathcal{F} = \{X \backslash \cup \mathcal{V} : \mathcal{V} \subset \mathcal{U} \text{ and } |\mathcal{V}| < \omega\}$. The intersection of any two members of $\mathcal{F}$ is again a member of $\mathcal{F}$ by DeMorgan's Laws, and since $\mathcal{U}$ has no finite subcover, all elements of $\mathcal{F}$ are nonempty closed subsets of X. Extend $\mathcal{F}$ to a filter, and let x be a cluster point of that filter. For each $U \in \mathcal{U}$, $X \backslash \cup \{U\} = X \backslash U \in \mathcal{F}$. Hence $x \in \cap \{X \backslash U : U \in \mathcal{U}\} = X \backslash \cup \mathcal{U}$. Hence $\mathcal{U}$ cannot be a cover of X, a contradiction. Hence every open cover of X must have a finite subcover, and the proof is complete. $\blacksquare$

Our next result is one of the most famous theorems in topology. It is a part of the preservation theorem, but we will state it separately because of the importance of the result.

Theorem 18.22 *(Tychonoff theorem) A nonempty product space is compact if and only if each factor space is compact.*

Proof. Suppose $X = \prod_{\alpha \in I} X_\alpha$ is a nonempty product space. Since the projection mappings are all continuous and onto, it follows from theorem 7.5 that if X is compact, then so must be each factor space X_α.

Suppose that each factor space is compact. We wish to show that X is compact. Suppose that $\mathcal{F}$ is an ultrafilter on X. Suppose $\alpha \in I$. We know that $\pi_\alpha(\mathcal{F})$ is an ultrafilter on X_α. Since X_α is compact, choose $x_\alpha \in X_\alpha$ such that $\pi_\alpha(\mathcal{F}) \to x_\alpha$. Do this for each $\alpha \in I$, and consider the point $x = (x_\alpha : \alpha \in I) \in X$. By theorem 16.37, $\mathcal{F} \to x$. Hence by the previous theorem, X is compact. ∎

This is the one piece of the preservation theorem that was not proved in part one. Even though there is almost nothing left to prove, for completeness sake, we now state the preservation theorem for compactness.

Theorem 18.23 *(Preservation theorem for compact spaces)*

1. *Subspaces of compact spaces need not be compact. However, closed subspaces of compact spaces are compact, and (almost conversely) compact subsets of Hausdorff spaces are closed.*

2. *A nonempty product space is compact if and only if each factor space is compact.*

3. *Any continuous image of a compact space is compact.*

Toward developing some intuition for the utility of compactness, compact spaces "act" like they are finite. Obviously, compact spaces may not be finite, but still they act like they are, and the ability to handle them as if they are finite is part of what makes them so useful. The tube lemma, which we saw earlier in this chapter, is a theorem of this sort. We further illustrate this idea with the following four theorems.

Theorem 18.24 *(Max-min theorem) If that X is compact, and $f : X \to \mathbb{R}$ is continuous, then there exist points x^* and x_* in X such that $f(x_*) \leq f(x) \leq f(x^*)$ for every $x \in X$.*

That is, every continuous real-valued function on a compact space has a maximum value and a minimum value.

Proof. This is exercise 5 from chapter 7. ∎

Theorem 18.25 *If X is a T_2-space, H is a compact subset of X, and $x \in X \backslash H$, then there exist disjoint open sets U, V in X with $H \subset U$ and $x \in V$.*

Proof. Suppose that X is a Hausdorff space, H is a compact subset of X, and $x \in X \backslash H$. For each $y \in H$, we can choose a pair U_y, V_y of disjoint open sets with $y \in U_y$ and $x \in V_y$. Now the collection $\{U_y : y \in H\}$ is a collection of open subsets of X that covers H. Since H is compact, we choose a finite set $y_1, y_2, ..., y_n$ of points of H such that $H \subset \bigcup_{k=1}^{n} U_{y_k}$. Let $U = \bigcup_{k=1}^{n} U_{y_k}$ and $V = \bigcap_{k=1}^{n} V_{y_k}$, and it is easy to check that U and V provide the separation we seek. ∎

Theorem 18.26 *If X is a T_2-space, and H and K are disjoint compact subsets of X, then there exist disjoint open sets U, V in X with $H \subset U$ and $K \subset V$.*

Proof. We leave this as an exercise. Use the theorem above, and then mimic the construction above. ∎

Theorem 18.27 *If X is a T_3-space, and H and K are disjoint subsets of X with H compact and K closed, then there exist disjoint open sets U, V in X with $H \subset U$ and $K \subset V$.*

Proof. We leave this as an exercise. Again, you can mimic the construction above. ∎

We showed in part one that compact Hausdorff spaces are normal (it would also follow from the theorems above). We also proved the Heine-Borel theorem, which characterizes the compact subsets of $\mathbb{R}$ as the closed bounded subsets. We won't prove those theorems again, but they should be stated here.

Theorem 18.28 *(See theorem 7.8) Every compact T_2-space is normal.*

Theorem 18.29 *(Heine-Borel theorem, see theorem 7.7) A subset of $\mathbb{R}$ is compact if and only if it is closed and bounded.*

We now have an interesting characterization of Tychonoff spaces.

Theorem 18.30 *Suppose X is a topological space. The following are equivalent.*

   1. X is a completely regular T_1-space.

   2. X can be embedded into a cube.

   3. X can be embedded into a compact T_2-space.

4. X can be embedded into a T_4-space.

Proof. 1 $\implies$ 2 because the embedding lemma gives us that any Tychonoff space X can be embedded into the product space $\prod_{f \in C^*(X)} I_f$ by the evaluation mapping, and $\prod_{f \in C^*(X)} I_f$ is a cube.

To see that 2 $\implies$ 3, use the Heine-Borel theorem. Any closed bounded interval is compact, and thus by the Tychonoff theorem, any cube is compact. Any cube is Hausdorff because Hausdorff is preserved by products. Hence any cube is a compact T_2-space.

For 3 $\implies$ 4, simply refer to the theorem above that any compact Hausdorff space is normal.

Finally to see that 4 $\implies$ 1, note that by Urysohn's lemma, any T_4-space is completely regular, and completely regular and T_2 are both hereditary properties. ∎

From the preservation theorem and the Heine-Borel theorem, we have many examples of compact spaces. For instance, for any ordinal κ the product space 2^κ is compact, any finite space is compact, any closed bounded subset of the reals is compact, any product of closed bounded subsets of the reals is compact, and any continuous image or closed subset of any of these types of spaces is compact. On the other hand, there are many noncompact spaces, for instance $\mathbb{R}$, $\mathbb{M}$, $\mathbb{S}$, Γ, ω_1, or any infinite discrete space.

Often, the full power of compactness is not needed to obtain the results we seek, and it is enough to know that each point sits inside a compact neighborhood. These spaces are called *locally compact* (see chapter 8).

Definition 18.31 *A space X is **locally compact** if and only if each point of X has a compact neighborhood in X.*

We showed in part one (see theorem 8.8), that for Hausdorff spaces this condition is equivalent to having a neighborhood base of compact sets. Many books will use the characterization below as the definition of locally compact spaces. To avoid any confusion between the two, we will always assume that we are working with Hausdorff spaces in the theorems to follow.

Theorem 18.32 *If X is a T_2-space, then X is locally compact if and only if each point of X has a neighborhood base in X consisting of compact sets.*

In terms of basic open sets, this results translates as the following theorem.

Theorem 18.33 *A T_2-space X is locally compact if and only if X has a base $\mathcal{B}$ such that, for each $B \in \mathcal{B}$, $\overline{B}$ is compact.*

Proof. We leave this as an exercise. ∎

We showed in part one that if X is a locally compact T_2-space, then if we add one additional point to obtain X^*, let X be open in X^*, and let neighborhoods of the one new point be complements of compact subsets of X, then the space X^* is a compact T_2-space. This space is called the Alexandroff one-point compactification of X. (See theorem 8.3.)

From the fact that every locally compact Hausdorff space can be embedded into a compact T_2-space, and that all subsets of T_4-spaces are Tychonoff, we have the following.

Theorem 18.34 *Every locally compact T_2-space is completely regular.*

The preservation theorem for locally compact spaces is a bit more complicated than for most properties. We really get no general theorems for subspaces, products or quotients, but we have a consolation prize in every category. We will also concern ourselves only with Hausdorff spaces for this result.

Theorem 18.35 *(Preservation theorem for locally compact spaces)*

1. *Subspaces of locally compact spaces need not be locally compact. However, any subspace of a locally compact Hausdorff space that can be written as the intersection of an open subset with a closed subset is locally compact. Conversely, any locally compact subset of a Hausdorff space is the intersection of an open subset with a closed subset.*

2. *Products of locally compact spaces need not be locally compact. However, any finite product of locally compact Hausdorff spaces is locally compact. Actually, a Hausdorff product space is locally compact if and only if each factor is locally compact and all but finitely many factors are compact.*

3. *Quotients of locally compact spaces need not be locally compact. However, open continuous images of locally compact Hausdorff spaces are locally compact. For closed mappings we have preservation if the fibers are compact. That is, if X is a locally compact Hausdorff space, $f : X \to Y$ is closed continuous onto and for each $y \in Y$, $f^{-1}(y)$ is compact, then Y is locally compact.*

Proof. We first show that the intersection of two locally compact subsets is again locally compact. Suppose A, B are locally compact subsets of a T_2 -space X, and suppose that $x \in A \cap B$. Choose neighborhoods U_A, U_B of x in X such that $A \cap U_A$ and $B \cap U_B$ are compact sets. Now $(A \cap U_A) \cap (B \cap U_B) = (A \cap B) \cap (U_A \cap U_B)$ is a neighborhood of x in the subspace $A \cap B$, and this set is also the intersection of two compact sets, and it is therefore compact.

We now prove 1. Suppose that X is locally compact. We first show that each open subspace of X is locally compact. Suppose that V is an open subspace of X. For any point $x \in V$, there exists a compact neighborhood U such that $x \in U \subset V$, and hence x has a compact neighborhood in V. Hence every open subspace of a locally compact T_2-space is locally compact.

Next, we show that every closed subspace of X is locally compact. Suppose that F is a closed subspace of a locally compact T_2-space X, and $x \in F$. Choose a compact neighborhood U of x in X, and note that $F \cap U$ is a closed subset of U, is a neighborhood of x in the space F, and hence is compact. Thus, every closed subset of a locally compact T_2-space is locally compact. Thus, we have that the intersection of any open subset and any closed subset of a locally compact T_2-space is again locally compact.

For the converse, suppose that X is a Hausdorff space and K is a locally compact subset of X. We will show that K is an open subset of $\overline{K}$, and this will provide us with the result since then there is an open set $U \subset X$ with $K = U \cap \overline{K}$.

Suppose that $x \in K$. Choose an open set $V \subset X$ with $x \in V$ such that $cl_K(K \cap V)$ is compact. We show that $V \cap \overline{K} \subset K$. Now, $cl_K(K \cap V)$ is a closed set in X, and so $cl_K(K \cap V) = cl_X(K \cap V)$. Suppose that $z \in V \cap \overline{K}$. For each open set $W \subset X$ with $z \in W$, we would have $V \cap W$ is a neighborhood of z, and thus $W \cap V \cap K \neq \emptyset$. Hence $z \in \overline{V \cap K}$. Thus $V \cap \overline{K} \subset cl_X(V \cap K) = cl_K(V \cap K) \subset K$, and $V \cap \overline{K}$ is a neighborhood of x in $\overline{K}$. Hence each point of K has a neighborhood in $\overline{K}$ which is contained in K, i.e., K is open in $\overline{K}$

Before proving part 2, we will prove part 3. Suppose that $f : X \to Y$ is an open continuous mapping onto Y, and X is locally compact. Suppose that $y \in Y$, and choose $x \in X$ with $f(x) = y$. Now since X is locally compact, we can choose an open set $U \subset X$ with $x \in U$ and $\overline{U}$ is compact. Since $y \in f(U) \subset f(\overline{U})$, and $f(U)$ is open, we see that $f(\overline{U})$ is a compact neighborhood of y in Y. Hence Y is locally compact.

The proof for the closed case is a little more involved. Suppose that $f : X \to Y$ is a closed continuous mapping onto Y, that X is locally compact, and that for each $y \in Y$, $f^{-1}(y)$ is compact. Suppose that $y \in Y$. For each $x \in f^{-1}(y)$, we can choose an open set U_x that contains x and has compact closure. By the compactness, we can find finitely many $U_{x_1}, U_{x_2}, ..., U_{x_n}$ that cover $f^{-1}(y)$. Let

$U = \bigcup_{k=1}^{n} U_{x_k}$, and note that $\overline{U} = \bigcup_{k=1}^{n} \overline{U_{x_k}}$. Hence $\overline{U}$ is compact. Since f is a closed mapping, we can choose an open set $V \subset Y$ with $y \in V$ and $f^{-1}(V) \subset U$. Now $\overline{V} = \overline{f(f^{-1}(V))} \subset \overline{f(U)} = f(\overline{U})$, and $f(\overline{U})$ is compact. Hence $\overline{V}$ is a compact neighborhood of y in Y. Thus Y is locally compact.

Now we prove part 2 If a nonempty product space $\prod_{\alpha \in I} X_\alpha$ is locally compact and Hausdorff, then, by 3, each factor space must be locally compact since projections are open continuous surjections. Suppose $x \in \prod_{\alpha \in I} X_\alpha$ and let W be a compact neighborhood of x in the product space. Since W must contain a basic open set about x, choose $\alpha_1, \alpha_2, ..., \alpha_n$ and open sets $U_1, U_2, ..., U_n$ in the respective factor spaces such that $x \in \bigcap_{k=1}^{n} \pi_{\alpha_k}^{-1}(U_k) \subset W$. For any $\alpha \notin \{\alpha_k : 1 \le k \le n\}$, we will have $X_\alpha = \pi_\alpha(W)$ which is a compact set. Hence only finitely many factors of $\prod_{\alpha \in I} X_\alpha$ can be non-compact.

For the converse, suppose that $x \in \prod_{\alpha \in I} X_\alpha$, and let

$$F = \{\alpha \in I : X_\alpha \text{ is not compact}\}.$$

For each $\alpha \in F$ choose U_α open in X_α with $\overline{U_\alpha}$ compact. For $\alpha \in I \backslash F$, let $U_\alpha = X_\alpha$, and now we see that $U = \prod_{\alpha \in I} U_\alpha$ is a basic open set containing x. Further, by the Tychonoff theorem, $\overline{U} = \prod_{\alpha \in I} \overline{U_\alpha}$ is compact. Hence $\prod_{\alpha \in I} X_\alpha$ is locally compact. ■

We now give examples to illustrate the negative parts of the theorem above.

Example 18.36 *We number as in the theorem.*

1. *Since any completely regular T_2-space is a subspace of a cube, and cubes are compact T_2-spaces, all we need is a non-locally compact Tychonoff space, $\mathbb{S}$ or $\mathbb{M}$ for instance.*

2. *Since we actually did give a characterization of when a product will be locally compact, no example is really needed here, but $\mathbb{R}^\omega$ would certainly provide a non-locally compact product of locally compact spaces.*

3. *The decomposition space obtained from $\mathbb{R}^2$ by identifying the x-axis to a point is not locally compact since there is no compact neighborhood of the x-axis. Since the x-axis is a closed set, the natural quotient mapping is closed and continuous.*

Remark 18.37 *Functions of the type mentioned in statement 3 are often called perfect mappings. That is, a mapping $f : X \to Y$ is **perfect** provided f is a* **closed continuous mapping onto** Y *and for each* $y \in Y$, $f^{-1}(y)$ **is compact.**

Hausdorff spaces that are quotients of locally compact spaces are called **k-spaces** , or sometimes *compactly generated spaces*. We will look at this class a bit more in the exercises.

Exercises

1. Prove that example 18.4 has the properties that we claimed.

2. (Introduction to the Souslin problem) A space X is said to satisfy the *countable chain condition* , or we say X is a *ccc-space* , provided that every pairwise disjoint collection of open subsets of X is countable. Prove that every separable space is a *ccc* -space. The Souslin problem deals with going the other way. In particular, the problem can be stated as the conjecture that every linearly ordered space that satisfies the countable chain condition must be separable.

   The problem arises naturally from the observation that any linearly ordered space without first or last element that is connected and separable must be homeomorphic to the real line. This can be proved by matching up the rationals with the countable dense subset inductively, making sure that we choose points of the countable dense set in the same order that the first n rationals were, and then extend continuously. The Souslin conjecture is that *ccc* can replace separable in this construction.

   A counterexample to the Souslin conjecture is called a Souslin line. That is, a Souslin line is a linearly ordered topological space that is *ccc*, connected, without first or last element, and is not separable. It has been shown that Souslin lines exist in some "small" models of set theory, for example Gödel's constructible universe. It has also been shown that certain "large" models do not have Souslin lines, for instance any model of Martin's axiom and the negation of the continuum hypothesis.

3. Prove that if X is separable, then $|C(X)| \leq \mathfrak{c}$.

4. Prove lemma 18.9, that if every open subspace of X is Lindelöf, then every subspace of X is Lindelöf.

5. Prove corollary 18.11, that every T_3 hereditarily Lindelöf space is perfectly normal.

6. Prove the remaining part of theorem 18.19, that a space X is hereditarily separable if and only if X has no uncountable left-separated subspaces.

7. Prove theorems 18.26 and 18.27, that disjoint compact sets can be separated by disjoint open sets in T_2-spaces, and disjoint closed sets, one of which is compact, can be separated by disjoint open sets in T_3-spaces.

8. Prove theorem 18.33, that a T_2-space X is locally compact if and only if X has a base $\mathcal{B}$ such that for each $B \in \mathcal{B}$ we have $\overline{B}$ is compact.

9. Prove that the rationals $\mathbb{Q}$, the Sorgenfrey line $\mathbb{S}$, and the Michael line $\mathbb{M}$ are all not locally compact.

10. A k-space is a T_2-space X such that for some locally compact space Y there is a quotient mapping $f : Y \to X$ that is onto X. Prove that a T_2-space X is a k-space if and only if for any $A \subset X$, A is closed in X if and only if $A \cap K$ is closed in K for every compact $K \subset X$. Hint: for the "if" direction, consider the space that is the disjoint union of all the compact subspaces of X. Prove that every first-countable Hausdorff space is a k-space.

Chapter 19

Stone-Čech Compactification

In this chapter we will discuss the Stone-Čech compactification of a topological space. A compactification of a space is a compact Hausdorff space into which the given space can be embedded as a dense subset. In particular, to describe a compactification we need a compact Hausdorff space and an embedding of our space into that compact Hausdorff space. Furthermore, any space that can be embedded into a compact Hausdorff space must be completely regular and T_2. Hence, *for this chapter, we assume that all spaces are Tychonoff spaces.*

Definition 19.1 *A **compactification** of a Tychonoff space X is a compact Hausdorff space K with a function $h : X \to K$ that is a homeomorphism of X onto the dense subspace $h(X)$ of K.*

We have seen in part one the construction of the Alexandroff one-point compactification X^* of a locally compact space X. This is a compactification, according to the definition above, if X is not itself compact. If X is actually compact itself, then the new point is isolated, and X is not dense in X^*. If we start with a noncompact space, this is the smallest compactification we could hope to have. We added only one point. We will now present what is, in a sense to be made clear shortly, the largest compactification. This construction has the advantage that it can be carried out for any Tychonoff space, rather than just for the locally compact ones.

Definition 19.2 *Suppose that X is a completely regular T_2-space. For each $f \in C^*(X)$, let I_f be the closed, bounded interval $[\inf f(X), \sup f(X)] \subset R$. By the embedding lemma, $e : X \to \prod_{f \in C^*(X)} I_f$ where $[e(x)]_f = f(x)$ is an embedding*

241

*of X into this product space, and the product space is compact by the Tychonoff
theorem. The* **Stone-Čech compactification** *of X is $\beta X = \overline{e(X)}$.*

Depending on the part of the world you inhabit, you may call this the Čech-Stone
compactification, but the name βX is universal.

We usually think of X as *being* a subset of βX, rather than thinking about
the embedding. Being a closed subset of a compact Hausdorff space, βX is a
compact Hausdorff space, and certainly X is dense in βX. Thus the construction
of βX does indeed give us a compactification of X. We want to look at some other
characterizations of βX. First we will prove the extension theorem.

Theorem 19.3 *If X is a Tychonoff space, $f : X \to K$ is continuous, and K is
a compact T_2-space, then there is a continuous extension $F : \beta X \to K$ such that
$F|X = f$.*

Proof. Suppose that we have such a function f, i.e. $f : X \to K$ where K is a
compact Hausdorff space, and f is continuous. Since K is also a Tychonoff space,
we will embed both X and K into cubes using the respective evaluation mappings.
The diagram below may be helpful to keep track of where the various functions
are defined.

$$
\begin{array}{ccccccc}
X & \xrightarrow{e} & e(X) & \subset & \beta X & \subset & \prod_{\gamma \in C^*(X)} I_\gamma \\
\downarrow f & & & & \downarrow F & & \downarrow H \\
K & \xrightarrow{e'} & e'(K) & \subset & \beta K & \subset & \prod_{g \in C^*(K)} I_g
\end{array}
$$

Since we want $F|X = F \circ e$, by the construction of βX, we will work in the
setting of the product space $\prod_{\gamma \in C^*(X)} I_\gamma$. First we define $H : \prod_{\gamma \in C^*(X)} I_\gamma \to$
$\prod_{g \in C^*(K)} I_g$ by taking $t \in \prod_{\gamma \in C^*(X)} I_\gamma$ and we define $H(t) \in \prod_{g \in C^*(K)} I_g$ by
$[H(t)]_g = [t]_{g \circ f}$. In other words, $\pi_g(H(t)) = \pi_{g \circ f}(t)$, and note that this does
make sense because for each $g \in C^*(K)$, we know $g \circ f$ is a bounded continuous
real-valued function on X, and $I_{g \circ f} \subset I_g$. Since for each $g \in C^*(K)$, we have
$\pi_g \circ H = \pi_{g \circ f}$, which is a continuous function, we know that H is continuous.

We want to show that H actually maps $e(X)$ into $e'(K)$. Suppose $x \in X$. For
each $g \in C^*(K)$, we have $[H(e(x))]_g = [e(x)]_{g \circ f} = g(f(x)) = [e'(f(x))]_g$, and thus
$H(e(x)) = e'(f(x)) \in e'(K) \subset \beta K$. Hence $H(e(X)) \subset \beta K$. Hence, by continuity,
$H(\beta X) = H(\overline{e(X)}) \subset \overline{H(e(X))} \subset \overline{\beta K} = \beta K$. Furthermore, since K is compact
and e' is continuous, $e'(K)$ is compact and $\beta K = e'(K)$.

Define $F : \beta X \to K$ by $F = (e')^{-1} \circ (H|\beta X)$, and we see that F is continuous, being the composition of two continuous functions. Moreover for each $x \in X$, $F(x) = (e')^{-1}(H(e(x))) = f(x)$. Thus F is the extension we seek. ∎

We will show in the next theorem that the extension property of theorem 19.3 characterizes βX among the compactifications of X.

Theorem 19.4 *If L is a compact Hausdorff space containing X as a dense subset, and for any compact Hausdorff space K and any continuous $f : X \to K$ there is a continuous extension $F : L \to K$ such that $F|X = f$, then L is homeomorphic to βX.*

Proof. Suppose L is a compactification of X having the extension property. Let us denote by $i_1 : X \to L$ the embedding of X into L so that $i_1(X)$ is dense in L. By theorem 19.3 there is a continuous extension $I_1 : \beta X \to L$ such that $I_1 \circ e = i_1$. Now $I_1(\beta X)$ is a dense compact subset of L, and thus $I_1(\beta X) = L$ since L is Hausdorff. Similarly, $e : X \to \beta X$ is continuous, and since L has the extension property, there is a continuous extension $I_2 : L \to \beta X$ such that $I_2 \circ i_1 = e$. Again, $I_2(L)$ is a dense compact subset of βX, and thus $I_2(L) = \beta X$.

Suppose that $z \in \beta X$, and choose a net $(x_\lambda : \lambda \in \Lambda)$ in X with $e(x_\lambda) \to z$. By continuity, $I_2 \circ I_1(e(x_\lambda)) \to I_2 \circ I_1(z)$. For each λ, $I_2(I_1(e(x_\lambda))) = I_2(i_1(x_\lambda)) = e(x_\lambda)$, and thus $I_2 \circ I_1(e(x_\lambda)) \to z$. Since βX is Hausdorff, limits are unique, and thus $z = I_2 \circ I_1(z)$. Similarly, $t = I_1 \circ I_2(t)$ for each $t \in L$. Thus we have that $I_1 \circ I_2$ is the identity mapping on L, and $I_2 \circ I_1$ is the identity mapping on βX. So $I_2 = I_1^{-1}$, making I_1 the homeomorphism we seek. ∎

Related to the extension property is the notion of C^*-embedding.

Definition 19.5 *A space X is C^*-embedded in a space Y if and only if X is. homeomorphic to a subspace of Y and for any $f \in C^*(X)$ there exists $\widehat{f} \in C^*(Y)$ such that $\widehat{f}|X = f$.*

This is a notion that we have actually discussed before in another context. The Tietze extension theorem (theorem 17.41) is really just the statement that a space X is normal if and only if all of the closed subspaces of X are C^*-embedded in X. We also can characterize βX using C^*-embedding.

Theorem 19.6 *A Tychonoff space X is C^*-embedded in a compactification L of X if and only if L is homeomorphic to βX.*

Proof. We first show that X is C^*-embedded in βX. Suppose $f \in C^*(X)$, and choose a closed bounded interval I_f such that $f : X \to I_f$. By the Heine-Borel theorem, I_f is compact, and so, by theorem 19.3, there is a continuous $F : \beta X \to I_f$ such that $F|X = f$. Clearly F is bounded, and thus $F \in C^*(\beta X)$.

Now suppose that X is C^*-embedded in a compactification L. Let $h : X \to L$ be the embedding of X into L with the property that if $f \in C^*(X)$, then there is $\widehat{f} \in C^*(L)$ such that $\widehat{f} \circ h = f$. To show that L is homeomorphic to βX, we will show that L has the extension property of theorem 19.3. Suppose that K is a compact Hausdorff space, and $f : X \to K$ is continuous.

Define $G : L \to \prod_{g \in C^*(K)} I_g$ as follows. For each $g \in C^*(K)$, the composition $g \circ f \in C^*(X)$. Since X is C^*-embedded in L, there exists $h_g \in C^*(L)$ such that $h_g \circ h = g \circ f$. For each $t \in L$, we define $G(t) \in \prod_{g \in C^*(K)} I_g$ by $[G(t)]_g = h_g(t)$.

For each $g \in C^*(K)$, $\pi_g \circ G = h_g$, which is continuous. So G is a continuous function. Suppose that $x \in X$. We see that

$$[G(h(x))]_g = h_g(h(x)) = g(f(x)) = [e(f(x))]_g$$

for each $g \in C^*(K)$, where $e : X \to \beta X$ is the evaluation map. Hence we have that $G(h(x)) = e(f(x))$ for all $x \in X$. In particular, by the continuity of G, $G(L) = G(\overline{h(X)}) \subset \overline{G(h(X))} \subset e(K) = e(K)$ since K is compact.

We define $\widehat{f} = e^{-1} \circ G$, and we see that $\widehat{f} : L \to K$, $\widehat{f}$ is continuous, and for $x \in X$, $\widehat{f}(h(x)) = e^{-1}(G(h(x))) = e^{-1}(e(f(x))) = f(x)$, as desired. Hence we see that L has the extension property, and thus, by theorem 19.4, L is homeomorphic to βX. ∎

At the beginning of this chapter, we commented that in some sense βX is the largest compactification of X. We want to make that more clear. To decide what is largest, we need an order, and the order we want to use is essentially that for two compactifications K_1 and K_2 of X, we will say $K_1 \geq K_2$ if there is a continuous function $f : K_1 \to K_2$ that leaves X fixed. Of course, here we are thinking of X as *being* a subset of the compactifications, and to be precise, we should probably refer to the embeddings. We will do this more carefully in the definition below, but it is good to keep the idea in mind of what we are trying to say. It is also worth noting that since functions do not increase cardinality, with this order we would have that $K_1 \geq K_2 \implies |K_1| \geq |K_2|$, and this is also a sense in which βX is the largest compactification of X.

Definition 19.7 *Suppose that X is a Tychonoff space, and that K_1 and K_2 are two compactifications of X. Let $h_1 : X \to K_1$ and $h_2 : X \to K_2$ be the embeddings* *of X as dense subsets of K_1 and K_2 respectively. We say $K_1 \geq K_2$ if and only if there is a continuous function $f : K_1 \to K_2$ such that for any $x \in X$, we have that $f(h_1(x)) = h_2(x)$.*

It may be helpful to look at a diagram again. Our function that witnesses the order should be the one going across the top.

$$
\begin{array}{ccc}
K_1 & \longrightarrow & K_2 \\
{}_{h_1}\nwarrow & & \nearrow_{h_2} \\
& X &
\end{array}
$$

To truly make this be an order relation on the class of all compactifications of X, we would have to interpret "equal" as being the same as "homeomorphic." With respect to this order, βX is the largest compactification of X.

Theorem 19.8 *If X is a Tychonoff space and K is any compactification of X, then $\beta X \geq K$.*

Proof. Suppose that K is a compactification of X, and $h : X \to K$ is the embedding of X as a dense subset of K. Note that h is continuous, and K is a compact Hausdorff space. Since βX satisfies the extension property of theorem 19.3, there is a function $f : \beta X \to K$ such that $f \circ e = h$. This is the definition of $\beta X \geq K$. ∎

We now turn our attention to one of the most interesting, thoroughly investigated, and still fascinating examples in topology. Many open questions still remain about this space.

Example 19.9 *(Stone-Čech compactification of the natural numbers) This example has historically been called $\beta \mathbb{N}$, but in the last twenty years it has become more common to call it $\beta \omega$. Since ω, the natural numbers with the discrete topology (i.e. the Euclidean topology), is a metric space, ω is certainly a completely regular T_2-space. Consequently, $\beta \omega$ is a compact T_2-space which contains ω as a countable dense subspace.*

We list several properties of $\beta \omega$. We will leave the proofs as exercises (but we will give some hints).

Proposition 19.10 $|\beta \omega| = 2^{\mathfrak{c}}$.

Proof. Hint: The product space $2^{\mathfrak{c}}$ is separable, where we think of $2 = \{0,1\}$ with discrete topology. Let $D \subset 2^{\mathfrak{c}}$ be a countable dense set, and find a function $f : \omega \to D$ that is one-to-one and onto. Note that f is continuous, and use the extension property. Further, $\beta\omega \subset \prod_{f \in C^*(\omega)} I_f$, and now show that $|C^*(\omega)| = \mathfrak{c}$, and thus $\left|\prod_{f \in C^*(\omega)} I_f\right| = \mathfrak{c}^{\mathfrak{c}} = 2^{\mathfrak{c}}$. ■

Proposition 19.11 $\beta\omega$ *is neither first countable nor second countable.*

Proof. Hint: How large can a first-countable separable space be? ■

Proposition 19.12 ω *is open in* $\beta\omega$.

Proof. Hint: Is ω locally compact? Is every locally compact space open in its Stone-Čech compactification? ■

Proposition 19.13 $\beta\omega \backslash \omega$ *is a compact* T_2-*space. This space is often called the* **Stone-Čech remainder of** ω.

The Stone-Čech remainder of ω is a very nasty compact Hausdorff space. People have spent entire careers studying the structure of this space. This includes looking at the structure of points, for instance, can a point have the property that the intersection of countably many neighborhoods is still a neighborhood?

Proposition 19.14 *Every separable compact Hausdorff space is a continuous image of* $\beta\omega$.

Proof. Hint: Use the extension property. ■

People who work with $\beta\omega$, and there are many such people, usually describe it as a space of ultrafilters. We will give this description now. It is, perhaps, the most natural way to construct a compactification. If we ask the question, "What keeps ω from being compact?" The answer must be that there are some ultrafilters that do not converge. Here we simply make sure that we give all of these ultrafilters limit points. Most of these propositions are easy exercises, and we will leave them to the reader.

Definition 19.15 *Let* $Z = \{p : p \text{ is a free ultrafilter on } \omega\}$, *and let* $B\omega = \omega \cup Z$. *For any* $U \subset \omega$, *let* $U^* = U \cup \{p \in Z : U \in p\}$.

Proposition 19.16 *If* U *and* V *are subsets of* ω, *then the following are true:*

1. $\varnothing^* = \varnothing$.

2. $\omega^* = B\omega$.

3. $\{n\}^* = \{n\}$, for each $n \in \omega$.

4. $(U \cup V)^* = U^* \cup V^*$.

5. $(U \cap V)^* = U^* \cap V^*$.

Proposition 19.17 *The collection $\mathcal{B} = \{U^* : U \subset \omega\}$ is a base for a topology on $B\omega$.*

From this point on, we will assume that $B\omega$ is the topological space having $\mathcal{B}$ as a base.

Proposition 19.18 *If $p \in Z = B\omega \backslash \omega$, then $\mathcal{B}_p = \{U^* : U \in p\}$ is a local base at p.*

Proposition 19.19 *ω is dense in $B\omega$.*

Proposition 19.20 *For each $U \subset \omega$, U^* is both open and closed in $B\omega$.*

Proposition 19.21 *$B\omega$ is Hausdorff.*

Proposition 19.22 *$B\omega$ is compact.*

Proposition 19.23 *If K is a compact T_2-space, and $f : \omega \to K$, then there exists a continuous $\widehat{f} : B\omega \to K$ such that $\widehat{f}|\omega = f$.*

If we combine propositions 19.19, 19.21, 19.22, and 19.23, then the following theorem is true because of theorem 19.4.

Theorem 19.24 *$B\omega$ is homeomorphic to $\beta\omega$.*

A similar construction to the above can be done to give another characterization of βX for any Tychonoff space X. The complication is that we don't want to use all ultrafilters, but rather the *z-ultrafilters*.

Definition 19.25 *A subset $A \subset X$ is called a **zero-set** if and only if there is a continuous $f : X \to [0,1]$ such that $A = f^{-1}(0)$.*

It is an easy exercise to show that the intersection of two zero-sets is again a zero-set, and that the union of two zero-sets is again a zero-set. Clearly zero-sets are always closed, and they are always G_δ-sets. In a discrete space, like ω for instance, all subsets are zero-sets.

Definition 19.26 *If X is a space and t is a nonempty collection of nonempty subsets of X, then we call t a **z-filter** provided the following are true:*

1. *all elements of t are zero-sets.*

2. *if $Z_1, Z_2 \in t$, then $Z_1 \cap Z_2 \in t$.*

3. *if $Z_1 \in t$, and Z is a zero-set in X with $Z_1 \subset Z$, then $Z \in t$.*

In colloquial terms, a z-filter is a just like a filter of zero-sets, except that the superset property only holds for zero-sets.

Definition 19.27 *A **z-ultrafilter** on a space X is a z-filter which is maximal in the sense that it is not contained in any strictly finer z-filter.*

The zero-sets and the subsets of ω coincide, and thus the ultrafilters on ω and the z-ultrafilters on ω coincide. The correct analogue of the filter description of $\beta\omega$ for general spaces is to attach to a copy of X all free z-ultrafilters. This construction is developed very nicely in chapter 6 of the book *Rings of Continuous Functions* by Leonard Gillman and Meyer Jerison.

Exercises

1. Prove that if X is a locally compact T_2-space, then X is open in βX. Prove proposition 19.12; ω is open in $\beta\omega$.

2. Suppose that X is a Tychonoff space. Prove that X is connected if and only if βX is connected.

3. (The space $C^*(X)$) Suppose X is a topological space.

   (a) We can define a metric on $C^*(X)$, since the functions all have range in $\mathbb{R}$, by declaring that $d(f, g) = \sup\{|f(x) - g(x)| : x \in X\}$. Prove that this is a metric. Compare the metric topology on $C^*(X)$ with the subspace topology which $C^*(X)$ inherits from the product space $\mathbb{R}^X$.

(b) $C^*(X)$ also has rich algebraic structure by doing arithmetic "pointwise," i.e., $(f+g)(x) = f(x)+g(x)$, $(fg)(x) = f(x)g(x)$ and $(kf)(x) = kf(x)\forall x \in X \forall k \in \mathbb{R}$. Prove that with these operations $C^*(X)$ is both a ring and a vector space over $\mathbb{R}$, and additionally for any $k \in \mathbb{R}$, and $f,g \in C^*(X)$, we have $k(fg) = (kf)g = f(kg)$. Prove that these three algebraic operations are continuous. Such a space is called an *algebra* over $\mathbb{R}$.

(c) Prove that $C^*(\omega)$, with the metric topology, is homeomorphic to the Banach space l_∞. The space l_∞ is the space of all bounded sequences of real numbers with the "sup norm," i.e.,

$$\|(x_n : n \in \omega)\| = \sup\{|x_n| : n \in \omega\}.$$

4. Prove that the order on compactifications, given in definition 19.7, is essentially a partial order (where we are assuming two spaces are equal if they are homeomorphic).

5. Prove proposition 19.10 that $|\beta\omega| = 2^{\mathfrak{c}}$.

6. (Sequences in $\beta\omega$) Prove proposition 19.11 that $\beta\omega$ is not first countable. This result is an easy cardinality argument, but much more is true. Prove that every infinite closed subset of $\beta\omega$ must contain copy of $\beta\omega$. Here is a hint: within the infinite closed set, construct a countably infinite set of points, $A = \{a_n : n \in \omega\}$, and a sequence of pairwise disjoint open sets $\{V_n : n \in \omega\}$ with $a_n \in V_n \forall n$. If $g : A \to [0,1]$, then define $g_1 : \omega \to [0,1]$ by $g_1(n) = g(a_i)$, if $n \in \omega \cap V_i$, and $g_1(n) = 0$, otherwise.

Now show that the extension of g_1 to $\beta\omega$ induces an extension to $\overline{A}$ of g. Hence $\overline{A}$ is βA, and since A is homeomorphic to ω, $\overline{A}$ is a copy of $\beta\omega$ inside the given closed set. Show that a nontrivial convergent sequence together with its limit point would be a countably infinite closed set. Now use this result to show that any convergent sequence in $\beta\omega$ must be eventually constant. This is the result that is colloquially referred to as "$\beta\omega$ has no convergent sequences."

7. Prove proposition 19.13 that $\beta\omega\setminus\omega$ is a compact Hausdorff space.

8. Prove proposition 19.14 that every separable compact Hausdorff space is a continuous image of $\beta\omega$.

9. Prove proposition 19.16 and use that result to prove proposition 19.17 that $\{U^* : U \subset \omega\}$ is a base for a topology on $B\omega$.

10. Prove proposition 19.18 that $\mathcal{B}_p = \{U^* : U \in p\}$ is a local base at p for each $p \in \beta\omega\backslash\omega$.

11. Prove directly (that is, without using the fact that $B\omega = \beta\omega$) propositions 19.19 and 19.20.

12. Prove directly (that is, without using the fact that $B\omega = \beta\omega$) propositions 19.21 and 19.22 that $B\omega$ is a compact Hausdorff space.

13. Prove directly (that is, without using the fact that $B\omega = \beta\omega$) proposition 19.23 that $B\omega$ has the extension property.

14. Suppose that Z_1, Z_2 are zero-sets in a space X. Prove that $Z_1 \cap Z_2$ and $Z_1 \cup Z_2$ are also zero-sets. Prove that any zero-set is a closed G_δ-set. Prove that in perfectly normal spaces the closed sets and the zero-sets coincide.

Chapter 20

Paracompact Spaces

The introduction of the notion of paracompactness was instrumental in the solution to two of the most difficult problems in topology from the first half of the twentieth century. One problem was to find a common generalization of metric space and compact space which could be used to explain the common results for these the two most powerful classes of topological spaces. The other problem was called the "general metrization problem." This problem is to describe, in terms that involve only the topology, necessary and sufficient conditions on a topological space for there to exist a metric that generates the given topology. The metrization problem we will take up in the next chapter. Our goal for this chapter is to introduce that common generalization—paracompactness. We prove some of the basic results. In particular, we will prove that all compact spaces and all metric spaces are paracompact.

Paracompactness is a covering property, that is, it a statement about open covers of topological spaces. Of course, compactness is the most common such property, and what we want to do here is generalize compactness in some natural way. In compactness, we say that every open cover has a finite subcover. Now we will replace the idea of a finite subcover with a condition that says that each point of the space *thinks* we have a finite subcover. We need several definitions to make this precise.

Definition 20.1 *A **cover** of a topological space is a collection of subsets whose union is the entire space. An **open cover** of a space X is a collection $\mathcal{U}$ of open subsets of X such that $\cup \mathcal{U} = X$. A **closed cover** of a space X is a collection $\mathcal{H}$ of closed subsets of X such that $\cup \mathcal{H} = X$.*

In a similar way, we use other adjectives which describe subsets to describe
covers made up of sets which are of the type described.

Definition 20.2 *If $\mathcal{A}$ is a collection of subsets of a space X, then we say $\mathcal{B}$ is a*
partial refinement *of $\mathcal{A}$ provided that $\mathcal{B}$ is a collection of subsets of X and for
each $B \in \mathcal{B}$ there exists $A_B \in \mathcal{A}$ with $B \subset A_B$. If a partial refinement $\mathcal{B}$ also
satisfies $\cup\mathcal{B} = \cup\mathcal{A}$, then we call $\mathcal{B}$ a* ***refinement*** *of $\mathcal{A}$.*

We use the adjectives open, closed, compact, and so on, to describe refinements in
the same way we used them to describe covers. So, for example, an open refinement
of a cover, would be a collection of open sets which is a partial refinement and also
covers.

Definition 20.3 *Suppose X is a topological space and $\mathcal{A}$ is a collection of subsets
of X.*

1. *We say $\mathcal{A}$ is a* ***pairwise disjoint*** *collection if and only if whenever $A, B \in \mathcal{A}$
   and $A \neq B$, $A \cap B = \varnothing$.*

2. *We say $\mathcal{A}$ is a* ***point finite*** *collection if and only if for each point $x \in X$,
   the collection*
$$\{A \in \mathcal{A} : x \in A\} \text{ is finite.}$$

3. *We say $\mathcal{A}$ is a* ***locally finite*** *collection if and only if for each point $x \in X$
   there exists a neighborhood U_x of x such that the collection*
$$\{A \in \mathcal{A} : U_x \cap A \neq \varnothing\} \text{ is finite.}$$

4. *We say $\mathcal{A}$ is a* ***discrete collection*** *if and only if for each point $x \in X$ there
   exists a neighborhood U_x of x such that*
$$|\{A \in \mathcal{A} : U_x \cap A \neq \varnothing\}| \leq 1.$$

5. *For a property P, we say $\mathcal{A}$ is a* ***σ-P collection*** *if and only if $\mathcal{A} = \bigcup \mathcal{A}_n$
   where each $\mathcal{A}_n$ satisfies property P. For example, a* ***σ-discrete*** *collection is
   a countable union of discrete collections.*

We could have defined a pairwise disjoint as a collection $\mathcal{A}$ such that for each
$x \in X$, $|\{A \in \mathcal{A} : x \in A\}| \leq 1$, and that would have made the definition above
appear more parallel. I think that would also obscure what the definition says,
but it is easy to see that this is equivalent. From the definitions it is clear that

discrete collections are locally finite, locally finite collections are point finite, and pairwise disjoint collections are point finite. Of course, in any T_1-space X, the collection $\{\{x\} : x \in X\}$ is a pairwise disjoint collection that would be locally finite only when the space is discrete. So there are certainly examples of pairwise disjoint collections (and thus point finite collections) that are not locally finite.

Example 20.4 *In the space $\mathbb{R}$, the real numbers with the usual topology, the collection $\{(n, n+1) : n \in \omega\}$ is a locally finite collection which is not discrete.*

Proof. We leave this as an exercise. ■

In order to avoid certain trivial counterexamples, we will assume that all paracompact spaces are T_2.

Definition 20.5 *A Hausdorff space X is called **paracompact** if and only if every open cover of X has an open locally finite refinement.*

Since a finite subcover would clearly be a locally finite refinement, we easily get the following theorem.

Theorem 20.6 *Every compact Hausdorff space is paracompact.*

There are other examples of paracompact spaces. We will see that all metric spaces are paracompact, for example. In the spirit of avoiding proofs of too many theorems about a class of spaces only to find that there are no spaces in the class or that all spaces are in the class, we will mention a couple of other examples now. The proofs that these examples have the properties that we claim will be easier once we have completed more of the results in this chapter.

Example 20.7 *The Michael line $\mathbb{M}$ is paracompact.*

Proof. We will leave this as an exercise. The easiest way to see this is to use the fact, which we will prove shortly, that all metrizable spaces are paracompact. Consequently every subspace of the real line is paracompact and neighborhoods of rational points in $\mathbb{M}$ contain Euclidean open sets. ■

Example 20.8 *The ordinal space ω_1 is not paracompact.*

Proof. We will see shortly that all countably compact, paracompact spaces are compact, and since ω_1 is countably compact and not compact, it cannot be paracompact. ■

Hence we have a nice example of a paracompact space that is neither metric nor compact. We also have a nice example that has many other strong properties, but it is not paracompact.

We want to prove several characterizations of paracompactness. These were all done by E. A. Michael (that's right, the Michael line guy) in a series of very beautiful papers published in the 1950's. Before proceeding to the first of Michael's characterizations of paracompactness, we need another definition and a lemma. Both of these items have been mentioned in chapter 17, but we will repeat them here.

Definition 20.9 *A collection $\mathcal{H}$ of subsets of a space X is called **closure preserving** if and only if for any subcollection $\mathcal{F} \subset \mathcal{H}$ we have $\cup \{\overline{F} : F \in \mathcal{F}\} = \overline{\cup \{F : F \in \mathcal{F}\}}$.*

Lemma 20.10 *If $\mathcal{H}$ is a locally finite collection of subsets of a space X, then $\mathcal{H}$ is closure preserving.*

Proof. This is an exercise in chapter 17. ∎

Theorem 20.11 *(Michael's theorem) Suppose X is a T_3-space. The following are equivalent.*

1. *X is paracompact.*

2. *Every open cover of X has a σ-locally finite open refinement.*

3. *Every open cover of X has a locally finite (but not necessarily open) refinement.*

4. *Every open cover of X has a locally finite closed refinement.*

Proof. That $1 \implies 2$ is trivial. Let us show that $2 \implies 3$. Suppose that every open cover of X has a σ-locally finite open refinement, and $\mathcal{U}$ is an open cover of X. We want to construct a locally finite refinement of $\mathcal{U}$. Choose an open refinement $\mathcal{V} = \bigcup_{n \in \omega} \mathcal{V}_n$ of $\mathcal{U}$ such that for each $n \in \omega$, $\mathcal{V}_n$ is a locally finite collection of subsets of X. For each $n \in \omega$, let $W_n = \cup \mathcal{V}_n$, and note that each W_n is an open subset of X.

We construct a locally finite cover $\{A_n : n \in \omega\}$ of X as follows. Let $A_0 = W_0$, and for $n > 0$, let $A_n = W_n \setminus \bigcup_{k<n} W_k$. Notice that $W_n \cap A_m = \varnothing$ if $m > n$, and $A_n \cap A_m = \varnothing$ if $n \neq m$. Now we easily see that $\bigcup_{n \in \omega} A_n = \bigcup_{n \in \omega} W_n = X$, and for each $x \in X$, if $x \in A_n$, then W_n is a neighborhood of x that intersects only those A_k's where $k \leq n$. Hence $\{A_n : n \in \omega\}$ is a locally finite cover of X.

Now we will cut up each A_n to get a refinement of $\mathcal{U}$. Let

$$\mathcal{F} = \{A_n \cap V : V \in \mathcal{V}_n \text{ and } n \in \omega\}.$$

It is clear that $\mathcal{F}$ is a refinement of $\mathcal{V}$, and thus $\mathcal{F}$ is a refinement of $\mathcal{U}$. It remains only to show that $\mathcal{F}$ is locally finite. Suppose $x \in X$. Choose $n \in \omega$ such that $x \in A_n$. For each $k \le n$, choose an open set R_k with $x \in R_k$ and $\{V \in \mathcal{V}_k : R_k \cap V \ne \varnothing\}$ is finite. Let $G = W_n \cap \bigcap_{k=0}^n R_k$, and note that G is an open neighborhood of x that intersects only finitely many members of $\mathcal{F}$.

Now we prove that $3 \implies 4$. Assume that every open cover of X has a locally finite refinement. First, notice that if $\{A_\alpha : \alpha \in I\}$ is a locally finite collection, then the same open neighborhood for each point will witness that $\{\overline{A_\alpha} : \alpha \in I\}$ is also locally finite. Suppose that $\mathcal{U}$ is an open cover of X. For each point $x \in X$, choose $U_x \in \mathcal{U}$ with $x \in U_x$. By regularity, choose an open set V_x with $x \in V_x \subset \overline{V_x} \subset U_x$. Now $\{V_x : x \in X\}$ is an open cover of X, and, by 3, we choose a locally finite refinement $\mathcal{F}$ of $\mathcal{V}$. Now $\mathcal{F}_c = \{\overline{F} : F \in \mathcal{F}\}$ is also locally finite and covers X. Moreover, if $F \in \mathcal{F}$, then we can choose $x \in X$ with $F \subset V_x$, and thus $\overline{F} \subset \overline{V_x} \subset U_x \in \mathcal{U}$. Hence $\mathcal{F}_c$ is a closed locally finite refinement of $\mathcal{U}$.

Finally, we prove $4 \implies 1$. Assume that every open cover of X has a locally finite closed refinement. Suppose that $\mathcal{U}$ is an open cover of X. By 4, choose a closed locally finite refinement $\mathcal{H}$ of $\mathcal{U}$. For each $x \in X$ choose an open neighborhood W_x of x such that $\{H \in \mathcal{H} : W_x \cap H \ne \varnothing\}$ is a finite set. Now $\{W_x : x \in X\}$ is an open cover of X, and, again using 4, we choose a closed locally finite refinement $\mathcal{P}$ of $\{W_x : x \in X\}$. For each $H \in \mathcal{H}$, we let $V_H = X \setminus \cup \{P \in \mathcal{P} : P \cap H = \varnothing\}$. By lemma 20.10, $\mathcal{P}$ is closure preserving, and thus we see that V_H is an open set containing H, for each $H \in \mathcal{H}$. Clearly, $\{V_H : H \in \mathcal{H}\}$ covers X.

We claim that $\{V_H : H \in \mathcal{H}\}$ is a locally finite collection of sets. For each $x \in X$, we choose an open neighborhood G_x of x that intersects only finitely many elements of $\mathcal{P}$. We list as $P_1, P_2, ..., P_n$ those elements of $\mathcal{P}$ that intersect G_x, and notice that $G_x \subset \bigcup_{k=1}^n P_k$. Since $\mathcal{P}$ is a refinement of $\{W_x : x \in X\}$, we see that for each k, $\mathcal{H}_k = \{H \in \mathcal{H} : P_k \cap H \ne \varnothing\}$ is a finite set. Now if $G_x \cap V_H \ne \varnothing$, then there must be a k such that $P_k \cap H \ne \varnothing$. Hence $H \in \bigcup_{k=1}^n \mathcal{H}_k$, which is the finite union of finite sets. This shows that $\{V_H : H \in \mathcal{H}\}$ is a locally finite open cover of X.

For each $H \in \mathcal{H}$, we choose $U_H \in \mathcal{U}$ such that $H \subset U_H$, and note that this is possible since $\mathcal{H}$ is a refinement of $\mathcal{U}$. Now $\mathcal{V} = \{V_H \cap U_H : H \in \mathcal{H}\}$ is clearly an open locally finite collection that refines $\mathcal{U}$, and since $H \subset V_H \cap U_H$ and $\mathcal{H}$ covers X, $\mathcal{V}$ is actually an open locally finite refinement of $\mathcal{U}$, as desired. ∎

Corollary 20.12 *Every T_3 Lindelöf space is paracompact.*

Proof. Any countable subcover of a given open cover of any space is certainly a σ-locally finite refinement. ∎

We are now ready to prove the famous theorem of Arthur Stone that metrizable spaces are paracompact. The proof for this result breaks naturally into two parts. We will first show that every open cover of a metrizable space has a σ-discrete closed refinement. Once we have that refinement, we will use collectionwise normality to "puff" the sets out to a σ-discrete open refinement. Finally, Michael's theorem will deliver the result.

Even though this is not a trivial construction, it is simpler than the proof given originally by A. H. Stone. The construction of the σ-discrete closed refinement follows a construction that was introduced three years after Stone originally proved the theorem. It is a very clever construction used by R. H. Bing in connection with Moore spaces.

Theorem 20.13 *(Stone's theorem) Every metrizable space is paracompact.*

Proof. Suppose X is a metrizable space, and let $\mathcal{U}$ be an open cover of X. Choose a metric d that generates the topology on X. Well order $\mathcal{U}$, and for each $x \in X$, let $F(x,\mathcal{U})$ be the first element $U \in \mathcal{U}$ such that $x \in U$. Suppose $U \in \mathcal{U}$ and $n \in \mathbb{N}$. We define

$$H(n,U) = \left\{ x : U = F(x,\mathcal{U}) \text{ and } B(x, \frac{1}{n}) \subset U \right\}.$$

For each $n \in \mathbb{N}$, let $\mathcal{H}_n = \{H(n,U) : U \in \mathcal{U}\}$. It is clear that for each $n \in \mathbb{N}$, $\mathcal{H}_n$ is a partial refinement of $\mathcal{U}$ since $H(n,U) \subset U$ for each U.

For the moment, fix $n \in \mathbb{N}$, and we will show that $\mathcal{H}_n$ is a discrete collection of closed sets. First, we show that $\mathcal{H}_n$ is discrete. Suppose $x \in X$, and let $V = F(x,\mathcal{U}) \cap B(x, \frac{1}{n})$. Note that V is an open neighborhood of x. Suppose that $U \in \mathcal{U}$ and U precedes $F(x,\mathcal{U})$ in the well ordering on $\mathcal{U}$. Note that $x \notin U$, and thus for each $z \in H(n,U)$, we have that $x \notin B(z, \frac{1}{n})$. Hence $z \notin B(x, \frac{1}{n})$, and thus $z \notin V$. Hence $V \cap H(n,U) = \emptyset$. Now if U follows $F(x,\mathcal{U})$, then $F(x,\mathcal{U}) \cap H(n,U) = \emptyset$, and again $V \cap H(n,U) = \emptyset$. Hence V is an open neighborhood of x that intersects at most one element of $\mathcal{H}_n$, and thus we see that $\mathcal{H}_n$ is a discrete collection.

Suppose that $U \in \mathcal{U}$. We want to show that $H(n,U)$ is a closed set. Suppose that $x \in X \backslash H(n,U)$. We consider cases. First, suppose that $x \notin U$; this means that $B(x, \frac{1}{n}) \cap H(n,U) = \emptyset$. Suppose that $x \in U$; either $U = F(x,\mathcal{U})$, or not. On the one hand, if $U \neq F(x,\mathcal{U})$, then $F(x,\mathcal{U})$ is an open neighborhood of x, and $F(x,\mathcal{U}) \cap H(n,U) = \emptyset$. On the other hand, suppose that $U = F(x,\mathcal{U})$. Since

$x \notin H(n,U)$, we must have $B(x, \frac{1}{n}) \cap (X \backslash U) \neq \emptyset$. Choose $y \in B(x, \frac{1}{n}) \cap (X \backslash U)$, and now $B(y, \frac{1}{n})$ is an open neighborhood of x that is disjoint from $H(n,U)$. Thus, in every case, we have an open neighborhood of x that is disjoint from $H(n,U)$. Thus we have that $\mathcal{H}_n$ is a collection of closed sets.

So we have established that $\mathcal{H} = \bigcup_{n=1}^{\infty} \mathcal{H}_n$ is a σ-discrete closed collection. Suppose $x \in X$, and choose $n \in \mathbb{N}$ with $B(x, \frac{1}{n}) \subset F(x, \mathcal{U})$. Now $x \in H(n, F(x, \mathcal{U}))$, and thus $\mathcal{H}$ covers X and is a σ-discrete closed refinement of $\mathcal{U}$.

Since X is collectionwise normal, for each $U \in \mathcal{U}$ and $n \in \mathbb{N}$, we can choose an open set $V(n, U) \subset X$ such that $H(n, U) \subset V(n, U)$, $\{V(n, U) : U \in \mathcal{U}\}$ is a discrete collection, and $V(n, U) \neq V(n, W)$ if $U \neq W$. For each $n \in \mathbb{N}$, let $\mathcal{W}_n = \{U \cap V(n, U) : U \in \mathcal{U}\}$. For each $U \in \mathcal{U}$ and $n \in \mathbb{N}$, $H(n, U) \subset U \cap V(n, U)$, and thus we see that $\bigcup_{n=1}^{\infty} \mathcal{W}_n$ covers X. Hence $\mathcal{W} = \bigcup_{n=1}^{\infty} \mathcal{W}_n$ is a σ-discrete open refinement of $\mathcal{U}$. By Michael's theorem, X is paracompact. ∎

We now turn to another theorem of Michael that gives additional characterizations of paracompactness. Before we do this, we need another definition.

Definition 20.14 *A collection $\mathcal{V}$ of subsets of a space X is **cushioned** in another collection $\mathcal{U}$ of subsets of X if and only if there is a function $t : \mathcal{V} \rightarrow \mathcal{U}$ such that for any $\mathcal{W} \subset \mathcal{V}$ we have $\overline{\cup \mathcal{W}} \subset \cup \{t(W) : W \in \mathcal{W}\}$.*

If $\mathcal{V}$ is cushioned in $\mathcal{U}$ and is also a refinement, then we call $\mathcal{V}$ a cushioned refinement of $\mathcal{U}$. If $\mathcal{V}$ is cushioned in $\mathcal{U}$ by some function $t : \mathcal{V} \rightarrow \mathcal{U}$, then it is easy to see that $\{\cup(t^{-1}(U)) : U \in \mathcal{U}\}$ is also cushioned in $\mathcal{U}$. In particular, if a collection has a cushioned refinement, then it has a cushioned refinement that is *precise* in the sense that the cushioning function is one-to-one. In particular, if a collection $\mathcal{V}$ is cushioned in a collection $\mathcal{U}$, and $\mathcal{U}$ is indexed as $\{U_\alpha : \alpha \in A\}$, then we can assume that $\mathcal{V}$ is indexed by $\{V_\alpha : \alpha \in A\}$ where the cushioning function t assigns $t(V_\alpha) = U_\alpha$. Note that we may have to name $V_\alpha = \emptyset$ if U_α is not in the range of t.

We easily see that if $\mathcal{V}$ is a closure-preserving closed refinement of $\mathcal{U}$, then $\mathcal{V}$ is cushioned in $\mathcal{U}$. This is true since if $V \in \mathcal{V}$, and we choose $U_V \in \mathcal{U}$ with $V \subset U_V$, then for any $\mathcal{W} \subset \mathcal{V}$ we have $\overline{\cup \mathcal{W}} = \cup \mathcal{W} = \cup \{W : W \in \mathcal{W}\} \subset \cup \{U_W : W \in \mathcal{W}\}$. Both closure-preserving and cushioned refinements are mechanisms for controlling the growth of closures of unions of subcollections. The ability to do this gives you a great deal of strength as we see from the following lemma.

Lemma 20.15 *If every open cover of a space X has a cushioned refinement, then X is collectionwise normal.*

Proof. Suppose a space X has the property that every open cover of X has a cushioned refinement. Let $\{F_\alpha : \alpha \in A\}$ be a faithfully indexed discrete collection of closed subsets of X. For each $\alpha \in A$, let $U_\alpha = X \backslash \bigcup_{\beta \neq \alpha} F_\beta$. Note that $\{U_\alpha : \alpha \in A\}$ is an open cover of X since $F_\alpha \subset U_\alpha$ for each α and points outside all the F_α's are in all the U_α's. Choose a precise cushioned refinement $\{H_\alpha : \alpha \in A\}$ of $\{U_\alpha : \alpha \in A\}$. For each $\alpha \in A$, let

$$V_\alpha = X \backslash cl(\cup \{H_\beta : \beta \neq \alpha\}).$$

We note that for each $\alpha \in A$, V_α is open, and if $x \in F_\alpha$, then $x \notin U_\beta$ for any $\beta \neq \alpha$ and thus $x \in V_\alpha$ by the cushioning. Now for $\alpha, \beta \in A$, with $\alpha \neq \beta$,

$$V_\alpha \cap V_\beta = X \backslash [cl(\cup \{H_\gamma : \gamma \neq \alpha\}) \cup cl(\cup \{H_\gamma : \gamma \neq \beta\})] = X \backslash X = \varnothing$$

Hence $\{V_\alpha : \alpha \in A\}$ is pairwise disjoint, as desired. ∎

In particular, this result gives us all the separation we want in T_1 -spaces where every open cover has a cushioned refinement.

Corollary 20.16 *If X is T_1 and every open cover of X has a cushioned refinement, then X is T_4.*

We now prove part 2 of Michael's theorem. There are more parts that Michael proved in the papers "A Note on Paracompact Spaces," "Another Note on Paracompact Spaces," and "Yet Another Note on Paracompact Spaces," which were all published in the *Proceedings of the American Mathematical Society* in the 1950's.

Theorem 20.17 *(Michael's theorem) Suppose X is a T_3-space. The following are equivalent.*

1. *X is paracompact.*

2. *Every open cover of X has a σ-closure preserving open refinement.*

3. *Every open cover of X has a closure preserving closed refinement.*

4. *Every open cover of X has a cushioned refinement.*

5. *Every open cover of X has a σ-discrete open refinement.*

Proof. That $1 \implies 2$ is clear from theorem 20.11 since any σ-locally finite open refinement would be a σ-closure preserving open refinement. It is clear that $3 \implies 4$ since a closed closure preserving refinement is cushioned as we noted above. Also $5 \implies 1$ is clear from theorem 20.11 since discrete collections are locally finite. It remains only to prove that $2 \implies 3$ and $4 \implies 5$.

We prove that $2 \implies 3$. We suppose that every open cover of X has a σ-closure preserving open refinement. Let $\mathcal{U}$ be an open cover of X. Choose an open refinement $\mathcal{G} = \bigcup_{n=1}^{\infty} \mathcal{G}_n$ of $\mathcal{U}$ where each $\mathcal{G}_n$ is closure preserving. Since X is regular, we may assume without loss of generality that $\{\overline{G} : G \in \mathcal{G}\}$ is also a refinement of $\mathcal{U}$. For each $n \in \mathbb{N}$, we let $H_n = \cup \mathcal{G}_n$, and we let

$$\mathcal{K}_n = \left\{ \overline{G} \backslash \bigcup\nolimits_{k<n} H_k : G \in \mathcal{G}_n \right\}.$$

Now $\mathcal{K} = \bigcup_{n=1}^{\infty} \mathcal{K}_n$ is a collection of closed sets which is a partial refinement of $\mathcal{U}$.

We show that $\mathcal{K}$ covers X. If $x \in X$, then we can find n, the first subscript for which $x \in H_n$, and then x is covered by $\mathcal{K}_n$.

To see that $\mathcal{K}$ is closure preserving, suppose that $\mathcal{S} \subset \mathcal{K}$, and let $x \in \overline{\cup \mathcal{S}}$. We need to show that $x \in \overline{S}$ for some $S \in \mathcal{S}$. Choose $m \in \mathbb{N}$ such that $x \in H_m$. Now H_m is an open neighborhood of x which is disjoint from every $S \in \mathcal{S} \cap \mathcal{K}_j$ where $j > m$. Hence

$$x \in cl(\cup \{S : S \in \mathcal{S} \cap \mathcal{K}_i \text{ and } i \leq m\}).$$

Since the closure of a finite union is the union of those finitely many closures, there exists $i_0 \leq m$ with

$$x \in cl(\cup \{S : S \in \mathcal{S} \cap \mathcal{K}_{i_0}\}).$$

However, $\mathcal{K}_{i_0}$ is a closure preserving collection since we subtracted the same open set from each member of $\mathcal{G}_{i_0}$ to create it. Hence there is some $S_0 \in \mathcal{S} \cap \mathcal{K}_{i_0}$ with $x \in \overline{S_0}$, and we are finished.

Now we prove that $4 \implies 5$. Suppose that every open cover of X has a cushioned refinement. Let $\mathcal{U}$ be an open cover of X. We well order $\mathcal{U} = \{U_\alpha : \alpha < \lambda\}$ for some ordinal λ. We will construct a σ-discrete open refinement of $\mathcal{U}$ by induction.

For each $\alpha < \lambda$, we let $U(\alpha, 1) = U_\alpha$, and

$$\mathcal{U}_1 = \{U(\alpha, 1) : \alpha < \lambda\} = \mathcal{U}.$$

Choose $\mathcal{C}_1 = \{C(\alpha, 1) : \alpha < \lambda\}$ to be a precise cushioned refinement of $\mathcal{U}_1$. For each $\alpha < \lambda$, let

$$U(\alpha, 2) = U(\alpha, 1) \backslash cl(\bigcup_{\beta < \alpha} C(\beta, 1)),$$

and let
$$\mathcal{U}_2 = \{U(\alpha, 2) : \alpha < \lambda\}.$$

For each $x \in X$, choose $\alpha = first\{\gamma : x \in U(\gamma, 1)\}$, and we see that, by the cushioning,
$$x \in U(\alpha, 1)\backslash \bigcup_{\beta < \alpha} U(\beta, 1) \subset U(\alpha, 2)$$

since $cl(\bigcup_{\beta < \alpha} C(\beta, 1)) \subset \bigcup_{\beta < \alpha} U(\beta, 1)$. Hence $\mathcal{U}_2$ is an open cover of X. We now choose $\mathcal{C}_2 = \{C(\alpha, 2) : \alpha < \lambda\}$ to be a precise cushioned refinement of $\mathcal{U}_2$. We continue by induction for $n \geq 1$, supposing $\mathcal{U}_n = \{U(\alpha, n) : \alpha < \lambda\}$ to be an open cover of X and $\mathcal{C}_n = \{C(\alpha, n) : \alpha < \lambda\}$ to be a precise cushioned refinement of $\mathcal{U}_n$, we define
$$U(\alpha, n+1) = U(\alpha, n)\backslash cl(\bigcup_{\beta < \alpha} C(\beta, n)).$$

As before, $\{\mathcal{U}(\alpha, n+1) : \alpha < \lambda\}$ is an open cover of X. We choose $\mathcal{C}_{n+1} = \{C(\alpha, n+1) : \alpha < \lambda\}$ to be a precise cushioned refinement of $\mathcal{U}_{n+1}$, and continue by induction to construct open covers $\mathcal{U}_n$, $n \in \mathbb{N}$, and cushioned refinements $\mathcal{C}_n$, $n \in \mathbb{N}$.

For each $x \in X$ and $n \in \mathbb{N}$, let $\delta(x, n)$ be the first $\alpha < \lambda$ such that $x \in C(\alpha, n)$, and notice that if $x \in C(\alpha, n+1) \subset U(\alpha, n+1)$, then $x \notin C(\beta, n)$ for any $\beta < \alpha$. Hence $\delta(x, n+1) \leq \delta(x, n)$ for every $n \in \mathbb{N}$. So for each $x \in X$, the sequence $(\delta(x, n) : n \in \mathbb{N})$ is a decreasing sequence of ordinals, and thus there is a number $m(x) \in \mathbb{N}$ such that $\delta(x, m(x)) = \delta(x, k)$ for all $k \geq m(x)$. For each $\alpha < \lambda$ and $n \in \mathbb{N}$, let
$$F(\alpha, n) = \{x : x \in C(\alpha, n), \alpha = \delta(x, n), n \geq m(x)\}.$$

Let $\mathcal{F}_n = \{F(\alpha, n) : \alpha < \lambda\}$ for each $n \in \mathbb{N}$. Suppose that $n \in \mathbb{N}$, and we will show that $\mathcal{F}_n$ is a discrete collection. Suppose that $x \in X$, and choose $k \in \mathbb{N}$ such that $k > \max\{m(x), n\}$. Let $W = X\backslash cl(\cup\{C(\beta, k) : \beta \neq \delta(x, k)\})$. Since $k - 1 \geq m(x)$,
$$x \in C(\delta(x, k), k - 1),$$

and thus $x \notin U(\beta, k)$ for $\beta > \delta(x, k)$. Hence
$$x \notin cl(\cup\{C(\beta, k) : \beta > \delta(x, k)\}),$$

and since $x \in U(\delta(x, k), k + 1)$, we see that
$$x \notin cl(\cup\{C(\beta, k) : \beta < \delta(x, k)\}).$$

So we see that W is an open neighborhood of x. For each β, $F(\beta, k) \subset C(\beta, k)$, and thus $W \cap F(\beta, k) = \varnothing$ if $\beta \neq \delta(x, k)$. Since $k > n$, $F(\beta, n) \subset F(\beta, k)$ for each β, and thus $W \cap F(\beta, n) = \varnothing$ if $\beta \neq \delta(x, k)$. Hence we see that $\mathcal{F}_n$ is a discrete family of sets. Let $\mathcal{F} = \bigcup_{n=1}^{\infty} \mathcal{F}_n$, then it is just a matter of looking at the definition to see that $\mathcal{F}$ covers X.

Furthermore, for each α, n, $F(\alpha, n) \subset C(\alpha, n) \subset U(\alpha, n) \subset U(\alpha, 1)$, and thus $\mathcal{F}$ is a σ-discrete refinement of $\mathcal{U}$. Actually, $\overline{F(\alpha, n)} \subset C(\alpha, n) \subset U(\alpha, n)$, and so we see that $\{\overline{F} : F \in \mathcal{F}\}$ is a σ-discrete closed refinement of $\mathcal{U}$. Now, by lemma 20.15, X is collectionwise normal, so just as we did in the proof of Stone's theorem, we can expand the σ-discrete closed refinement to a σ-discrete open refinement using the collectionwise normality. This completes the proof. ■

Corollary 20.18 *Every paracompact space is collectionwise normal.*

Proof. This will follow from the theorem above and lemma 20.15 once we know that paracompact spaces are regular. We leave it as an exercise to show that if every open cover of a T_2-space X has an open locally finite refinement, then X is regular. ■

In the previous theorem and in the proof of Stone's theorem, we constructed a σ-discrete closed refinement of our given open cover, and then we built a σ-discrete open refinement using collectionwise normality. It is natural to wonder whether the existence of a σ-discrete closed refinement would be enough to give us paracompactness. As we see in the next example, this property is actually weaker than paracompactness.

Definition 20.19 *A space X is called **subparacompact** if and only if every open cover of X has a σ-discrete closed refinement.*

Example 20.20 *The Moore plane Γ is subparacompact but not paracompact.*

Proof. We will leave it as an exercise to show that every open cover of Γ has a σ-discrete closed refinement. Here is a hint: look at the subspace $\mathbb{R} \times [\frac{1}{n}, \infty)$, and note that it is a metric space.

We also leave it as an exercise to show that any point finite open cover of a separable space must be countable. Hence if Γ were paracompact, then Γ would be Lindelöf, and we know that is not the case. ■

As we have seen in the proofs of previous results, subparacompact collectionwise normal spaces are paracompact.

Theorem 20.21 *A T_1-space X is paracompact if and only if X is subparacompact and collectionwise normal.*

Proof. It is a good exercise to prove just this result directly and not in the context of some more complicated construction. We leave it as an exercise. ∎

We will now prove that every countably compact paracompact space is compact. First notice that if we choose one point from each set of a discrete collection, then the resulting set of points is a closed discrete set.

Lemma 20.22 *If X is a T_1-space, $\{F_\alpha : \alpha \in A\}$ is a faithfully indexed discrete collection of subsets of X, and $x_\alpha \in F_\alpha$ for each $\alpha \in A$, then $\{x_\alpha : \alpha \in A\}$ is a closed discrete set.*

Proof. Left to the reader. ∎

Lemma 20.23 *If X is T_1 countably compact space, then every discrete collection of subsets of X is finite.*

Proof. This is an easy consequence of lemma 20.22. ∎

Lemma 20.24 *If every open cover of a space X has a countable refinement, then X is Lindelöf.*

Proof. Left to the reader. ∎

Theorem 20.25 *Every paracompact countably compact space is compact.*

Proof. It is clear from the previous lemmas that every open cover of a paracompact (or even subparacompact) countably compact space must have a countable refinement. Hence we have a countably compact Lindelöf space, and it is clear that such spaces are compact. ∎

We now move to the preservation theorem for paracompactness.

Theorem 20.26 *(Preservation theorem for paracompactness)*

1. *Subspaces of paracompact spaces need not be paracompact. However, F_σ-subspaces of paracompact spaces are paracompact. In particular, closed subspaces of paracompact spaces are paracompact.*

2. *Products of, even two, paracompact spaces need not be paracompact. However, if X is paracompact and K is compact and Hausdorff, then $X \times K$ is paracompact.*

3. *Quotients of paracompact spaces need not be paracompact. However, closed continuous images of paracompact spaces are paracompact.*

Proof. We will give examples below to illustrate the negative results in the theorem. Here we prove the positive results.

For 1, suppose that X is paracompact, and $F = \bigcup_{n=1}^{\infty} F_n$ is a subspace of X, where F_n is closed for each $n \in \mathbb{N}$. Suppose that $\mathcal{U}$ is an open cover of F. For each $U \in \mathcal{U}$, choose an open set $V_U \subset X$ such that $U = F \cap V_U$. For each $n \in \mathbb{N}$, the collection $\mathcal{V}_n = \{X \backslash F_n\} \cup \{V_U : U \in \mathcal{U}\}$ is an open cover of X. For each n, by the paracompactness of X, choose an open locally finite refinement $\mathcal{W}_n$ of $\mathcal{V}_n$. Let $\mathcal{H}_n = \{F \cap W : W \in \mathcal{W}_n \text{ and } W \cap F_n \neq \varnothing\}$ for each $n \in \mathbb{N}$. Now in the space F, $\bigcup_{n=1}^{\infty} \mathcal{H}_n$ is a σ-locally finite open refinement of $\mathcal{U}$. Hence F is paracompact.

The result that if X is paracompact and K is compact, then $X \times K$ is paracompact is such an instructive exercise in the use of the Tube lemma, that we will leave it to the reader as an exercise which really must be done.

For part 3, suppose that $f : X \to Y$ is a closed continuous mapping of X onto Y, and suppose that X is paracompact. We first show that every open cover of Y has a closed, closure preserving refinement. Suppose $\mathcal{U}$ is an open cover of Y. We see by the continuity that $\{f^{-1}(U) : U \in \mathcal{U}\}$ is an open cover of X. Since X is paracompact, we can find a closed closure preserving refinement $\mathcal{H}$ of $\{f^{-1}(U) : U \in \mathcal{U}\}$. Now since f is closed, continuous, and onto, $\{f(H) : H \in \mathcal{H}\}$ is a closed closure preserving refinement of $\mathcal{U}$. Now since Y is a closed continuous image of a T_1-space, Y is T_1. To complete the proof we need only show that Y is regular. This follows from the fact that every open cover has a closed, closure preserving refinement. Hence Y is paracompact. ∎

Example 20.27 *These examples show where preservation fails.*

1. *Since every Tychonoff space is a subspace of a compact Hausdorff space, it is clear that subspaces of paracompact spaces need not be paracompact. In particular, ω_1 is an open nonparacompact subspace of the compact T_2-space $\omega_1 + 1$. Actually, it is true that if every open subspace of a given space is paracompact, then so is every subspace. We will leave this result as an exercise.*

2. *The Sorgenfrey line is Lindelöf, and thus paracompact, but $\mathbb{S} \times \mathbb{S}$ is not normal, and so $\mathbb{S} \times \mathbb{S}$ cannot be paracompact. Also the Michael line $\mathbb{M}$ is*

paracompact, but the product of $\mathbb{M}$ *with the metric space* $\mathbb{R}\backslash\mathbb{Q}$ *is not normal, and thus not paracompact.*

3. *Since discrete spaces are metric, and thus paracompact, every space is the one-to-one continuous image of a paracompact space. So certainly continuous images fail to preserve paracompactness. We have seen earlier that the open continuous image of a metric space may fail to be Hausdorff and thus may fail to be paracompact.*

We will close out the chapter on paracompact spaces with one more theorem which gives additional characterizations of paracompactness. These are of a rather different type than those given earlier, and they begin to have a little flavor of generalized metric space conditions. To be able to state this theorem, we need a couple of definitions.

Definition 20.28 *Suppose* X *is a space and* $\mathcal{G}$ *is a collection of subsets of* X*. For any* $A \subset X$*, we define the* **star** *of* A *in* $\mathcal{G}$ *as*

$$st(A,\mathcal{G}) = \cup\{G \in \mathcal{G} : A \cap G \neq \varnothing\}.$$

In the case when $A = \{x\}$*, we write* $st(x,\mathcal{G})$ *instead of* $st(\{x\},\mathcal{G})$*. That is,*

$$st(x,\mathcal{G}) = \cup\{G \in \mathcal{G} : x \in G\}.$$

Definition 20.29 *Suppose* X *is a space and* $\mathcal{U}$ *and* $\mathcal{V}$ *are covers of* X*.*

1. *We say that* $\mathcal{V}$ *is a* **barycentric refinement** *of* $\mathcal{U}$ *if and only if*

$$\{st(x,\mathcal{V}) : x \in X\}$$

is a refinement of $\mathcal{U}$*. Barycentric refinements are sometimes called pointwise star refinements.*

2. *We say that* $\mathcal{V}$ *is a* **star refinement** *of* $\mathcal{U}$ *if and only if*

$$\{st(V,\mathcal{V}) : V \in \mathcal{V}\}$$

is a refinement of $\mathcal{U}$*.*

An interesting exercise, which is instructive in understanding these kinds of refinements, is given by the following lemma.

Lemma 20.30 *Suppose X is a space, and $\mathcal{U}, \mathcal{V}, \mathcal{W}$ are covers of X. If $\mathcal{W}$ is a barycentric refinement of $\mathcal{V}$, and $\mathcal{V}$ is a barycentric refinement of $\mathcal{U}$, then $\mathcal{W}$ is a star refinement of $\mathcal{U}$.*

Proof. We leave this as an exercise. ∎

Spaces in which every open cover has an open star refinement were called *fully normal* by Tukey, who showed that metric spaces had this property. The real crux of the original proof of Stone's theorem was to show that fully normal is equivalent to paracompact. Now that we have Michael's characterizations, this result is much easier.

Theorem 20.31 *Suppose X is a T_1-space. The following are equivalent.*

1. *X is paracompact.*

2. *Every open cover of X has an open barycentric refinement.*

3. *Every open cover of X has an open star refinement.*

Proof. For $1 \implies 2$, suppose that X is paracompact and $\mathcal{U}$ is an open cover of X. Let $\mathcal{F}$ be a locally finite closed refinement of $\mathcal{U}$. For each $F \in \mathcal{F}$, choose $U_F \in \mathcal{U}$ with $F \subset U_F$. Now for each $x \in X$, let

$$W_x = \left(\bigcap_{x \in F} U_F \right) \setminus \cup \{H \in \mathcal{F} : x \notin H\}.$$

It is clear that $\mathcal{W} = \{W_x : x \in X\}$ is an open cover of X.

Suppose $x \in X$, and choose $F \in \mathcal{F}$ with $x \in F$. If $y \in X$ and $x \in W_y$ for some y, then from the definition of W_y we must have $y \in F$, and thus $W_y \subset U_F$. Hence $st(x, \mathcal{W}) \subset U_F$. Thus we see that $\mathcal{W}$ is an open barycentric refinement of $\mathcal{U}$.

That $2 \implies 3$ is immediate from lemma 20.30.

To show that $3 \implies 1$, suppose that X is a T_1-space and that every open cover of X has an open star refinement. We show that every open star refinement is a cushioned refinement. Suppose that $\mathcal{U}$ is an open cover of X and $\mathcal{V}$ is an open star refinement of $\mathcal{U}$. For each $V \in \mathcal{V}$, choose $U \in \mathcal{U}$ with $st(V, \mathcal{V}) \subset U_V$. Suppose $\mathcal{H} \subset \mathcal{V}$ and $x \in cl(\cup\mathcal{H})$. Choose $V \in \mathcal{V}$ with $x \in V$, and since V is open, $V \cap (\cup\mathcal{H}) \neq \emptyset$. So there is $H \in \mathcal{H}$ with $V \cap H \neq \emptyset$, and thus $x \in V \subset st(H, \mathcal{V}) \subset U_H$. Hence $cl(\cup\mathcal{H}) \subset \cup\{U_H : H \in \mathcal{H}\}$. So the function $V \longmapsto U_V$ provides the function to establish that $\mathcal{V}$ is cushioned in $\mathcal{U}$. Now X is regular by corollary 20.16, and by Michael's theorem, X is paracompact. ∎

Exercises

1. Prove that the example 20.4 has the properties claimed. That is, prove that in the space $\mathbb{R}$ with the usual topology $\{(n, n+1) : n \in \omega\}$ is a locally finite, pairwise-disjoint collection that is not a discrete collection.

2. Prove that the Michael line $\mathbb{M}$ is paracompact (example 20.7). Notice that this would give you the fact that $\mathbb{M}$ is normal, which was a difficult exercise earlier in the text.

3. Prove directly that every paracompact space is regular. That is, prove that if every open cover of a T_2-space X has an open locally finite refinement, then X is regular.

4. Prove that every open cover of the Moore plane Γ has a σ-discrete closed refinement. There is a hint given in example 20.20.

5. Prove theorem 20.21. That is, prove that for T_1-spaces paracompactness is equivalent to subparacompactness plus collectionwise normality.

6. Prove that if X is a separable space and $\mathcal{U}$ is a point-finite open collection in X, then $\mathcal{U}$ is countable. Prove lemma 20.24 that if every open cover of X has a countable refinement, then X is Lindelöf. Prove that every separable paracompact space is Lindelöf. So, for separable T_3 spaces, paracompactness is equivalent to the Lindelöf property.

7. Prove lemma 20.22. That is, prove that if $\{F_\alpha : \alpha \in A\}$ is a faithfully indexed discrete collection of subsets of a space X, and $x_\alpha \in F_\alpha$ for each $\alpha \in A$, then $\{x_\alpha : \alpha \in A\}$ is a closed discrete set in X.

8. Prove that if every open subspace of a space X is paracompact, then every subspace of X is paracompact. Prove that every perfect paracompact space is hereditarily paracompact.

9. Prove lemma 20.30. That is, prove that if $\mathcal{W}$ is a barycentric refinement of $\mathcal{V}$, and $\mathcal{V}$ is a barycentric refinement of $\mathcal{U}$, then $\mathcal{W}$ is a star refinement of $\mathcal{U}$.

Chapter 21

Metrization

A topological space X is called *metrizable* if there is some metric on X that is compatible with the topology on X. From the time that topology became recognizable as a discipline, a major question was to determine which spaces were metrizable. At first glance this seems like a strange question. Indeed, doesn't the definition of the word tell us which spaces are metrizable? The answer to that is that the definition is not really a statement about the *topology*. It is a statement about the relationship between some external structure, namely some function from $X \times X$ into $\mathbb{R}$, and the topology. What was wanted was to give an *internal* characterization; that is, a characterization which talks only about the set X and the open subsets of X which will tell when there is a metric that generates these open sets. This problem was called the general metrization problem . (By the way, in some parts of the world, the word metrization is spelled "metrisation.")

Before taking on the metrization problem, which will consume the balance of this chapter, we state the preservation theorem.

Theorem 21.1 *(Preservation theorem for metrizable spaces)*

1. *All subspaces of metrizable spaces are metrizable.*

2. *Countable products of metrizable spaces are metrizable. Nontrivial uncountable products are not metrizable.*

3. *Quotients of metrizable spaces need not be metrizable.*

Proof. The restriction of a metric to a subspace is a metric on the subspace. We showed in the project in chapter 6 that the countable product of spaces with bounded metrics is metrizable. Every metric space is metrizable by a bounded metric, and so the result follows. Nontrivial uncountable products will not even be first countable. We have seen quotients of subspaces of the plane $\mathbb{R}^2$ which are not even Hausdorff. ∎

For the separable case, the metrization problem was answered in the 1920's. Urysohn proved that a T_3 separable space is metrizable if and only if it has a countable base. This was the project in chapter 6. We will state the theorem again here, but we will not risk spoiling the fun for the students by giving away the proof.

Theorem 21.2 *(Urysohn metrization theorem) Suppose X is a T_1-space. The following are equivalent.*

1. *X is separable and metrizable.*

2. *X is regular and second countable.*

The solution to the general problem was more difficult to obtain, and it was also more difficult to determine when it was obtained. What I mean by that is that the question of being *internal* was not so easy to decide. Are sequences of open sets internal? Well, they do involve a function from the natural numbers into the topology, and that seems like an external object. Furthermore, people wanted a condition that was easy to check. When Nagata, Smirnov, and Bing proved their amazing theorems, however, people were satisfied that the general metrization theorem had been solved.

In this chapter, we will give several approaches to trying to solve this problem, and we will give several metrization theorems, culminating in the Nagata-Smirnov-Bing theorem. We then use the Nagata-Smirnov-Bing theorem to establish a few more results that are especially relevant to Moore spaces.

Definition 21.3 *A **development** for a topological space X is a sequence $(\mathcal{G}_n : n \in \omega)$ of open covers of X such that for each point $x \in X$ the collection $\{st(x, \mathcal{G}_n) : n \in \omega\}$ is a neighborhood base at x. If a space X has a development, then we call X a **developable space**. A regular T_1-space which is developable is called a **Moore space**.*

Example 21.4 *If X is a metric space and for each $n \in \omega$ we let*

$$\mathcal{G}_n = \left\{ B(x, \frac{1}{2^n}) : x \in X \right\},$$

then it is easy to see that $(\mathcal{G}_n : n \in \omega)$ *is a development for* X. *Hence every metrizable space is a Moore space.*

Proof. Using the triangle inequality, it is easy to check that for each $n \in \omega$, $st(x, \mathcal{G}_{n+1}) \subset B(x, \frac{1}{2^n})$. ■

We will need several facts about stars and developments, which are easy once you think about them for a few minutes. We will list them as a series of lemmas, and most of the proofs will be left as exercises.

Lemma 21.5 *If* $\mathcal{U}, \mathcal{V}$ *are covers of* X, *and* $\mathcal{V}$ *is a refinement of* $\mathcal{U}$, *then for each* $x \in X$ *we have* $st(x, \mathcal{V}) \subset st(x, \mathcal{U})$.

Proof. Suppose $\mathcal{U}, \mathcal{V}$ are covers of X, $\mathcal{V}$ refines $\mathcal{U}$, and $x \in X$. Let $y \in st(x, \mathcal{V})$. There is an element $V \in \mathcal{V}$ with $x \in V$ and $y \in V$. Now for some $U \in \mathcal{U}$ we have $V \subset U$. Thus $y \in U$ and $x \in U$, and thus $y \in U \subset st(x, \mathcal{U})$. Hence $st(x, \mathcal{V}) \subset st(x, \mathcal{U})$. ■

Lemma 21.6 *If a space* X *has a development* $(\mathcal{G}_n : n \in \omega)$ *and for each* $n \in \omega$, $\mathcal{W}_n$ *is an open refinement of* $\mathcal{G}_n$, *then* $(\mathcal{W}_n : n \in \omega)$ *is also a development for* X.

Proof. We leave this as an exercise. ■

Lemma 21.7 *If a space* X *has a development* $(\mathcal{G}_n : n \in \omega)$, *then* X *has a development* $(\mathcal{W}_n : n \in \omega)$ *that has the property that for each* $n \in \omega$, $\mathcal{W}_{n+1}$ *is a refinement of* $\mathcal{W}_n$.

Proof. Suppose we have a development $(\mathcal{G}_n : n \in \omega)$ for X. We will construct a sequence $(\mathcal{W}_n : n \in \omega)$ of open covers of X by induction. Let $\mathcal{W}_0 = \mathcal{G}_0$ and for $n \in \omega$ we let
$$\mathcal{W}_{n+1} = \{W \cap G : W \in \mathcal{W}_n \text{ and } G \in \mathcal{G}_{n+1}\}.$$
Clearly, $(\mathcal{W}_n : n \in \omega)$ is the development we seek. ■

Notice that this last result says that if we have a development, then we can assume that our development generates a sequence of stars at each point that is a decreasing neighborhood base at that point.

Our next result says that stars are symmetric.

Lemma 21.8 *If* $\mathcal{U}$ *is a cover of* X *and* $x, y \in X$, *then* $x \in st(y, \mathcal{U})$ *if and only if* $y \in st(x, \mathcal{U})$.

Proof. We leave this as an exercise. ■

Lemma 21.9 *If a space X has a development $(\mathcal{G}_n : n \in \omega)$, then $\bigcup_{n\in\omega} \mathcal{G}_n$ is a base for X.*

Proof. We leave this as an exercise. Here is a hint: if $x \in G \in \mathcal{G}_n$, then $G \subset st(x, \mathcal{G}_n)$. ∎

Theorem 21.10 *Every Lindelöf Moore space is metrizable.*

Proof. If $(\mathcal{G}_n : n \in \omega)$ is a development for a Lindelöf space X, then for each $n \in \omega$ find a countable subcover $\mathcal{W}_n \subset \mathcal{G}_n$. Now, by lemma 21.6, $(\mathcal{W}_n : n \in \omega)$ is also a development for X. Hence by lemma 21.9, X has a countable base. So by the Urysohn metrization theorem, X is metrizable. ∎

We now list a couple of examples.

Example 21.11 *Here we have an example of a space that is, and one that is not, a Moore space.*

1. *The Moore plane Γ is a Moore space.*

2. *The Sorgenfrey line $\mathbb{S}$ is not a Moore space.*

Proof. Part 2 is clear since $\mathbb{S}$ is Lindelöf and not second countable. To construct a development for the Moore plane, for each $n \in \omega$ take $\mathcal{G}_n$ to be the collection consisting of all ordinary open balls in $\mathbb{R}^2$ whose radii are $\frac{1}{2^n}$ and centers are at least $\frac{1}{2^n}$ above the x-axis together with all tangent balls to the x-axis (including the tangent point) with radii equal to $\frac{1}{2^n}$. ∎

One of the earliest intrusions of set theory into topology came from Burton Jones' famous result, which we give now.

Theorem 21.12 *(Jones' theorem) Assuming the continuum hypothesis to be true, then every separable normal Moore space is metrizable.*

Proof. By Jones' lemma, theorem 17.31, if X is a separable normal space, then for every closed discrete subset S of X we have $|\mathcal{P}(S)| \leq \mathfrak{c}$. Hence by Cantor's result, we have that $|S| < \mathfrak{c}$. Assuming the continuum hypothesis (CH), this means that every closed discrete subset of X would be countable. If we can show that every Moore space is subparacompact; that is, every open cover has a σ-discrete closed refinement, then we would have that every separable normal Moore space is Lindelöf, and thus metrizable, if CH holds.

The construction we use here is essentially the same as in the proof of Stone's theorem. Suppose $\mathcal{U}$ is an open cover of a Moore space X. Choose a development $(\mathcal{G}_n : n \in \omega)$ for X with $\mathcal{G}_{n+1}$ a refinement of $\mathcal{G}_n$ for each n. Well order $\mathcal{U}$, and for each $x \in X$, let $F(x, \mathcal{U})$ be the first element of $\mathcal{U}$ which contains x. For each $n \in \omega$ and $U \in \mathcal{U}$, let

$$H(n, U) = \{x : U = F(x, \mathcal{U}) \text{ and } st(x, \mathcal{G}_n) \subset U\}.$$

Let $\mathcal{H}_n = \{H(n, U) : U \in \mathcal{U}\}$, and just as in the proof of Stone's theorem, we see that $\bigcup_{n \in \omega} \mathcal{H}_n$ is a σ-discrete closed refinement of $\mathcal{U}$. Hence all Moore spaces are subparacompact, and the result now follows. ∎

Jones knew this theorem in 1934, but the paper didn't appear until 1937 because he spent that time trying to prove that the continuum hypothesis was true. We know now that CH can neither be proved nor disproved from the axioms of Zermelo-Frankel and the axiom of choice.

Jones' theorem marks the introduction of the normal Moore space conjecture which is the conjecture that every normal Moore space is metrizable .

Problem 21.13 *(Normal Moore space problem) Is every normal Moore space metrizable?*

This problem has almost been solved, and we will have more to say about it at the end of this chapter. The problem itself has certainly been the most important and ardently pursued problem in general topology since 1937.

The result, which we included in the proof of Jones' theorem, that every Moore space is subparacompact probably deserves to be written down separately. We will call it a corollary since it is really a corollary to the proof of Jones' theorem.

Corollary 21.14 *Every Moore space is subparacompact. That is, every open cover of a Moore space has a σ-discrete closed refinement.*

In the 1920's, the terminology of Moore spaces had not become common, but our next theorem can be easily couched in those terms.

Theorem 21.15 *(Alexandroff-Urysohn metrization theorem) A T_1-space X is metrizable if and only if X has a development $(\mathcal{G}_n : n \in \omega)$ with the property that if $U, V \in \mathcal{G}_{n+1}$ and $U \cap V \neq \varnothing$, then there is $W \in \mathcal{G}_n$ with $U \cup V \subset W$.*

Remark 21.16 *A development as in the Alexandroff-Urysohn theorem is a called a* **regularly refining development***. Notice that this condition forces $\mathcal{G}_{n+1}$ to refine $\mathcal{G}_n$ for each $n \in \omega$.*

Proof. First suppose that (X, d) is a metric space. For each $n \in \omega$, we let $\mathcal{G}_n = \left\{ B(x, \frac{1}{2^n}) : x \in X \right\}$. From example 21.4, $(\mathcal{G}_n : n \in \omega)$ is a development for X. Suppose $U, V \in \mathcal{G}_{n+1}$ and $U \cap V \neq \varnothing$. Choose points $x, y, z \in X$ such that $U = B(y, \frac{1}{2^{n+1}})$, $V = B(z, \frac{1}{2^{n+1}})$, and $x \in U \cap V$. Suppose $p \in V$, and we compute $d(x, p) \leq d(x, z) + d(z, p) < \frac{1}{2^{n+1}} + \frac{1}{2^{n+1}} = \frac{1}{2^n}$. Hence $V \subset B(x, \frac{1}{2^n}) \in \mathcal{G}_n$. Similarly $U \subset B(x, \frac{1}{2^n})$, and the proof of the "only if" is complete.

Now suppose we have a regularly refining development $(\mathcal{G}_n : n \in \omega)$ for the T_1-space X. The development we will actually use to construct a metric is every other term of the original. For each $n \in \omega$, let $\mathcal{H}_n = \mathcal{G}_{2n+1}$. Now since, at each point x, the sequence $(st(x, \mathcal{G}_n) : n \in \omega)$ is decreasing, $(\mathcal{H}_n : n \in \omega)$ is also a development for the same topology on X. We now have the slightly stronger refining property that if $U, V, W \in \mathcal{H}_{n+1}, U \cap V \neq \varnothing$, and $V \cap W \neq \varnothing$ then there exists $S \in \mathcal{H}_n$ such that $U \cup V \cup W \subset S$.

We first define a distance function (called a semimetric) such as we can define for any development, which is not quite a metric, and then we will extend that function to the metric we seek. Let $t : X \times X \to [0, \infty)$ be defined by

$$t(x, y) = \begin{cases} 0, & \text{if } y \in st(x, \mathcal{H}_n) \text{ for all } n \\ \frac{1}{2^n}, & \text{if } n = first\{k : y \notin st(x, \mathcal{H}_k)\}. \end{cases}$$

Notice that by lemma 21.8 the stars are symmetric, and thus, for any points $x, y \in X$, we have $t(x, y) = t(y, x)$. Also if $x \neq y$, then since $(\mathcal{H}_n : n \in \omega)$ is a development, there exists some $m \in \omega$ with $x \notin st(y, \mathcal{H}_m)$, and so $t(x, y) \neq 0$. Hence $t(x, y) = 0$ if and only if $x = y$. Also note that if $t(x, y) < \frac{1}{2^n}$, then $t(x, y) \leq \frac{1}{2^{n+1}}$. In particular, if $t(x, y) < \frac{1}{2^n}$, then for $k = first\{j : y \notin st(x, \mathcal{H}_j)\}$ we must have $k \geq n+1$, and so $y \in st(x, \mathcal{H}_n)$. On the other hand, if $y \in st(x, \mathcal{H}_n)$, then $k = first\{j : y \notin st(x, \mathcal{H}_j)\} > n$, and thus $t(x, y) < \frac{1}{2^n}$. Hence we have that for each $x \in X$ and $n \in \omega$, $st(x, \mathcal{H}_n) = \left\{ y \in X : t(x, y) < \frac{1}{2^n} \right\}$. (This set is what we would call the ball at x of radius $\frac{1}{2^n}$ if t were a metric. Actually, t satisfies the properties of a metric except for the triangle inequality.)

We now define a function based on t which also satisfies the triangle inequality. Let $d : X \times X \to [0, \infty)$ be defined by

$$d(x, y) = \inf \left\{ \sum_{n=1}^{k} t(x_{n-1}, x_n) : x_0 = x, x_k = y, x_i \in X \forall i \right\}.$$

Notice that several things are easily true. First $d(x, y) = d(y, x)$ for all $x, y \in X$, and $d(x, y) = 0$ if and only if $x = y$. Furthermore, $d(x, y) \leq t(x, y)$.

We show that d satisfies the triangle inequality. Suppose that $x, y, z \in X$, and suppose that $d(x, z) > d(x, y) + d(y, z)$. So we would have that $d(x, z) - d(x, y) >$

$d(y, z)$. By the definition of infimum, we can find $y_0, y_1, ..., y_m \in X$ with $y_0 = y$ and $y_m = z$ such that $\sum_{i=1}^{m} t(y_{i-1}, y_i) < d(x, z) - d(x, y)$, and hence $d(x, y) < d(x, z) - \sum_{i=1}^{m} t(y_{i-1}, y_i)$. Again using the definition of infimum, we choose $x_0, x_1, ..., x_n$ with $x_0 = x$ and $x_n = y$ such that $\sum_{j=1}^{n} t(x_{j-1}, x_j) < d(x, z) - \sum_{i=1}^{m} t(y_{i-1}, y_i)$. Now we see that $d(x, z) > \sum_{j=1}^{n} t(x_{j-1}, x_j) + \sum_{i=1}^{m} t(y_{i-1}, y_i)$, which is an element of the set for which $d(x, z)$ is the infimum, a contradiction. Thus d satisfies the triangle inequality, and so d is a metric on X.

We now show that the topology generated by d is the topology that is given on X. In order to do this, we will first establish the relationship between d and t a little more. We show, by induction on k, the following:

If $n \in \omega$, $x, y \in X$, and there exist points $x_0, x_1, ..., x_k$

such that $x_0 = x$, $x_k = y$, and $\sum_{i=1}^{k} t(x_{i-1}, x_i) < \dfrac{1}{2^{n+2}}$,

then $t(x, y) < \dfrac{1}{2^n}$.

Clearly this is true when $k = 1$. Suppose $k > 1$ and the claim is true whenever $l < k$. We will show that the claim is true at k. Suppose $n \in \omega$, $x, y \in X$, and there exist points $x_0, x_1, ..., x_k$ such that $x_0 = x$, $x_k = y$, and

$$(*) \quad t(x, x_1) + t(x_1, x_2) + \cdots + t(x_{k-1}, y) < \frac{1}{2^{n+2}}.$$

We may assume, without loss of generality, that

$$t(x, x_1) + t(x_1, x_2) + \cdots + t(x_{k-1}, y) \geq \frac{1}{2^{n+3}}$$

since otherwise we could choose the largest n causing this $(*)$ to be true, and then $t(x, y)$ would be even smaller than $\frac{1}{2^n}$. Let

$$j = \max \left\{ i : t(x, x_1) + t(x_1, x_2) + \cdots + t(x_{i-1}, x_i) < \frac{1}{2^{n+3}} \right\}.$$

Since $j < k$, by the induction hypothesis,

$$t(x, x_j) < \frac{1}{2^{n+1}}.$$

Since

$$t(x, x_1) + t(x_1, x_2) + \cdots + t(x_j, x_{j+1}) \geq \frac{1}{2^{n+3}},$$

we know that

$$t(x_{j+1}, x_{j+2}) + t(x_{j+2}, x_{j+3}) + \cdots + t(x_{k-1}, y) < \frac{1}{2^{n+3}}.$$

Again using the induction hypothesis, we now know that

$$t(x_{j+1}, y) < \frac{1}{2^{n+1}}.$$

Since t takes only positive values, we also know that

$$t(x_j, x_{j+1}) < \frac{1}{2^{n+2}} < \frac{1}{2^{n+1}}.$$

Hence there exist $U, V, W \in \mathcal{H}_{n+1}$ such that $\{x, x_j\} \subset U$, $\{x_j, x_{j+1}\} \subset V$, and $\{x_{j+1}, y\} \subset W$. By the regular refining, there is $S \in \mathcal{H}_n$ with $U \cup V \cup W \subset S$. Hence $y \in S \subset st(x, \mathcal{H}_n)$, and thus $t(x, y) < \frac{1}{2^n}$. This completes the inductive step, and establishes the claim by the principle of mathematical induction.

From this claim and the definition of infimum,

$$\text{if } d(x, y) < \frac{1}{2^{n+2}}, \text{ then } t(x, y) < \frac{1}{2^n},$$

and hence

$$st(x, \mathcal{H}_{n+2}) \subset B(x, \frac{1}{2^{n+2}}) \subset st(x, \mathcal{H}_n) \text{ for each } n \in \omega \text{ and each } x \in X.$$

Thus we see that the topology generated by the metric d is the topology on X, and the proof is complete. ∎

We can now obtain as a corollary the uniform metrization theorem, so called because of the fact that it naturally arises in the study of uniform spaces.

Corollary 21.17 *(Uniform metrization theorem) A T_1-space X is metrizable if and only if X has a development $(\mathcal{G}_n : n \in \omega)$ such that for each n, $\mathcal{G}_{n+1}$ is a star refinement of $\mathcal{G}_n$.*

Proof. It is clear that a star refinement is a regular refinement in the sense of the Alexandroff-Urysohn metrization theorem, and so the "if" part follows. To see that a metric space must have such a development, consider the $\frac{1}{3^n}$ balls. The remainder of the details are left as an exercise. ∎

This result is also sometimes called the normal sequence theorem since a development as in corollary 21.17 is sometimes called a compatible normal sequence.

We now prove the Nagata-Smirnov-Bing metrization theorem. This remarkable theorem was proved simultaneously and independently by these three men in three different parts of the world. As a result of this amazing coincidence, all three are credited with solving the general metrization problem. Nagata and Smirnov proved the same result. Bing proved a slightly different, but easily equivalent, result. All three used the idea of paracompactness in an essential way.

Theorem 21.18 *(Nagata-Smirnov-Bing metrization theorem)*

1. *(Nagata-Smirnov metrization theorem) A T_3-space X is metrizable if and only if X has a σ-locally finite base.*

2. *(Bing metrization theorem) A T_3-space X is metrizable if and only if X has a σ-discrete base.*

Proof. We first prove the Nagata-Smirnov metrization theorem, and we will then use this result to establish the Bing metrization theorem.

Proof of the Nagata-Smirnov theorem. Suppose that (X, d) is a metric space. Using paracompactness, for each $n \in \omega$, let $\mathcal{B}_n$ be an open locally finite refinement of the open cover $\{B(x, \frac{1}{2^n}) : x \in X\}$. We claim that $\mathcal{B} = \bigcup_{n \in \omega} \mathcal{B}_n$ is a base for X. Suppose that U is an open set and $x \in U$. Choose $n \in \omega$ such that $B(x, \frac{1}{2^n}) \subset U$, and since $\mathcal{B}_{n+1}$ is a cover of X, there exists $B \in \mathcal{B}_{n+1}$ with $x \in B$. Let $y \in B$. Choose $z \in X$ such that $B \subset B(z, \frac{1}{2^{n+1}})$. Now we see that $d(x, y) \leq d(x, z) + d(z, y) < \frac{1}{2^{n+1}} + \frac{1}{2^{n+1}} = \frac{1}{2^n}$. Hence $y \in B(x, \frac{1}{2^n}) \subset U$. Thus $x \in B \subset U$, and we see that $\mathcal{B}$ is a base for X. Hence X has a σ-locally finite base.

Now suppose that X is a regular T_1-space and $\mathcal{B} = \bigcup_{n \in \omega} \mathcal{B}_n$ is a base for X where for each n, $\mathcal{B}_n$ is a locally finite collection of open sets. For each open cover $\mathcal{U}$ of X, $\{B \in \mathcal{B} : \exists U \in \mathcal{U} \text{ with } B \subset U\}$ is a σ-locally finite open refinement of $\mathcal{U}$. Hence X is paracompact, and therefore X is normal.

Suppose $m \in \omega$ and $U \in \mathcal{B}_m$. For each $n \in \omega$, find a continuous function $f_{n,U} : X \to [0, 1]$ by Urysohn's lemma such that

$$f_{n,U}(\cup\{\overline{V} : V \in \mathcal{B}_n \text{ and } \overline{V} \subset U\}) = 0 \text{ and}$$
$$f_{n,U}(X \backslash U) = 1.$$

Note that this is possible since these are clearly disjoint sets and $\cup\{\overline{V} : V \in \mathcal{B}_n \text{ and } \overline{V} \subset U\}$ is closed since $\mathcal{B}_n$ is locally finite.

For $m, n \in \omega$, define $d_{mn} : X \times X \to [0, \infty)$ as follows:

$$\text{for } x, y \in X, \ d_{mn}(x, y) = \sum \{|f_{n,U}(x) - f_{n,U}(y)| : U \in \mathcal{B}_m\}.$$

Notice that since $\mathcal{B}_m$ is locally finite, it is also point finite, and thus $f_{n,U}(x) = f_{n,U}(y) = 1$ for all but finitely many $U \in \mathcal{B}_m$. Hence the sum in the definition of $d_{mn}(x, y)$ is actually a sum of terms of which all but finitely many are 0. Hence d_{mn} is well defined.

Our plan is to add up all of these distance functions, so we need to truncate them. For $m, n \in \omega$, define $\rho_{mn} : X \times X \to [0, \infty)$ by

$$\rho_{mn}(x, y) = \min\{d_{m,n}(x, y), 1\}$$

for each pair $(x, y) \in X \times X$.

Since $\omega \times \omega$ is countable, index

$$\{\rho_{mn} : (m, n) \in \omega \times \omega\} \ \text{ as } \ \{\delta_k : k \in \mathbb{N}\}.$$

Define $d : X \times X \to [0, \infty)$ as follows:

$$\text{for each } x, y \in X, \ d(x, y) = \sum_{k=1}^{\infty} \frac{\delta_k(x, y)}{2^k}.$$

Since the terms of this series are dominated by a geometric series, it follows that this series converges for each $(x, y) \in X \times X$, and thus d is well defined.

We show that d is a metric. It is clear that for any $x \in X$, $d(x, x) = 0$, and $d(x, y) = d(y, x) \geq 0$ for all $x, y \in X$. Suppose $x \neq y$. Since X is T_1, there exists some $U \in \mathcal{B}_m$ with $x \in U$ and $y \in X \backslash U$. Since X is regular, there exists $V \in \mathcal{B}_n$ with $x \in V \subset \overline{V} \subset U$. Now $f_{n,U}(x) = 0$, and $f_{n,U}(y) = 1$. So $d_{mn}(x, y) \geq 1$, and thus $d(x, y) \neq 0$. The triangle inequality holds for each d_{mn} because of the triangle inequality for absolute value of real numbers, and thus the triangle inequality holds for each ρ_{mn} and consequently for each δ_k. Thus the triangle inequality holds for d. Hence we see that d is a metric on X.

Now we show that d is compatible with the topology on X. Suppose that $U \subset X$ is open, and $x \in U$. Choose $B \in \mathcal{B}$ with $x \in B \subset U$, and choose $m \in \omega$ with $B \in \mathcal{B}_m$. By regularity, choose $V \in \mathcal{B}$ with $x \in V \subset \overline{V} \subset B$, and choose $n \in \omega$ with $V \in \mathcal{B}_n$. Notice that $f_{n,B}(x) = 0$. Choose $k \in \mathbb{N}$ such that $\delta_k = \rho_{mn}$.

We claim that $B(x, \frac{1}{2^k}) \subset U$. Suppose that $d(x, y) < \frac{1}{2^k}$. Note that this implies that $\delta_k(x, y) < 1$, and so $d_{mn}(x, y) < 1$. Hence $|f_{n,B}(x) - f_{n,B}(y)| < 1$, and since $f_{n,B}(x) = 0$, we must have $f_{n,B}(y) \neq 1$. Thus $y \in B \subset U$, and the claim follows.

We complete the proof that the balls form neighborhood bases at the points of X, by showing that each ball is a neighborhood of its center. Suppose $x \in X$, and $\varepsilon > 0$. Suppose $k \in \mathbb{N}$. We shall construct an open set V_k with $x \in V_k$ and $V_k \subset B_{\delta_k}(x, \frac{\varepsilon}{2})$. Choose $(m, n) \in \omega \times \omega$ with $\delta_k = \rho_{mn}$. Since $\mathcal{B}_m$ is locally finite, choose an open set W with $x \in W$ and $\mathcal{A}_m = \{B \in \mathcal{B}_m : W \cap B \neq \varnothing\}$ is a finite set. Let $j = |\mathcal{A}_m|$, and for each $U \in \mathcal{A}_m$, by continuity of $f_{n,U}$, choose an open set V_U with $x \in V_U$ and

$$f_{n,U}(V_U) \subset (f_{n,U}(x) - \frac{\varepsilon}{2j}, f_{n,U}(x) + \frac{\varepsilon}{2j}).$$

Let $V_k = W \cap (\bigcap_{U \in \mathcal{A}_m} V_U)$. Note that V_k is open, and $x \in V_k$. Now suppose that $y \in V_k$. We see that

$$\delta_k(x,y) = \sum \{|f_{n,U}(x) - f_{n,U}(y)| : U \in \mathcal{A}_m\} < \sum_{U \in \mathcal{A}_m} \frac{\varepsilon}{2j} = \frac{\varepsilon}{2}.$$

We construct V_k in this way for each $k \in \mathbb{N}$. Choose $N \in \mathbb{N}$ such that $\sum_{i=N+1}^{\infty} \frac{1}{2^i} < \frac{\varepsilon}{2}$, and let $V = \bigcap_{k=1}^N V_k$. Note that V is open, and $x \in V$. If $y \in V$, then

$$d(x,y) = \sum_{k=1}^{\infty} \frac{\delta_k(x,y)}{2^k} < \sum_{k=1}^{N} \frac{\delta_k(x,y)}{2^k} + \frac{\varepsilon}{2}$$
$$< \frac{\varepsilon}{2}(\sum_{k=1}^{N} \frac{1}{2^k}) + \frac{\varepsilon}{2} < \frac{\varepsilon}{2} + \frac{\varepsilon}{2} = \varepsilon.$$

Hence $V \subset B(x,\varepsilon)$, and thus $B(x,\varepsilon)$ is a neighborhood of x. This completes the proof of the Nagata-Smirnov theorem.

Proof of the Bing theorem. Since any metrizable space is paracompact, we know that for any metrizable space X, and each $n \in \omega$, we can choose $\mathcal{B}_n$ to be a σ-discrete open refinement of the cover of X consisting of all the balls of radius $\frac{1}{2^n}$. Now $\bigcup_{n \in \omega} \mathcal{B}_n$ is a σ-discrete base for X. Thus every metrizable space has a σ-discrete base.

Conversely, discrete collections are locally finite. So if a T_3-space X has a σ-discrete base, then X has a σ-locally finite base, and thus X is metrizable by the Nagata-Smirnov theorem. This completes the proof. ∎

From this theorem we easily get the result that paracompact Moore spaces are metrizable.

Theorem 21.19 *A T_3-space X is metrizable if and only if X is a paracompact Moore space.*

Proof. The "only if" part we already know. Suppose X is a paracompact Moore space, and let $(\mathcal{G}_n : n \in \omega)$ be a development for X. For each $n \in \omega$, let $\mathcal{V}_n$ be an open locally finite refinement of $\mathcal{G}_n$. By lemma 21.6, we know that $(\mathcal{V}_n : n \in \omega)$ is also a development. Hence by lemma 21.9, $\bigcup_{n \in \omega} \mathcal{V}_n$ is a σ-locally finite base for X. Thus by the Nagata-Smirnov theorem, X is metrizable. ∎

In 1951, Bing showed that all Moore spaces are subparacompact (Lemma 21.14), and that all collectionwise normal subparacompact spaces are paracompact (theorem 20.21). Actually, if you look in Bing's paper, you won't see the word "subparacompact," since that word was coined by Dennis Burke in 1969, but Bing did prove these results. Combining these theorems, he obtained the first really important result on the normal Moore space problem after the time of Jones' original paper.

Theorem 21.20 *(Bing's metrization criterion) A space X is metrizable if and only if X is a collectionwise-normal Moore space.*

To close out this chapter (and the book), we will say a few more things about the normal Moore space problem. First, a *Q-set* is an uncountable subset A of $\mathbb{R}$ with the property that, in the subspace topology on A, every subset of A is a G_δ-set. In his 1951 paper, Bing showed that if we take the subspace of the Moore plane Γ consisting of all points above the x-axis and a Q-set on the x-axis, then we obtain a separable normal nonmetrizable Moore space. The problem was that it was not known whether or not Q-sets existed. In 1964, Bob Heath showed that if there is a separable normal nonmetrizable Moore space, then there must exist a Q-set in the real line. In 1969, Jack Silver and Frank Tall showed that if we assume the axiom $MA + \neg CH$, then there do exist Q-sets in the real line. It was already known that $MA + \neg CH$ is consistent with ZFC. Hence the separable case of the normal Moore space problem was completely settled. It is consistent with ZFC and independent of ZFC that there exist separable normal non-metrizable Moore spaces. This means that problem 21.13 can neither be proved in the affirmative or in the negative in the case that the spaces involved are separable. Of course, the existence of the counterexample, in certain models, which is separable, shows that it is consistent that the answer to problem 21.13 is, "no." The consistency of a positive answer has been put into a strange kind of philosophical never-never land by a beautiful result of Peter Nyikos.

Definition 21.21 *If λ is a cardinal, then the **"usual measure"** on the product space $2^\lambda = \prod_{\alpha<\lambda}\{0,1\}$ is the measure μ which assigns $\mu(B) = \frac{1}{2^{|F|}}$ if B is a basic open set in 2^λ, and $F = \{\alpha : \pi_\alpha(B) \neq \{0,1\}\}$.*

The tool that Nyikos used is the following.

Axiom 21.22 *(Product measure extension axiom, PMEA) For any cardinal λ, the usual measure on 2^λ can be extended to a $\mathfrak{c}$-additive measure on all subsets of 2^λ.*

What we mean by a $\mathfrak{c}$-additive measure is that if $\mathcal{H}$ is a collection of sets of measure 0, and $|\mathcal{H}| < \mathfrak{c}$, then $\cup\mathcal{H}$ is also a set of measure 0. We will now give Nyikos' proof that $PMEA$ implies that every normal Moore space is metrizable. To borrow the words which were used when this result was presented at the Spring Topology Conference in 1978, "It is a thing of rare beauty and stunning simplicity."

Theorem 21.23 *(Nyikos theorem) Assuming PMEA, every normal first countable space is collectionwise normal.*

Proof. Assume $PMEA$. Suppose that X is a normal first countable space, and let $\{C_\alpha : \alpha < \lambda\}$ be a discrete collection of closed subsets of X. For each $A \subset \lambda$, use the normality to choose disjoint open sets U_A, V_A such that $\cup\{C_\alpha : \alpha \in A\} \subset U_A$ and $\cup\{C_\alpha : \alpha \notin A\} \subset V_A$. Using $PMEA$, find a $\mathfrak{c}$-additive measure μ which extends the usual measure on 2^λ and is defined for all subsets of 2^λ. For each $\alpha < \lambda$, let $\mathcal{B}_\alpha = \{f \in 2^\lambda : f(\alpha) = 1\}$. Notice that $\mu(2^\lambda) = 1$, $\mu(\mathcal{B}_\alpha) = \frac{1}{2}$, and $\mu(\mathcal{B}_\alpha \backslash \mathcal{B}_\beta) = \frac{1}{4}$ for $\alpha, \beta < \lambda$ and $\alpha \neq \beta$. Also, each $f \in 2^\lambda$ is the characteristic function of the set $A_f = \{\alpha : f(\alpha) = 1\}$.

For each $\alpha < \lambda$ and each $p \in C_\alpha$, let $\{U_n(p) : n \in \omega\}$ be an open neighborhood base at p. Let $\mathcal{A}(p, n) = \{f \in 2^\lambda : U_n(p) \subset U_{A_f}$ or $U_n(p) \subset V_{A_f}\}$. For each fixed p, $\cup\{\mathcal{A}(p, n) : n \in \omega\} = 2^\lambda$ since for any f, U_{A_f} and V_{A_f} are open sets and one of them contains C_α. Now since $\mu(2^\lambda) = 1$, there exists $n_p \in \omega$ such that $\mu(\bigcup_{i=0}^{n_p} \mathcal{A}(p, i)) > \frac{7}{8}$. Choose $\gamma_p \in \omega$ such that $U_{\gamma_p}(p) \subset \bigcap_{i=0}^{n_p} U_i(p)$, and note that $\mu(\mathcal{A}(p, \gamma_p)) > \frac{7}{8}$. We do this for all $\alpha < \lambda$ and $p \in C_\alpha$. Now we have that $\mu(\mathcal{A}(p, \gamma_p) \cap \mathcal{A}(q, \gamma_q)) > \frac{3}{4}$ for all $p, q \in X$. Suppose that $\alpha \neq \beta$, $p \in C_\alpha$, and $q \in C_\beta$. Since $\mu(\mathcal{B}_\alpha \backslash \mathcal{B}_\beta) = \frac{1}{4}$, there must exist $f \in (\mathcal{B}_\alpha \backslash \mathcal{B}_\beta) \cap \mathcal{A}(p, \gamma_p) \cap \mathcal{A}(q, \gamma_q)$. Since $f(\alpha) = 1$ and $f(\beta) = 0$, we know that $p \in U_{A_f}$ and $q \in V_{A_f}$. Since $f \in \mathcal{A}(p, \gamma_p)$, $U_{\gamma_p}(p) \subset U_{A_f}$, and similarly, $U_{\gamma_q}(q) \subset V_{A_f}$. Thus $U_{\gamma_p}(p) \cap U_{\gamma_q}(q) = \varnothing$. Now for each $\alpha < \lambda$, we let $U_\alpha = \cup\{U_{\gamma_p}(p) : p \in C_\alpha\}$, and the collection $\{U_\alpha : \alpha < \lambda\}$ provides the desired separation. ∎

Since Moore spaces are clearly first countable, from Bing's criterion it would follow that all normal Moore spaces are metrizable.

It would seem that the normal Moore space problem is successfully solved. This is not quite the case. What about the consistency of $PMEA$? It would follow from the consistency of a strongly compact cardinal, but those are very large cardinals. They are so large that if you take all sets which are of "rank" less than a strongly compact in the inductive construction of the universe, then that class would be a model of ZFC. The thing wrong with that is that if ZFC could prove the consistency of strongly compact cardinals, then, by looking at the sets of rank less than a strongly compact, it could be used to prove its own consistency, and that would violate Gödel's incompleteness theorem. On the other hand, most set theorists believe in the consistency of strongly compact cardinals, and therefore in the consistency of $PMEA$. So the normal Moore space problem is solved, almost.

Exercises

1. Complete the proof of example 21.4 that all metric spaces are developable.

2. Prove lemma 21.6. That is, prove that if $(\mathcal{G}_n : n \in \omega)$ is a development for a space X, and for each $n \in \omega$, $\mathcal{W}_n$ is an open refinement of $\mathcal{G}_n$, then $(\mathcal{W}_n : n \in \omega)$ is also a development for X.

3. Prove lemma 21.8. That is, prove that for any cover $\mathcal{U}$ of X, for $x, y \in X$ it is true that $x \in st(y, \mathcal{U})$ if and only if $y \in st(x, \mathcal{U})$.

4. Prove lemma 21.9. That is, prove that if $(\mathcal{G}_n : n \in \omega)$ is a development for a space X, then $\bigcup_{n \in \omega} \mathcal{G}_n$ is a base for X.

5. Prove corollary 21.14. That is, prove that every Moore space is subparacompact.

6. Complete the proof of the uniform metrization theorem, corollary 21.17. Show that if $(\mathcal{G}_n : n \in \omega)$ is a development for X such that for each n, $\mathcal{G}_{n+1}$ is a star refinement of $\mathcal{G}_n$, then whenever $U, V \in \mathcal{G}_{n+1}$ and $U \cap V \neq \varnothing$, there is $W \in \mathcal{G}_n$ with $U \cup V \subset W$. Also prove that if X is a metric space and $n \in \omega$, then $\left\{ B(x, \frac{1}{3^{n+1}}) : x \in X \right\}$ is a star refinement of $\left\{ B(x, \frac{1}{3^n}) : x \in X \right\}$.

7. Show that the construction in example 21.11 does give a development for the Moore plane Γ. Show that the Michael line $\mathbb{M}$ is not a Moore space. Show that the ordinal space ω_1 is not a Moore space.

8. Prove that every continuous image of a compact metrizable space is metrizable (and compact).

9. Prove that every countable first countable T_3-space is metrizable. In particular, the rational numbers with the Sorgenfrey topology is a metrizable space. Describe a compatible metric.

10. Prove that any compact Hausdorff space in which each point is a G_δ-set is a first countable space.

11. (Spaces with G_δ-diagonal) A space X is said to have G_δ-diagonal if the diagonal $\Delta = \{(x, x) : x \in X\}$ is G_δ-set in the product space $X \times X$. Prove that a space X has G_δ-diagonal if and only if there is a sequence $(\mathcal{G}_n : n \in \omega)$ of open covers of X such that for each $x \in X$, we have $\bigcap_{n \in \omega} st(x, \mathcal{G}_n) = \{x\}$. Prove that any compact Hausdorff space with G_δ-diagonal is metrizable.

12. (Locally metrizable spaces) A space X is called locally metrizable if each point of X has a metrizable neighborhood in X. Prove that the ordinal space ω_1 is locally metrizable. Prove that every locally metrizable paracompact space is metrizable.

Bibliography

[1] P. Alexandroff and P. Urysohn, *Une Condition Nécessaire et Suffisante pour qu'une Classe (L) soit une Classe (D)*, C. R. Acad. Paris 177 (1923), 1274–1276.

[2] A. V. Arhangel'skiĭ, *Mappings and Spaces*, Russian Math. Surveys 21 (1966), 115–162.

[3] R. Baire, *Sur les Fonctions de Variables Réelles*, Ann. di Mat. 3 (1899), 1–123.

[4] Zoltan Balogh, *A Small Dowker Space in ZFC*, Proc. Amer. Math. Soc. 124 (1996), 2555–2560.

[5] S. Banach, *Sur les Opérations dans les Ensembles Abstraits et leurs Applications aux Equations Intégrales*, Fund. Math. 3 (1922), 7–33.

[6] R. H. Bing, *Metrization of Topological Spaces*, Canad. J. Math. 3 (1951), 175–186.

[7] L. E. J. Brouwer, *Beweis des Jordanschen Kurvensatz*, Math. Ann. 69 (1910), 169–175.

[8] D. K. Burke, *On Subparacompact Spaces*, Proc. Amer. Math. Soc. 23 (1969), 655–663.

[9] D. Burke and G. Gruenhage, *The Mathematics of Zoltan T. Balogh*, Publ. Math. Debrecen 63 (2003), 5–17.

[10] G. Cantor, *Über Unendliche, Lineare Punktmannigfaltigkeiten*, series of papers in Math. Ann. 15 (1879), 1–7; 17 (1880), 355–388; 20 (1882), 113–121; 21 (1883), 51–58; 21 (1883), 545–591; 22 (1884), 453–488.

[11] E. Čech, *On Bicompact Spaces*, Ann. Math. 38 (1937), 823–844.

[12] P. J. Cohen, *The Independence of the Continuum Hypothesis I*, Proc. Nat. Acad. Sci. USA 50 (1963), 1143–1148.

[13] P. J. Cohen, *The Independence of the Continuum Hypothesis II*, Proc. Nat. Acad. Sci. USA 51 (1964), 105–110.

[14] S. W. Davis, G. M. Reed, and M. L. Wage, *Further Results on Weakly Uniform Bases*, Houston J. Math. 2 (1976), 57–63.

[15] J. Dieudonné, *Une Généralisation des Espaces Compacts*, J. de Math. Pures et Appl. 23 (1944), 65–76.

[16] C. H. Dowker, *On Countably Paracompact Spaces*, Canad. J. Math. 3 (1951), 219–224.

[17] James Dugundji, *Topology*, Allyn and Bacon, Boston, 1966.

[18] Ryszard Engelking, *General Topology*, Heldermann Verlag, Berlin, 1989.

[19] Ben Fitzpatrick, *The Students of R. L. Moore: Preliminary Report*, preprint, 2000.

[20] Theodore W. Gamelin and Robert Everist Greene, *Introduction to Topology*, Saunders, New York, 1983.

[21] Leonard Gillman and Meyer Jerison, *Rings of Continuous Functions*, Springer–Verlag, New York, 1976.

[22] K. Gödel, *The Consistency of the Axiom of Choice and the Generalized Continuum Hypothesis with the Axioms of Set Theory*, Uspehi Mat. Nauk 3 (1948), 96–149.

[23] Paul R. Halmos, *Naive Set Theory*, Springer–Verlag, New York, 1974.

[24] F. Hausdorff, *Grundzüde der Mengenlehre*, Leipzig, 1914.

[25] R. W. Heath, *Screenability, Pointwise Paracompactness and Metrization of Moore Spaces*, Canad. J. Math. 16 (1964), 763–770.

[26] R. W. Heath and W. F. Lindgren, *Weakly Uniform Bases*, Houston J. Math. 2 (1976), 85–90.

[27] John G. Hocking and Gail S. Young, *Topology*, Addison-Wesley, Reading, 1961.

[28] Miroslav Hušek and Jan van Mill, *Recent Progress in General Topology*, North-Holland, Amsterdam, 1992.

[29] Thomas Jech, *Set Theory*, Academic Press, New York, 1978.

[30] F. Burton Jones, *Concerning Normal and Completely Normal Spaces*, Bull. Amer. Math. Soc. 43 (1937), 671–677.

[31] F. Burton Jones, *The Moore Method*, American Mathematical Monthly 84 (1977), 273–278.

[32] John L. Kelley, *General Topology*, Van Nostrand, New York, 1955.

[33] Kenneth Kunen, *Set Theory, An Introduction to Independence Proofs*, North-Holland, Amsterdam, 1980.

[34] K. Kunen and J. E. Vaughan, *Handbook of Set-Theoretic Topology*, North-Holland, Amsterdam, 1984.

[35] K. Kuratowski, *Sur l'Opération $\overline{A}$ de l'Analysis Situs*, Fund. Math. 3 (1922), 182–199.

[36] K. Kuratowski, *Topology*, Academic Press, New York, 1966.

[37] K. Kuratowski and A. Mostowski, *Set Theory*, North-Holland, Amsterdam, 1976.

[38] E. Lindelöf, *Sur Quelques Points de la Théorie des Ensembles*, C. R. Acad. Paris 137 (1903), 697–700.

[39] David J. Lutzer, *On Generalized Ordered Spaces*, Dissertationes Math. 89 (1971).

[40] E. A. Michael, *A Note on Paracompact Spaces*, Proc. Amer. Math. Soc. 4 (1953), 831–838.

[41] E. A. Michael, *Another Note on Paracompact Spaces*, Proc. Amer. Math. Soc. 8 (1957), 822–828.

[42] E. A. Michael, *Yet Another Note on Paracompact Spaces*, Proc. Amer. Math. Soc. 10 (1959), 309–314.

[43] E. A. Michael, *The Product of a Normal Space and a Metric Space Need Not Be Normal*, Bull. Amer. Math. Soc. 69 (1963), 375–376.

[44] J. Donald Monk, *Introduction to Set Theory*, McGraw-Hill, New York, 1969.

[45] E. H. Moore and H. L. Smith, *A General Theory of Limits*, Amer. J. Math. 44 (1922), 102–121.

[46] R. L. Moore, *On the Foundations of Plane Analysis Situs*, Trans. Amer. Math. Soc. 17 (1916), 131–164.

[47] R. L. Moore, *Foundations of Point Set Theory*, New York, 1932. (reprinted: Amer. Math. Soc. Colloq. Publ. vol. XIII, Providence, 1976)

[48] R. L. Moore, *Concerning Simple Continuous Curves*, Trans. Amer. Math. Soc. 21 (1920), 333–347.

[49] R. L. Moore, *Concerning the Cut Points of Continuous Curves and of other Closed and Connected Point Sets*, Proc. Nat. Acad. Sci. USA 9 (1923), 101–106.

[50] R. L. Moore, *An Extension of the Theorem that no Countable Point Set Is Perfect*, Proc. Nat. Acad. Sci. USA 10 (1924), 168–170.

[51] James R. Munkres, *Topology, 2nd Ed.*, Prentice-Hall, Upper Saddle River, 2000.

[52] A. Mysior, *A Regular Space Which Is Not Completely Regular*, Proc. Amer. Math. Soc. 81 (1981), 852–853.

[53] J. Nagata, *On a Necessary and Sufficient Condition of Metrizability*, J. Inst. Polyt. Osaka City Univ. 1 (1950), 93–100.

[54] P. J. Nyikos, *A Provisional Solution to the Normal Moore Space Problem*, Proc. Amer. Math. Soc. 78 (1980), 429–435.

[55] A. J. Ostaszewski, *On Countably Compact, Perfectly Normal Spaces*, J. London Math. Soc. 14 (1976), 501–516.

[56] G. M. Reed and P. L. Zenor, *Metrization of Moore Spaces and Generalized Manifolds*, Fund. Math. 91 (1976), 203–209.

[57] Judith Roitman, *Introduction to Modern Set Theory*, John Wiley & Sons, New York, 1990.

[58] Joseph J. Rotman, *The Theory of Groups*, Allyn and Bacon, Boston, 1965.

[59] Mary Ellen Rudin, *Countable Paracompactness and Souslin's Problem*, Canad. J. Math. 7 (1955), 543–547.

[60] Mary Ellen Rudin, *Souslin's Conjecture*, Amer. Math. Monthly 76 (1969), 1113–1119.

[61] Mary Ellen Rudin, *A Normal Space X for Which X × I Is Not Normal*, Fund. Math. 73 (1971), 179–186.

[62] Mary Ellen Rudin, *Lectures on Set Theoretic Topology*, CBMS, American Mathematical Society, Providence, 1975.

[63] Walter Rudin, *Homogeneity Problems in the Theory of Čech Compactifications*, Duke J. Math. 23 (1956), 409–419.

[64] Walter Rudin, *Principles of Mathematical Analysis*, McGraw–Hill, New York, 1964.

[65] W. Sierpiński, *General Topology*, University of Toronto Press, Toronto, 1956.

[66] George F. Simmons, *Topology and Modern Analysis*, McGraw-Hill, New York, 1963.

[67] Ju. M. Smirnov, *On Metrization of Topological Spaces*, Uspehi Mat. Nauk 6 (1951), no. 6, 100–111. (English translation: Amer. Math. Soc. Transl. Ser. 1, 8 (1962), 63–77.)

[68] R. H. Sorgenfrey, *On the Topological Product of Paracompact Spaces*, Bull. Amer. Math. Soc. 53 (1947), 631–632.

[69] M. Souslin, *Problème 3*, Fund. Math. 1 (1920), 223.

[70] Lynn A. Steen and J. Arthur Seebach, Jr., *Counterexamples in Topology*, 2nd ed. Springer-Verlag, New York, 1978.

[71] A. H. Stone, *Paracompactness and Product Spaces*, Bull. Amer. Math. Soc. 54 (1948), 977–982.

[72] M. H. Stone, *Applications of the Theory of Boolean Rings to General Topology*, Trans. Amer. Math. Soc. 41 (1937), 375–481.

[73] F. D. Tall, *Set-theoretic Consistency Results and Topological Theorems Concerning the Normal Moore Space Conjecture and Related Problems*, Thesis, Univ. of Wisconsin 1969, Dissertationes Math. 148 (1977)

[74] S. Tennenbaum, *Souslin's Problem*, Proc. Nat. Acad. Sci. USA 59 (1968), 60–63.

[75] H. Tietze, *Beiträge zur Allgemeinen Topologie I*, Math. Ann. 88 (1923), 290–312.

[76] S. Todorčević, *Forcing Positive Partition Relations*, Trans. Amer. Math. Soc. 280 (1983), 703–720.

[77] J. W. Tukey, *Convergence and Unifromity in Topology*, Ann. Math. Studies 2, Princeton, 1940.

[78] A. Tychonoff, *Über die Topologische Erweiterung von Räumen*, Math. Ann. 102 (1930), 544–561.

[79] P. Urysohn, *Sur la Métrisation des Espaces Topologiques*, Bull. Intern. Acad. Pol. Sci. Sér. A (1923), 13–16.

[80] P. Urysohn, *Über die Mächtigkeit der Zusammenhängenden Mengen*, Math. Ann. 94 (1925), 262–295.

[81] P. Urysohn, *Zum Metrisationsproblem*, Math. Ann. 94 (1925), 309–315.

[82] Stephen Willard, *General Topology*, Addison-Wesley, Reading, 1970.

Index